Systems Engineering

Other McGraw-Hill Communications Books of Interest

*For more information about other McGraw-Hill materials,
call 1-800-2-MCGRAW in the United States. In other
countries, call your nearest McGraw-Hill office.*

Systems Engineering

Architecture and Design

Walter R. Beam

Professor of Systems Engineering
George Mason University
Fairfax, Virginia

McGraw-Hill Publishing Company

New York   St. Louis   San Francisco   Auckland   Bogotá
Caracas   Hamburg   Lisbon   London
Madrid   Mexico   Milan   Montreal
New Delhi   Oklahoma City   Paris
San Juan   São Paulo   Singapore
Sydney   Tokyo   Toronto

To Teka

Library of Congress Cataloging-in-Publication Data

Beam, Walter R.
  Systems engineering: architecture and design

    Includes index.
    1. Systems engineering.   I. Title.
TA168.B35   1990   89-13463
620'.0042—dc20   CIP
ISBN 0-07-004259-4

234567890   DOC/DOC   98543210

ISBN 0-07-004259-4

The editors for this book were Daniel A. Gonneau and Jim Halston, and the production supervisor was Dianne Walber.

Printed and bound by R. R. Donnelly and Sons.

*For more information about other McGraw-Hill materials,
call 1-800-2-MCGRAW in the United States. In other
countries, call your nearest McGraw-Hill office.*

Contents

Chapter 6. Process-driven Design Objectives   190

Chapter 7. Synthesis and Analysis in System Design   216

Chapter 8. Other Design Considerations   250

Chapter 9. Managing System Architecture and Design   302

PREFACE

This book was developed as a first professional text in the application of architecture and design practice in systems design. It is intended for readers with engineering or scientific backgrounds who have also had design experience. What its readers will have in common is a strong interest, or urgent involvement, in systems developments so large or complex as to challenge any traditional engineer or scientist's knowledge base and organizational abilities.

The material, in the form of extensive notes, was used in successive academic years, for an elective graduate course in systems engineering at George Mason University. Relatively few universities yet offer systems engineering programs. Accordingly the book was designed for use by its reader with or without guidance from a professor or professional mentor.

Systems engineers are, of necessity, technical generalists. Though this book is largely lacking in mathematical expressions, this should not suggest that a systems engineer will not need to use them. A wealth of applied mathematics and science must be correctly applied to competent systems engineering practice. This book, however, is not about applied math and science. Rather it deals with the practice of systems engineering, which is not intrinsically mathematical. Rather, it is organizational, judgmental, logical, goal-oriented, and admittedly must often be subjective. We attempt to teach these basics, within a framework of actual systems engineering, architecture, and design examples.

A systems engineer will often need information on specialized technologies. If one is lucky enough to be part of a comprehensive, first-rate, systems design organization, there should be numbers of knowledgeable specialists, some at hand and others on call. Otherwise one must turn to the library, but certainly not under the title "systems engineering," for which even the U.S. Library of Congress showed only 213 references in its 1989 holdings. In most cases systems engineers will seek information on principles, capabilities, and limitations of a technology, rather than on highly specialized details or unproven research hypotheses. It is difficult to find recent professional journal articles which explain the fundamentals in succinct understandable terms. In many technical fields, survey or reprint publications, or some of the better trade magazines, may better serve these needs.

A fair proportion of systems engineering usually involves application of mature technologies, sometimes so long out of the research lab that they are no longer discussed in new academic textbooks. Though older textbooks are still found in many libraries, they often represent a world of smokestack industries without transistors, integrated circuits, personal computers, or foreign technical competition. Contemporary handbooks prepared by active, experienced designer/authors are often good sources of routine state-of-the-art information. Though most engineers do not like to be thought of as "handbook engineer," design data must be obtained in whatever form it is available. Data from subsystem, parts, or materials suppliers will supplement the generic handbook descriptions with up-to-date materials, components, or subsystems data.

The book's introductory Chapter 1 defines systems architecture and design and their place within the convenient system life cycle framework. Chapter 2 provides perspective, first through historic building architectures, then by examples from modern computer architectures. In Chapter 3, the reader learns how to recognize architectural elements and to distinguish architectural differences and variants.

Chapter 4 begins coverage of system design with description of the design process and other aspects such as simulations and prototyping which are common to design efforts for many kinds of systems. In Chapters 5 and 6, special foci of design efforts are discussed with a viewpoint which we hope will make the material more interesting. Though most real system design effort must deal simultaneously with issues such as system cost, performance, weight, and reliability, we isolate these and similar design factors to better illustrate their application. For example, the discussion of reliability deals with system examples for which the utmost in reliability is required. Chapter 5 deals with design factors which derive from user and product objectives. Chapter 6 views other design issues—cost and process-driven factors, which drive the design to also meet the needs of designer or builder.

The remaining chapters deal with special "how-to" aspects of system design. Chapter 7 explores the matter of innovation in systems engineering, with a "better mousetrap" example illustrating processes suitable for synthesis of new systems. Chapter 8 deals with a variety of design issues, including available and desirable computer-based aids for system architectural and design evolution. Such matters as testing, development and use of design rules, and competitive aspects of systems engineering are dealt with. Chapter 9 examines management aspects of systems engineering or design, a challenging matter because it often requires close coordination of innovative intellectual activities of many designers. Because systems engineering is so broad in scope, responsive management is important at every level.

The appendix describes the evolution, over a five-year period, of a commercially successful manufacturing data system in which the author was centrally involved. Although in this case the system was relatively small by contrast to a rail system or mainframe computer installation, its complexity continued to evolve with its market, requiring repeated system decisions which tested the original architectural basis.

By drawing examples from a variety of commercial and military system areas, we can deal with systems architecture and design on a broad basis. My experience, first in electronic devices, later in computers and large military systems, suggests to me that many system truths are universal. Learning about fields outside my own has usually provided ideas which travel well.

With a broad viewpoint, the accusation of shallowness is probably inevitable. Getting specific about a few large systems which incorporated interesting design features would have provided too narrow a set of examples and all of them obsolete. Instead, examples are mostly fragmentary. Many deal with modern systems with which most readers will be familiar as users. This theme is followed up in each chapter by practical exercises dealing with architecture and design of familiar systems such as residential systems, automotive systems, and entertainment systems. These assignments have no "school solutions." The problems and responses, when fleshed out by the reader from his or her experience, will be strongly personalized. Readers should also be able to assess the quality or innovativeness of their response without assistance, or with review by a mentor or colleague.

I appreciate the assistance provided by the George Mason University graduate students in systems engineering who have used this material. Thanks are also due to the administration at GMU, for encouraging development of the material. Stephen Silverman, President of Highland Technology, Inc., who was a developer and is the manufacturer of Shoplink manufacturing networks, graciously permitted publication of the appendix material.

Walter R. Beam

Introduction

A major objective of this book is to expose readers to the basics and some of the methodologies of good system design. A fair part of our treatment might be termed philosophical rather than mechanistic. This sort of generality is required by the wide variety in design of various kinds of systems: computers, power stations, aircraft, or rail systems, to name a few. Dealing with architectural issues is atypical of most books on design; however, attention to architecture should be an early step in the design of any large or complex system.

1.1 System architecture

The term *architecture*, long used to describe generic aspects of building design, has been applied to technical systems only for about a generation. We will define an architecture, for our purposes, as comprising design elements which are characteristic of a system, and of a class of systems to which it belongs. A class of systems, for present purposes, may be distinguished by their functions, performance range, cost, or other attributes.

Architecture is thus, by definition, a matter of repetition among members of the class, and often repetition within a single member.

Architectural elements may be visual in that appearance of systems in a class is similar. Manufactured kitchen storage units in the United States have, since the early 1900s, followed a modular architecture. Although different product groups may have significantly different surface treatments and use different materials (metal, wood), the modular dimensions and the makeup of a typical set of products are nearly identical.

The architectural elements may also be functional in that each member of the class operates in similar or identical manner, despite physical differences. Many different computers developed since the mid-1960s by IBM, for example, have used a common "System/370" architecture which offered a large degree of software compatibility across computer generations, and in computers of vastly different processing capacities. While this architecture's age has been showing for many years, it has accomplished what it was intended for: It has allowed computer applications developed early in the computer era to remain executable.

The important architectural elements of modern systems may consist largely of standardized interfaces. This lets the different parts of a system relate to each other and to the outside world in predefined ways having known feasibility and well-understood characteristics. Even small electrical appliances exhibit many architecturally standardized elements: For example, the power cord is almost always terminated by a plug designed to fit mechanically into a standard 110-V AC power receptacle. In the United States, the appliance itself will be designed to operate using a nominally 110-V, 60-cycle alternating-current electric power source. More complex systems, such as transportation systems or communications systems, typically have several or many interfaces, both internally and externally, which represent architectural elements.

Whereas interfaces are one sort of architectural element, the overall system structure represents another. Highway systems are characterized by *network* structures, whereas public health and emergency services systems are generally *nodal*, a collection of police stations, firehouses, or hospitals, each containing aggregated resources to serve a large geographic area. In contrast, aircraft and many other vehicles have a *monolithic*, i.e., one-piece, nature. All the essential parts are collected and connected to form a single, portable structure. Tractor/semitrailer "rigs" or railway trains typically consist of several kinds of major elements, each having special functions, which can be connected together in different orders and combinations.

Architecture, thus, is most often a measure of a system's commonality with others of its class, rather than of uniqueness. This is not to say that one cannot create new architectures. Departures from traditional design may, if imitated and improved, become recognized as architectural innovations. However, major innovative experiments in system architecture often impose major risks. An attempt to change more than a few architectural elements in a new system may involve enormous additional design costs, special manufacturing techniques and facilities, and training of designers and builders who must deal with unfamiliar structures, materials, processes, etc.

As an example, modern residential architecture was spearheaded by the late Frank Lloyd Wright. It used unpainted laminated natural-wood beams. These could not be fabricated, handled, and joined in the rough manner possible with the nailed-up "balloon framing" which has been traditional in residential construction in the United States for a century. Balloon framing in turn represented an economical replacement for pegged-together log frames, which in the early 1800s replaced stacked-log "cabin" constructions of our early settlers. Acceptance of Wright's design concepts, esthetically pleasing as they were, continues to be slow even today because of the preponderance of traditional construction in which the visually unappealing framing members are totally covered by painted wall panels.

Why is there a need to address architecture in design of other kinds of systems? Although many systems are yet designed with little initial attention to architectural concepts, this is a reasonable first step in a *top-down* approach to system design . . . a simple concept which has been widely accepted as the most effective way to address large, costly, and/or complex systems. Without the unification which a sound architecture can bring, details of system design and integration are far more likely to lead to inconsistencies and incompatibilities late in the system development process. Unless a new system's departure from conventional architecture and design and production practices are understood in advance, its consequence may be failure late in development or system production, yielding unacceptable, uneconomic, or even disastrous results.

Architectural design is an essential precursor to detailed system design, which we will refer to as *engineering design*. In practice, it is difficult to separate architectural and design activities uniquely. New ground broken in architecture initially takes the form of innovative design and is recognized as architecture only later.

Not all innovative design is architecturally sound or useful. We can observe this, for example, from motion pictures of experimental aircraft during the early years of aviation. Reviewing early developments in what has become a highly evolved system field is good practice for the performing system architecture and design. The early work usually reveals steps of progress toward the important concepts, while much later work represents mainly refinements.

Though by no means the sole architects of the airplane, the Wright brothers made many fundamental contributions whose longevity suggest that they should be called *architectural* innovations. Soundness of many of Wilbur and Orville Wright's flight-control concepts, and airfoil design principles, is demonstrated by their adaptation, use, and rediscovery for aircraft designed 8 decades (and many design generations) later. However, the Wrights' location of the pitch-control device (elevator) forward of the wing in their 1903 airplane evolved, through others' designs and their own 1911 work, into an aft-mounted empennage (tail assembly). Is this proof that they were originally off-base? By no means. Successful aircraft designs by the innovative aeronautical designer Bert Rutan again placed horizontal control surfaces (canards) forward of the main wing, like those of the Wrights, and also render his aircraft designs (if correctly constructed) incapable of stalls or spins.

Prior to the Wrights', most "flying machines" were modeled on birds' body structures, the most successful flying objects in all creation. The Wright's architecture was well suited to the materials from which the machine had to be built (wood and cloth, not feathers), to the means of propulsion (a rotating propellor, not flapping of the wings), and to a human as controller and passenger, rather than as source of motive power. Although digital computers have evolved from mechanical calculating machines rather than from a concept of the human brain, the brain offers a capability which many would-be computer designers have attempted to copy, thus far without any real success. A main reason for their failure may be the almost total lack of knowledge about the true connection architecture of animals' brains. Mimicking details such as behavior of individual brain cells (neurons) in response to stimuli, a popular research area for at least 25 years, has failed to produce anything remotely resembling "brainpower."

1.2 Engineering design

Engineering design principles and disciplines are acknowledged to have special value for dealing with large and complex endeavors. In recognition of this, for example, the term *software engineering* has been applied to methodologies designed to formalize and to make more rigorous the development of large and complex software. Observe that in this case there is little or no involvement with "engines," or with any hardware other than commercially available computers.

Despite avowed commitment to engineering design as a discipline, no two engineering designers will likely agree on precisely what collection of moods, methodologies, and standards, taken together, would constitute good engineering design. Some hints may, however, be drawn from a brief examination of a typical present-day undergraduate engineering curriculum: It (1) contains significant amounts of "hands-on" laboratory experimentation, (2) is heavy in mathematics, (3) requires students to document their study or experimental objectives, studies, and findings, and (4) features mathematical modeling of most of the static and dynamic effects or processes met with. Surprising, perhaps, that though versed in graphic and tabular representations of the things they deal with, most graduate engineers rank rather poorly in verbal and written expression, as compared to graduates in the sciences or humanities.

By comparison with most other intellectual pursuits, in good engineering design a great deal of attention is paid to documentation; indeed, great attention is paid to detail by the traditional engineer no matter how tedious it may seem to others. The old maxim "a chain is no stronger than its weakest link" comes often to mind. Engineers think of complex systems as forming close-coupled chains or networks of connected elements, while laypersons may view the same systems as collections of eclectic parts.

Traditionally educated engineers, after widely ranging careers in detailed engineering, often evolve to be highly qualified in system design. Many find transition to the systems engineering role unpleasant, when they must relinquish attention to detail to others to merely manage processes in which they once participated. Once established at the system engineering level, however, the previously learned skills will find new application in the new and broader domain.

A central element in engineering design skill is an ability to conceptualize. The design engineer must often determine, from indirect evidence, how some device operates, or why it fails. This is where easy conceptualization is most valuable. Designers operate in different and personal ways. Some may imagine themselves small enough to climb inside a tiny mechanism, or conceive what is occurring in a sealed combustion chamber through applying, intuitively, some combination of the basics of physics, chemistry, electronics, mathematics, and other knowledge. Almost all skilled designers (engineers, artists, architects, and others) make extensive use of graphic aids, as simple as sketchpads or as complex as supercomputers, to assist visualization or apply theory. They will often make lists of alternative possibilities which might explain or solve some engineering problem.

Every designer has limits which are set by his or her experiences. Those who can readily incorporate experience of others as well, by reading and personal interaction, clearly enjoy an advantage. This is especially so in complex system designs for which no single individual is likely to have personal experience in all aspects of the problems.

At the system level, as noted above, the primitive elements of design are frequently grander in scope or scale than they are in traditional engineering. A good designer may be able to understand only the basic principles involved in the system details but will be practiced at asking questions of all available specialists. Careful note will be made of the system elements' basic characteristics and limitations, for these may be central to success or failure.

Traditionally, engineering design, like most other intellectual creations, is heavy in solo activity. Design of very small systems, or of system components, may indeed be the work of a single individual or a small group led by a skilled designer. Most patented inventions emerge from sole inventors. This is no mystery: If necessary, considerations and compromises could be attacked without need for discussion and compromise, or the elimination of personal biases and avoiding of bureaucratic processes, they can usually be resolved more rapidly and resolutely. For most large or complex systems, no one individual has the knowledge, or the time, required to carry out a complete design.

A major factor in good system design—one we shall revisit later—is the intelligent subdivision, or *partitioning* of the system, or merely its design, into a discrete set of parts. This decomposition should insofar as reasonable satisfy the criterion that, after its definition, each part should be designable independently of the rest. This of course requires a very intelligent subdivision, by the system designer. Within the limitations imposed by the real needs of the system and its users, there should furthermore be no unnecessary constraints on subsystem or component design. Clearly this decomposition is a challenging task, especially so since it must usually be performed at a time when the system exists only on paper.

1.3 Architecture and design within the system life cycle framework

The term *system life cycle* usually refers to that sequence of steps involved in evolution and application of an important or expensive system. There can be no fixed definition of these steps, which depend on (1) roles carried out by developer(s) and user(s), (2) system complexity, (3) the processes by which the system is acquired by or for its user(s), (4) the number of duplicate systems to be constructed, and other factors.

The listing of life cycle steps below is intended only to be representative of large or complex systems designed by a system developer to meet the requirements of particular users. Many examples of this sort of system development situation may be found in systems acquired by the U.S. Department of Defense and National Aeronautics and Space Administration. In these instances, the client pays the costs of system engineering and production, since in most cases the developer is unable to find a customer for duplicate systems. However, since a proposal describing the system is usually required by the client prior to ordering the development, the most important parts of the system architecture and high level design may be carried out speculatively, at the expense of the developer. Where these costs are so high as to discourage competition, multiple contracts may be issued to guarantee competition in early design.

Many complex systems are designed using the system manufacturer's resources and even may be built speculatively, for sale to customers not committed at time development begins; automobiles, consumer equipment, and commercial computer systems fall in this category. Since the design or manufacturing firm is paying the full cost of system design, it will often be scheduled to satisfy that organization's availability of personnel and other resources. However, since no income is generated until after development has been completed, pressure for rapid development may be greater than when a client is paying for the development. Ergo, most commercial developments require no more than 5 years from initiation to system deliveries, while governmentally funded system developments may stretch this to 8 to 12 years or even more.

There are also many systems whose design, installation, and testing are carried out largely, or even exclusively, by the systems' end users. This usually occurs where the users are the most knowledgeable group and where no outside contractor has comparable expertise. Unfortunately, too many organizations which have made the decision to perform their own systems architecture and engineering work have discovered unrecognized shortfalls in their expertise, or biases which led to eccentricities in the resulting design. A typical example was a large commercial organization (a former client of this author) which determined, on the basis of existing in-house skills (and over our objections), to use Fortran as the programming language for a large communication-connected database information system. The results were far more expensive than expected and were delayed in arriving. The system initially performed far less well than had been promised.

1.3.1 Exemplary system life cycles

The steps below are typical of large system developments for the *U.S. Department of Defense* (DoD). Notwithstanding periodic attacks by the Congress and the press, DoD system developments are, in light of the higher risks which are undertaken, at least as successful as comparable high-risk developments carried out by other federal agencies, state or local governments, or even most business organizations.

1. Perception of need The system's end users must first recognize need for a system capability. Users may be stimulated by weaknesses in an existing system performing similar functions or by knowledge of technology which makes a greatly improved system feasible.

2. Requirements definition This is usually in the form of documents prepared by the user or a surrogate. A formal definition of requirements is essential as a basis for asking developers to come up with system solutions. To permit as broad a range of solutions as possible, perhaps including "off-the-shelf" systems, requirements should not be constrained in unessential ways.

3. Draft Request For Proposal (RFP) This document may be prepared by, or for, the user as a "talking paper" for discussions with potential system contractors. It is intended to stimulate contractor suggestions or caveats (perhaps "It can't be built. . ."). A good draft RFP has an additional benefit of stimulating thinking by potential system contractors about how to meet new or especially challenging user needs. The final RFP, on which cost estimates will be based will often reflect insights obtained from responses to a draft. Sometimes the user will insert constraints into the final RFP intended to favor certain contractors, based on responses to the draft. Though not unknown, this is illegal for government contract activities, and is a poor business practice in any situation. In any event it is essential that prospective developers be appraised of tentative or incomplete aspects of a draft RFP, and the contractual status (if any) of such documents.

4. Comments on the draft RFP These are prepared, perhaps informally, by some of the potential contractors. These responses to the draft RFP may point out requirements believed difficult or costly to achieve. A prospective developer may use this opportunity to argue its perceived advantages over the competition.

5. Final RFP, including a Statement of Work This is prepared by the user or a surrogate. The Request For Proposal (RFP), in its final form, may include detailed information directing contractors how to be responsive. This is necessary since alternative offers must be comparable in sufficient ways to allow intelligent selection. The Statement of Work, which is part of the RFP, lists the actual work the user expects the contractor(s) to perform and should describe and number the "deliverables" (items which must be delivered to fulfill the contract). An RFP should also describe the criteria by which proposals will be evaluated and compared, in particular those aspects of the proposal which will be scored, and a basis for scoring.

6. Proposal development Each prospective system contractor must develop a proposal. The time permitted for proposal development may be brief; yet this step often must include selection of system architecture and carrying out top-level design. A contractors' most capable system designers are usually assigned this task, which should result in a proposal describing:

- The selected "approach"
- System architecture and critical design details
- The system's major partitioning and identification of subcontractors (if any) who will develop or supply parts or subsystems
- A work breakdown and work schedule
- Project management plans
- Cost proposal

7. Source selection The user or a trusted surrogate selects a contractor, or in some instances "downselects", i.e. reduces the numbers of contractors who will be involved in the next phase. Proposals are evaluated and ranked on the basis of the criteria mentioned in the RFP. One or more contractors are selected. A contract or contracts are negotiated and awarded to the successful developer(s).

8. Completion of system design Here the system contractor, now almost certain that the project will be funded to completion, completes the system design. Though the proposal presumedly defines the system, much additional detailed design is typically required, and responsibilities for performing and managing the rest of the work must be assigned and negotiated.

9. Final partitioning of system Needed changes in major and minor system subdivisions are made. The system should at this point be partitioned into subsystems and components which can be developed and manufactured independently, or which can simply be purchased. Few high-level system decisions should occur after this point if the system engineering job is done well and user needs are relatively stable.

10. Definition of subsystem specifications and system tests The system contractor must flesh out subsystem definitions to the degree necessary for compliance with the proposed system design. A test plan ensuring successful integration of the overall system should be developed and defined at this point. Subcontracts can now be negotiated and awarded based on these "final" specifications.

11. Development of hardware and software components Hardware or software developments or modifications must be carried out before the system itself can be integrated. Scheduling here is critically important, lest any part of the system which lags the rest cause a delay in system delivery. Often several classes of developments are required:

- Components not available off-the-shelf
- Major hardware subsystems
- Systems software
- Interface hardware, including cables/plug assemblies
- Application software
- Documentation (by system contractor and subcontractors)

The system contractor is least likely to be involved with the developments at the top of this list, and most likely to be involved with those at the bottom because of their total-system dependency.

12. Integration of subsystems Much of this will be done by the system contractor, but large subsystems performing special functions (e.g., antenna arrays) may be integrated by a major subcontractor with special expertise. As developmental components or subsystems become available in the form of prototypes or early-production items, they can be brought together for system integration. This is the most severe test of the quality of the systems engineering. Any or all of the following may be involved:

- Hardware which has been purchased or developed for use in the system is tested against specifications.
- Software which must function on system hardware is operated in a simulated or test environment; detected "bugs" are fixed.
- Testing of complete subsystems of assemblages is carried out, often by using an ad hoc test system to simulate the operation of subsystems which are not yet available.

- Documentation is completed and reviewed for correctness and consistency.
- The system is tried out on users and adjusted if necessary.

13. Full-system integration and testing These are always done by the system contractor when there is one. This step may not be feasible until after the system is delivered to the user's site, in fact. The subsystems of the system are connected as intended and the overall system is tested in an environment closely representing its actual operating environment. (A system's environment may involve features of both the physical environment in which it operates and the functional environment which exercises it.)

This process may involve a series of steps, for instance:

- Mechanical, electrical, and logical integration and interface testing
- Software integration and testing
- Full-system simulated operational testing
- Actual operational tests involving users who will operate the system

14. Preparation of user training material and operational documentation This usually involves extensive writing. It may also include construction of simulators to model the environment and/or operation of the real system. Course curricula, graphic aids, etc., may be part of the requirement. This activity is time-consuming and should be initiated as early as is feasible; clearly it is difficult to complete until the system can be examined in complete form.

15. Installation (fielding) of system The system is installed at user site(s), usually by its builder but sometimes with user involvement.

16. Initial operational testing is usually done by user personnel, with assistance from the system contractor and sometimes with the help of specialty contractors. A formal period or plan of operational testing may be the basis for verifying that the system meets its contractual requirements. Deviations from these requirements become problems for the system contractor; some may be passed to subcontractors for solution.

17. System operation and maintenance Even after the system is put into operation, it must be monitored for malfunctions or incorrect operation. The system contractor's period of responsibility will usually be defined in his contract, and it may extend well beyond delivery. Warrantees are common in the acquisition of most complex products and systems, but must be called out specifically in contracts.

18. System acceptance This is a formality which follows demonstration that requirements have been met, sometimes even without full compliance. The system contractor then receives the final portion of the payment, previous portions having been awarded on completion of defined parts of the work.

Following this point, duplicate systems (if contracted for) may now be assembled, tested, and delivered according to the contract schedule. Less exhaustive testing processes may suffice for these duplicate systems. Beyond this point in the life cycle, the system is in operational use, though its capabilities may be evolved further over its lifetime. The original system contractor (if a good initial job was done, and the follow-on bid is competitive) or other organizations may be awarded contracts for continued maintenance support, the provision of spare parts or expendable supplies, or for other needs.

19. System change requirements These may evolve as technology makes possible a better system, as the users' needs change, or merely as users find out things they didn't know when setting initial system requirements. Major change may be found necessary. Obsolescent portions of a system may need to be replaced. Any of these circumstances may involve hardware and/or software modification or addition. Specifications for these changes may be developed by the user but more likely by a contractor.

20. Contractor bid on system hardware and/or software changes Depending on the scope of required changes, contracts ranging from informal "support contracts" to new competitive system acquisition programs may be required.

21. Change development Equipment or software required for the change is purchased or developed by the contractor. Required level of system design and integration effort will usually depend on how much of the system is affected and to what degree.

22. Modification installation and testing Modifications may be installed by the contractor and initially tested, perhaps during periods when the system is not otherwise used. The user usually accepts the work after the contractor's test results meet contract specifications.

23. Operational testing of enhancements Implicit testing continues during the system's lifetime. In software-intensive systems it is not uncommon for a software error in the original design to go undetected years after the initial installation. Each round of system improvements leads to a new set of design and implementation errors.

At this point, one may loop to step 19 repeatedly. Most large systems evolve periodically throughout their working lifetime to meet user needs and/or to remain cost-effective. Changes may be made on a regular schedule (if there is incentive for such), or irregularly. A slowdown in system evolution often precedes system replacement, after it is determined that the existing system "isn't worth fixing."

Much essential detail has been omitted in this listing. For example, during system operation, there must be maintenance plans or strategies, including provisions for spare parts or "expendable" items (fuel, ammunition, printer paper, computer tapes, etc.). Maintenance and training for it can be major activities, especially if maintenance personnel change assignments frequently as they do in military service.

EXERCISES

1. A system life cycle is one example of a complex cycle of events whose order may vary slightly but whose initiation, end, and objectives are relatively consistent. Describe three other complex cycles from the natural or cultural world; compare these, according to aspects you believe are important, with the system life cycle described in this chapter. (An objective is to recognize and describe relationships within a complex cycle.)

2. Examine several books on classical or modern building architectures, from an available library. (You may, for example, find a book describing traditional building architecture characteristic of your state or region.) Distill from this material what you see as the essential elements of one or more distinct architectures. What features uniquely characterize each type of architecture? Identify any other features of the architecture so common in other architectures as to be poor identifiers.

3. Architectures may have strong and weak, good and bad points. Consider a common type of housing architecture in your area, for example "colonial." Identify and explain strong and weak points, from perspectives of: weather tightness, structural rigidity, soundproofing, and convenience of access. Make certain that the aspects you cite are characteristic of the architecture, in its common usage.

4. Consider major variables in modern automotive architectures: for example front- versus rear-wheel drive, unit-body versus body/frame construction, body style, and other salient architectural features.
(a) Identify four combinations of these elements which have been selected by a manufacturer as its principal architecture. (One example is the air-cooled rear-engine, rear-drive unit-body two-door sedan design first made popular by Porsche in the Volkswagen.)
(b) Compare these combinations in respect to their influence on system characteristics: cost, handling, useful lifetime, economy, noisiness, and any other factors which come to mind.

5. Examine the engineering design of a complex mechanical or electromechanical product in your home or office, one which you can open up to see the key working parts. (Suggested examples: toilet, gas or electric range, typewriter, mechanical chiming clock.)
(a) Identify six or more design features of the product which you believe required significant design effort by the designer(s) of the product.
(b) Identify several design features which you suspect or know were borrowed from earlier designs, and why you think so.

6. For a product meeting selection criteria in Exercise 5, identify three design features which you believe could be improved upon by redesign.

(a) Explain what aspect you believe needs improvement, and why. Use simple sketches where they will be helpful.

(b) Suggest improvement: what should be changed, what kind and how much improvement might be expected, and what would be involved in manufacture, e.g., change of shape, add part(s), use greater precision, use a different material.

BIBLIOGRAPHY

A selection of additional sources is suggested following each chapter. Only books are cited—though some of them are reprint collections and others are Proceedings of annual conferences. We acknowledge the importance of periodicals in technical progress. However, the effort to locate a book is no more than that for an article, and the rewards are often greater.

Many of the works cited can be found in a metropolitan library, or most in the library of a university having an engineering school. Most *handbooks* are revised periodically. Though we attempted to identify the latest editions available at the time of this writing, later ones may have become available at the time of reading. As in all science and technology, the obsolescence rate of books (or articles) depends on the rate of change of technology and products. In rapidly changing fields such as semiconductors or computers, the reader must seek out up-to-date publications.

For Chapter 1, books cited are light reading and should set the stage for system architecture and design.

A. J. Pulos, *American Design Ethic: A History of Industrial Design to 1940*, MIT Press, Cambridge, MA, 1983. Industrial design is the design of useful articles. Good industrial designs have many of the desirable features we describe in later chapters.

D. B. Steinman, *The Builders of the Bridge: The Story of John Roebling and His Son*, Arno Press, New York, 1972. Building the Brooklyn Bridge (1883) was one of the major systems engineering efforts in the nineteenth-century world. Some others were the creation of the huge ship *Great Eastern* by Isambard Kingdom Brunel (1859); London's *Crystal Palace*, an enormous iron and glass exhibition hall of 1851; and Gustav Eiffel's iron tower (1889) in Paris.

L. Wescott and P. Degen, *Wind and Sand*, Abrams, New York, 1983. (Available, e.g., from National Park Service, Kitty Hawk, NC). One of many descriptions of the Wright's pioneering work, this description of their Kitty Hawk activities from 1900 to 1911 employs their own words and photographs.

System Architecture

2.1 Historic basis of system architecture

We owe to the first system builders, those who constructed the large edifices and memorials of antiquity, the origination of the concepts of architecture. When faced with design and construction tasks requiring the labors of an army of builders, traditional methods of "hand-crafting" or "design-while-build" were not applicable.

2.1.1 Classical traditions

Our knowledge of ancient architecture comes only from those structures which have survived to the present. The oldest such structure on a grand scale is believed to be the pyramid-tomb of King Zoser of Egypt, at Sakkara, built about 3000 B.C. It was an evolution by the architect Imhotep, from the earlier practice of burying important individuals beneath an arrangement of stones or adobe blocks in the shape of a bench—the mastaba. Zoser's tomb was evolved from such a flat-topped pile, into a stepped pyramid (Figure 2.1). A principal architectural element of later pyramids were already evident: cut stone of uniform size.

The pyramids evolved over a few hundred years, first with less, later with greater steepness than in the eventual standard. Steep-sided pyramids were found to be less than completely successful. A large one, built just south of Sakkara, developed cracks during construction and was made less steep from that level to the top. The most familiar pyramids are the three giants at Giza (Figure 2.2), which embody the final design evolution.

Not to be overlooked, in connection with pyramid architecture, is that not only the edifice itself but the *method of building it* comprised a system. Uniform size blocks, each weighing about two tons, were cut from quarries far away, boated on the Nile and sledged to the site. It is believed that an earthen ramp was constructed beyond the extent of the pyramid

Figure 2.1. A photograph of the first stone pyramid built in Egypt, tomb of King Zoser. The designer was Imhotep, also known to history as a physician. Two humans at the base are dwarfed by the 200-foot high structure. (Photo by the author)

Figure 2.2. The Egyptian tomb pyramids at Giza, only examples remaining of the ancient "seven wonders of the world." The Great Pyramid, at extreme right, displays the modular stones from which it was constructed. (Photo by the author)

structure, to carry the upper blocks into position. While there are some blocks which differ, the vast majority of the 2.3 million pieces of stone in Cheops' Great Pyramid were of the standard size—a size which could readily be handled by 40-man teams of off-season farmers who provided labor.

After the pyramids of northern Egypt were built, the royal city was moved to Thebes (Luxor) several hundred miles south. Pyramids were replaced by elaborate excavated and hidden tombs, in unsuccessful attempts to foil grave robbers. Architectural development continued, however, in the form of stone temples in wide variety. Height was achieved using columns of stone. Some of the chief architectural elements were in the shapes of the columns, their use as a framework for tall enclosed spaces, the massive entrance gateways of characteristic sloped-back shape, the use of history-inscribed obelisks as focal ornaments, and multiple identical statues of pharaohs and other deities. Well-preserved ruins of two huge temples and other structures at Luxor, Egypt, reveal the consistency with which architectural developments were applied. The ancient Egyptian monumental architecture provided vast areas of stone which were usually inscribed with historic or religious scenes in low relief.

Fifteen hundred years later, the Greeks reached a high level of architectural development, in construction of temples and monuments, some of which have been well preserved to the present. For apparently the first time, regular "orders" of architecture evolved, known best by the evolving complexity of column capitals: the simple Doric, scroll-like Ionic, and elaborate acanthus-leaved Corinthian. These were accompanied by evolutions elsewhere. Means were invented to lighten the massive weight and hence increase the span of the ceiling blocks. The sloping roofs diverted rain (little of which fell in Egypt) and posed new construction needs. More subtle inventions of the Greeks included *entasis*, the slight curving of columns inward toward the top which gives a somewhat "softer" look and increases monumentality. Building foundations were slightly domed (a few inches in a hundred feet) for visual effect—no doubt adding considerably to implementation problems.

The next noted architects were the Romans. Their construction seems less elegant than that of the Greeks, for they often used stuccoed brick as a readily available substitute for stone.

The ancient Romans added significant inventions: the spherical vaulted dome (Figure 2.3) enabled large open spaces to be spanned by a relatively lightweight roof, albeit one requiring massive walls to sustain the vaults' out-thrusting forces. The *Romanesque*, or circular, arch replaced the earlier stone lintel across uprights, or the earlier corbeled arch which was formed by stacking layers of stones on the walls with ever-smaller gaps until a final layer closed over the top (Figure 2.4). Much later the more graceful Gothic arch and vaults of the same pattern appeared.

Figure 2.3. The Pantheon in Rome, capped by a poured concrete dome. Ihich was erected by the Roman emperor Hadrian in 123 A.D. This building has been the inspiration for many monumental domed structures, down through history. The concrete dome is covered by protective lead sheet. (Photo by the author)

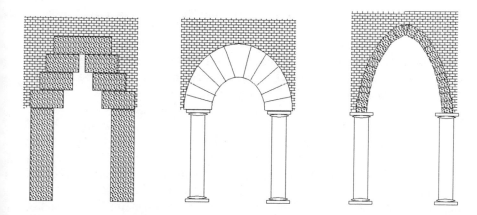

Figure 2.4. Arches have evolved from the corbeled arch at left, which appears in many prehistoric areas including Mayan buildings in Central America. The round Romanesque arch (center) and pointed Gothic arch (right) use the pressure between the blocks to support the arch. Each is also the basis for *vaults*, which are merely arches extended normal to the paper to form roofed areas. Arches and vaults produce large outward forces on their supports which are resisted by the mass of the masonry or by *buttresses*.

Architecturally, the fall of the Roman Empire signaled "dark ages," followed in Europe by the rise of Gothic cathedral construction. These buildings, higher and lighter in weight for their height than any before built, supported outward forces of their ogival-vaulted roofs not by the mass of the highly windowed exterior walls but by extensions referred to as "flying buttresses." That the cathedral architects' work was in part experimental is testified to by the large numbers of collapses which occurred with the more daring designs.

2.1.2 Evolution of architecture through advances in materials and technology

Ancient stone structures achieved stability entirely by the mass of the material itself. Many such structures failed to survive because of the earthquakes prevalent in the circum-Mediterranean region. Mortar, which was known from ancient times, merely improved the bedding of irregular stones on one another. The first great technological advance in stone construction came with the invention of concrete by the Romans, enabling construction of the Pantheon which has survived in almost perfect condition from the second century A.D. Technological progress in permanent construction was made, of course, when the characteristics of vaults and arches were better understood and the Gothic (pointed) arch evolved.

Ignoring the advances made in wooden construction, where the trussed arch evolved along with a system of fitting and doweling wood building frames, the next major advances were made possible by mass-producible cast-iron building members and later rolled steel beams. Some of the early demonstrations of the power of the new material were in the *Eiffel Tower* in Paris, the *Crystal Palace* (a lacy glass-enclosed exposition pavilion which was destroyed in World War II), and the *Great Eastern*, which was the largest ship of its day and the first modern "ocean liner." The massive dome of the U.S. Capitol Building is of cast iron.

Steel, stronger than iron and in the form of continuous drawn wire, made possible longer bridge spans than ever before, through the ingenuity of John A. Roebling who constructed the Brooklyn Bridge and a smaller predecessor. Along with each material-technology advance there was usually required a manufacturing system: forms for the poured concrete domes, smelters and rolling mills for the steel, and enormous foundries for the casting. It should not go unnoticed by the system designer of today that innovations at the architecture level often required changes in materials, production methods, and other contributing technologies.

Perhaps the most ubiquitous technology of modern times is electronics, without which modern communications, control, and computer-based systems simply could not exist.

2.1.3 Modern repetitive or modular building systems

The essence of modern building architecture, more than any other feature, is repetition. Windows, walls, and even entire clusters of buildings often make use of the same repeated elements. This is of course beneficial if mass production methods are available for fabrication. Such repetition also greatly decreases the amount of detailed design since only a single "repeat" of an iterated construction detail needs to be designed. However, the architect must be responsible for seeing that the repeated elements fit together accurately.

Role of architectural standards Architectural standards play an important role in reducing the cost and improving the quality of modern buildings. Many if not most of these standards have evolved in ad hoc fashion, the result of an innovation by one designer whose work was copied by others. There are, for example, standard building-block dimensions, standard window dimensions, and standard wood-trim cross-sectional shapes. Dimensions (such as the 16- or 24-inch spacing between vertical wooden "studs") have become standardized through their adoption in municipal building codes. Innovators continue to eschew standards and, along with many failures, succeed in generating new standards.

Subsystems in building architectures In the earliest constructions, edifices provided little more than a sheltered area with access through doors and sometimes lighting through ceiling or wall areas. Modern buildings contain a wide range of subsystems: water supply and drain; electrical supply and distribution; voice and data communication networks; heating, ventilating, and air conditioning; security alarms; and more. Traditionally, each subsystem has been designed and installed, and is maintained, with little or no relationship to other subsystems. Electrical supply and distribution is the most significant exception, since it must provide power for certain other subsystems.

In very few buildings are these subsystems intimately connected with the architecture. Water pipe and electrical conduit runs are located where found most convenient at installation time, within the inevitable hollow wasted spaces of most structures. In recent years, some innovations have been introduced such as reliance on waste heat from lighting as part of the heating resources of a building, incorporation of solar energy collectors in roof or wall surfaces, or unified control of energy use.

2.2 Computer architecture

Though building architects over the ages have pioneered many generic and lasting architectural principles, within the past 40 years the architects and designers of computer systems have carried the arts and sciences of system integration to high levels.

That computers are complex systems there is no question. Even the least expensive of personal computers comprises tens of thousands of identifiable parts, each with a specific function. The largest present-day computers are much smaller than those of a decade or two ago because technology has enabled miniaturization. This is fortunate in a second respect. Physically larger computers have larger internal time delay than smaller ones, because the speed of signals moving through the computer cannot exceed the velocity of light. One can confidently predict that future supercomputers will be smaller than today's.

While some minor portion of the architecture of a computer is connected with the structures housing it, most deals with functional operations and the interfaces by which it communicates internally or with peripheral units and external systems.

2.2.1 Central and peripheral concepts: von Neumann's stored-program architecture

Soon after the emergence of the first electronic computers, attention was directed largely to what was called the central processor. Though in today's personal computers the corresponding portion is always on a single integrated circuit chip, this structure is and was the most important portion of the computer system. In the initial description of a stored-program computer by Dr. John von Neumann, there were five major elements: arithmetic and logical unit, control, memory, input subsystem, and output subsystem. The first two, or sometimes the first three of these make up the central processor, and the others are referred to as peripherals. Later peripheral additions have included storage and display/control subsystems. At present, most computer systems also have extensive connections to outside users and to other computer systems through a communications subsystem. Accordingly, many or most computer systems can be described by the block diagram of Figure 2.5. (A *block diagram* is typically a symbolic representation, at some selected level of detail, of a system's parts and their interconnections. The outlines of blocks may, in some variations, be given shapes or sizes indicative of their role or importance of the function. Likewise interconnections may be represented by narrow or broad lines to represent single or parallel connections. Block diagrams containing more than 10 to 20 blocks lose tutorial impact but are of value in operating and maintenance manuals.)

This figure, describing a major subdivision of parts and/or function, already portrays an architecture, the von Neumann architecture. A single memory holds both data and programs, a single arithmetic and logical unit performs all manipulations on data (and perhaps on memory address

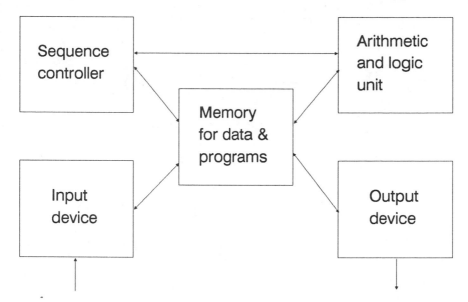

Figure 2.5. John von Neumann's concept of a *stored program computer* has influenced the architecture of most computers designed since 1945.

values as well). Input and output communicate traditionally with the computational unit; however, in most modern computers there is an alternative route into or out of the memory (direct memory access).

From the standpoint of those who will write the first programs for use by such a computer, the most important aspect of its architecture is implicit in the computing unit and control. The term *instruction set architecture* is intended to describe the aspects of the machine with which the programmer communicates. To understand this, a more detailed view (Figure 2.6) is required. The computational unit consists of a group of small memory units called registers plus a binary adder. Each register can contain one or perhaps two data units of the size contained in the larger memory. Since binary logical operations are logically very similar to addition or subtraction, the adder doubles as a "logic unit" by such artifices as disconnecting the "carry bit" so that the addition becomes an exclusive OR operation.

The set of registers is central to the programming of most computers. Whereas the memory is relatively remote and slow of access, the registers are nearby and much speedier. Most computers set aside certain registers for certain purposes. We describe here the architecture of the Intel 8080 microprocessor. This first commercial 8-bit microprocessor was the direct progenitor of 8088s, 80286s, 80486s, and other Intel processors which, along with Motorola's 6800, 68000, 68020, 68030, etc., dominate the personal computer market. Its architecture (Figure 2 .7) is

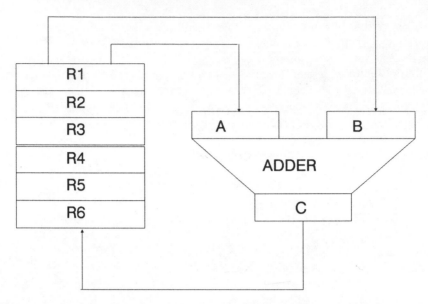

Figure 2.6. Typical arithmetic and logical unit (ALU) of an electronic digital computer. The same circuitry performs both arithmetic (+ , –) and boolean (AND, OR, NOT, XOR) functions.

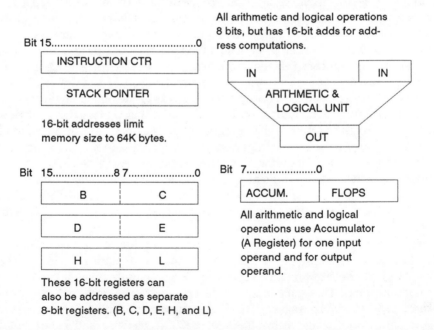

Figure 2.7. Register architecture of the Intel 8080 8-bit microprocessor chip. Although this chip might be considered an historic relic, its design strongly influenced later chips produced by the same firm.

typical of that of 8-bit "engines." Architecture of 16- or 32-bit processors is derived from 8-bit processors, principally by doubling, and doubling again, the bit lengths of their registers.

Register A (which holds an 8-bit value) has a traditional role as the *accumulator*. Results of all additions, subtractions, and logical operations are placed in it by the *adder*. It likewise must hold the input operand of every one-operand operation, or one of the inputs of a two-operand operation. The other 8-bit registers B, C, D, E, H, and L, lack the accumulator function, though each plays special roles in certain computer instructions. What's more, B-C, D-E and H-L, for certain instructions, are treated as *16-bit registers*, the lower 8 bits being in C, E, or L. If one needs to transfer a string of ten 8-bit values from one location (say, starting at address 1234) to another (say, starting at 2345), one first places the address 1234 in the "HL register" using a single instruction, puts the address 2345 in the DE register using another instruction, likewise the length (10) in BC. A single instruction (which the programmer refers to as LDIR, if writing a primitive assembly language program) effects the 10-byte transfer, which copies the data, leaves the next following addresses (1244 and 2355) in HL and DE, and zero in BC.

An important characteristic of the 8080 and other microprocessors is that they are *single memory address* computers. That is, no instruction contains more than one 16-bit memory address value. When registers are addressed, instead of memory, only *3* bits are required to refer to any of the (7) addressable 8-bit registers. Thus it is possible for a single instruction only 8 bits in length to specify a move of data, from any one of the seven 8-bit registers, to any other. If memory is to be addressed, the 8-bit op-code is followed by a 16-bit memory location value. It certainly would be feasible to provide in an 8-bit op-code a set of 3 bits defining a particular one of the 7 registers to (or from) which the memory word is to be transferred. However, the 8080 allows transfers only between memory and the A register—a limitation in circuit complexity imposed by the technology prevalent when it was developed. (The term *word*, in early computers, always referred to the fixed length of instructions and all registers. It now most often refers to the largest-size data unit which can be moved between memory and registers by a single LOAD or STORE instruction. Modern computers may have instructions whose length is many "words," may be capable of transferring different-size quantities of data in a memory transfer, and have registers which can be ganged to hold longer data units. In short, "word length" has lost much of its importance as the measure of computer capability.)

Three other registers usually play unique roles in the operation of most processors: the *instruction pointer*, which contains the memory address of the next instruction due for execution, the *stack pointer*, which holds the address of a special location in memory, and the *flag register* which contains a number of 1-bit status indicators or "flags," whose states are set by the operation of the ALU and the value of the result it produces during an arithmetic or logical operation.

The instruction pointer (IP) is the key to moving around in the program. If the programmer wants to switch to another location in the program, he or she includes a branch or jump instruction, the only effect of which is to place in the instruction pointer the address of the desired location. When the next instruction is due for execution, it is "fetched" from that location. The IP is like the other registers in most ways, but it is automatically incremented in value after each portion of the instruction is brought (to the control) from memory.

The stack pointer (SP) is unique also because of the way it is used. Certain instructions, usually referred to as PUSH and POP, cause data to be transferred between some (other) specified register and memory. A PUSH causes the SP to be decremented and the contents of the specified register to be transferred to the memory location to which it then points. A POP causes the contents of the memory location to which SP is "pointing" to be transferred to the designated register and the SP to be incremented. This procedure allows a series of data in registers to be "saved" (PUSHed) into separate memory locations, without need to define each memory location, then returned (POPped) later to desired registers. The effect is identical to the operation of a stack of data receptacles with the scheme that the last data in is the first data out (LIFO stack).

Stack implementation is also invaluable in operations which temporarily transfer processor operation to a program subroutine. A CALL instruction causes the address of the instruction following the CALL to be placed on the stack. At the end of the subroutine, the RETurn instruction causes the value "on top of" the stack to be placed in the instruction pointer, and the program returns to its original course. This stack concept, now used even in very small computers, was a relatively late arrival in processor architecture. So profound was its effect that, for some years following its introduction by Burroughs in its 5500 computer, similar designs were referred to as stack architectures.

The status register (sometimes termed the flag register or the *program status word*) contains a variety of 1-bit data items associated with the operation of the ALU. If, for example, an addition or subtraction causes a carry bit to be produced at the most significant bit of the output, this is transferred to the carry flag. It can thus be used to extend the size of numbers the processor can add, the carry from one such addition being added as a unit increment to one of the next-most-significant operand data. Many other "flags" can be defined. A common feature is the zero flag or zero flop, which signifies that the most recent arithmetic or logical operation produced a zero result. (The term flop, which is itself a contraction of the name of a device, flip-flop, can refer to any computer circuit which will remain in a binary 1 or 0 condition until it is reset.) If the high-order bit of the result is a binary 1, the sign flop will be set, for it is common particularly in small processors to represent negative integers using those bit combinations whose high bit is 1. Flags may in addition indicate when the processor is operating in a special mode; for example, in some processors the supervisory mode allows additional instructions to be executed that would not otherwise be accessible. (This is characteristically reserved for use by operating system software, and not by application programs.)

Microprogrammable processors, introduced in the 1960s, added programmability to the control unit—not for the computer users but for its designers. The *microprogram* is no more than a further breakdown of the (previously) "hard-wired" sequence of actions carried out by the control unit in executing a computer instruction. This makes it feasible to change not only the instruction set but also the definitions of many of the registers—without hardware changes. This feature can enable a processor to *emulate* another one having different register usage and a different instruction set. Though a majority of recently designed computers are microprogrammed, this feature is generally hidden from everyone except the computer's designers.

So much for the internals of the processor. What about the important input and output? You may be surprised to learn that a processor's view of input and output is a relatively primitive one. In most cases it simply accepts a data unit (which is commonly of 8 bits or more) from one of a number of *input ports*, or likewise transmits such a data unit to some *output port*, which may be the same port as the output port, but serving in the opposite role. The mechanics of manipulation of the devices attached to the port are matters for the software (to deal with), not the processor hardware. The only exception may be a direct memory access (DMA) "channel," a simple subsystem which can transfer a requested quantity of data units between the external input or output devices and a group of adjacent memory locations. Typically, the starting address and length (in whatever size of data unit is transferred to or from memory) are transferred to the DMA circuits, which contain registers to hold them. The actual transfers take place by "cycle steal," or by briefly inhibiting the processor's access to the memory while the I/O device takes its turn. Since the processor usually generates transfers to or from memory at less than one-half the rate at which the memory is able to service these requests, the DMA scheme impedes the processor operations but little.

One final architectural element found on most modern processors is an *interrupt* scheme. This is an arrangement whereby an input or output device needing attention by the processor may signal its needs. Upon completing the instruction then in process, the processor usually saves only the status register and the address of the next instruction and jumps to an address whose value was earlier stored in the I/O device. At that address is an *interrupt handler* responsible for saving (e.g., on the stack) the values of any registers it uses and restoring them after completion of its work. With an interrupt scheme, combined with DMA, a processor has many different ways to handle input or output: by having the program poll the input or output device(s) periodically to determine their needs for service; by use of the DMA; or by having input or output devices interrupt program operations and use a separate program subroutine to "process" the interrupt.

This level of description of a computer is usually the *most* detailed level which could be characterized as architecture and of interest to users. The essential lower-level detail will generally be referred to as design features, understandable mainly by circuit designers.

There are many essential elements of a computer design which do not relate directly to programming the computer. For example, some computers use a single, highly paralleled, set of data- and signal-carrying wires called a bus as a means to interconnect the processor, memory, input/output, display, and storage devices. A computer relying on this unifying element may be referred to as having *bus* (or single-bus) *architecture*.

If a computer is designed to use a small, high-speed memory along with a larger slower memory, so that large segments of stored data and program are transferred into the fast memory when any part of a segment is requested, it is said to have virtual memory architecture. The small high-speed memory is referred to as a *cache*; hence this architecture is sometimes referred to as cache architecture. The actual or *physical* memory is much smaller than the addressable memory size, or *logical* memory. All of memory is logically divided into large blocks, which are referred to as *pages*. A particular program may require one or more of these pages to be present in *physical* memory in order to be run. If a program of higher priority than one currently stored in the physical memory is able to run, often as a result of having received new inputs, one or more pages of the lower-priority program may be "swapped out" and the higher-priority program run. If physical memory contains P pages and on average each program requires N pages to execute, there could concurrently be under execution in the machine only about P/N programs, even though the machine might have several processors sharing the memory and themselves able to execute programs.

Computer speed can be increased if following instructions are "fetched" from memory to the processor while the current instruction is being executed, particularly if the memory's data rate is high. This may be referred to as instruction pre-fetch architecture and is common today, even in some microcomputers.

Our purpose in these brief descriptions is primarily to disclose features of the sort to which the term architecture has been applied. Rather than being small changes in arrangement, each is capable of producing a significant improvement in computer capability. It is important—and interesting—to note that the enormous reductions in computer cost and increases in performance which have been achieved through advances in semiconductor circuits are usually viewed as *technological*, rather than architectural, advances. Even the replacement of large, low-speed, and costly magnetic memory devices by semiconductors, which took place in the early 1970s, was a technological, rather than architectural advance.

2.2.2 Multiprocessors and other non-von Neumann architectures

Computer systems are excellent, if somewhat specialized, examples for study of system architecture. The conventional (von Neumann) architecture has been expanded upon in a variety of ways, none of which has as yet achieved as broad an area of application.

Since a von Neumann computer is capable, in principle, of dealing with any problem which can be programmed, the usual basis for architectural innovation is the desire to solve some problem, or class of problems, more rapidly, or less expensively. Throughout the history of computer evolution, technological improvements (of semiconductor circuits, for a large part, but also of magnetic disk storage density and speed) have contributed most to the factor of nearly 1000 in performance improvements achieved. If at some given stage of technological development, however, a computer designer needed to build a machine which outperformed others having similar component technology, the sole route available was architectural change.

The first significant architectural change was the shared-memory multiprocessor, where the single (ALU and control) "processor" in the von Neumann structure is replaced by two, four, or more such units. If all processors have access to a common memory, then within the capability of that memory to serve all their needs concurrently, a multiprocessor architecture can handle two, four, or more different programs simultaneously. To achieve the high memory access rates required, it is common to use an interleaved memory in which the two or more least-significant bits of the address refer to independent memory "banks." The advantages of an n-way multiprocessor over n independent machines are (1) that programs requiring large amounts of memory can be executed along with others requiring small amounts of memory, and (2) that peripheral input-output devices can be shared by (or share) all of the processors.

Multiprocessors allow several different programs to execute concurrently. They will allow one program which requires a large amount of memory to access most of the memory normally shared between all executing programs. They do *not*, however, solve the problem posed by a program which requires very long times to execute on a conventional computer.

One direction of architectural evolution toward the solution of that problem was through computers able to perform *vector operations* at high speed—that is, to repeat one arithmetic or logical combination, for a series of data pairs. "Pipelined" architectures represent a popular approach to vector processing. The single adder of a traditional computer must be recycled repeatedly, even when merely multiplying two numbers. Each *partial product* is added to the sum of previously computed partial products. A pipelined vector ALU typically contains *many adders*. The first of these handles what would be the first addition, in a multiplication sequence carried out by a traditional ALU. The next adder handles the second addition for a first pair of operands to be multiplied, while the first adder is handling the first step for a *second* pair of operands. By the time the first pair of numbers reach the far end of the pipeline, their product has been determined. Thus a process which would ordinarily be repeated, using one adder circuit, is carried out concurrently, in the pipelined processor.

Were there only two numbers to be multiplied together, a practical pipelined (vector) processor will often be even slower than a conventional unit-processor because of additional preparatory ("overhead") operations. If there are 20, 40, or 100 sets of numbers to be multiplied together, then all of the adder circuits in the pipeline processor will be simultaneously busy, and for an extended period. The first product may take longer to emerge than in a conventional machine, but following ones could arrive several times more rapidly than in a single-adder system. Control Data Corporation, and more recently Cray Computer Corporation, led in the early development of computers with vector processing architecture, in most cases embedded in a multiprocessor arrangement containing one or a few vector processors, plus a few conventional "processing engines."

These architectural variants have for the most part been found in *general purpose computers*, that is, in computers capable of efficient performance performing almost any kind of problem. Where a computer is required to handle only a single problem—as is the case in many military systems, for example—*application specific architectures* have been developed. They have the objectives of achieving substantially greater processing power, memory, and/or input data rates.

One large class of architectures is referred to as parallel processors. A great variety of structural arrangements of memory banks, processors, and interconnection schemes have been proposed over the years, and a relatively limited number of designs have actually been built. Still fewer have been built in quantity, except for special-purpose military *signal processors.* (A signal processor receives inputs from one or more sources of digital signals, and processes inputs, usually in "real time," to produce a desired output. Examples are found in modern radar and sonar processing.) Parallel processors are usually designed to divide portions of a single program into parts which can be executed simultaneously. Since there is frequently little *inherent* parallelism in *most* parts of any particular problem, the task of creating parallelism usually falls to the programmer, or sometimes to language translation software which converts users' programs to executable form. A highly paralleled processor might contain 16, 256, or more processors, each with its own memory and perhaps access to a common memory. A *control processor* able to communicate rapidly with the other processors may be included in the architecture. Given sufficient economic incentive—and talented programmers—such a system can achieve great execution speeds. With the advent of powerful single-chip processors such as (at time of writing) the Intel 80386 and Motorola 68030, parallel-processor arrays can be built for affordable costs. No doubt the search for more broadly useful parallel processor architectures will continue. More onus is, however, needed on *software* which can make effective use of large and costly processor arrays for solving large problems.

Very high speed computers have thus far been developed for one principal type of application: a single problem which would require prohibitive amounts of time to process, on other available computers. These processors have been very expensive, reflecting the special components and design requirements as well as the small market (until recently, mostly governmental).

Some computer experts believe that many, perhaps most, future high-speed computers will take the form of compact groupings of microprocessors and be far more affordable than traditional supercomputers. Their broad-scale realization, however, will also require revolutions in software, to handle effectively a breadth of work including:

1. *logically independent tasks*, uncoupled operations in which each processor executes a separate program using unique input data
2. *parallel tasks*, tightly coupled operation in which a single application utilizes all processors at the same time on different components of common input data
3. *eclectic tasks* which may change rapidly between coupled and uncoupled processor operations

Some parallel-processor arrays, in particular, those designed to solve specific problems, have structures which reflect their applications. For example, NASA's *Massively Parallel Processor* (Figure 2.8), a single-instruction stream, multiple-data-stream (SIMD) architecture, has processors arranged in a rectangular array. Each processor/memory unit is connected directly with four neighboring processors. This structure is

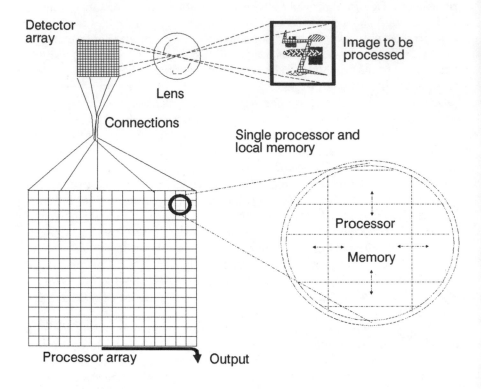

Figure 2.8. Concept of massively parallel processor, built for NASA image processing. The planar array of many thousands of processors has 1-bit word length.

intended for manipulation of two-dimensional imagery, each processor, for example, receiving as its input a photodetector signal which indicates if the intensity of a particular rectangular picture element (pixel).

A form of multiprocessor array which has recently been made commercially available using powerful microprocessors is termed a *Hypercube* structure. As depicted in Figure 2.9, each of 2^n processors and its associated memory can be envisioned as having a location in an n-space. It is connected directly only to $n - 1$ "closest" neighbors, those whose binary addresses differ from its own by a single bit. Connection to additional processors is by relay processor-to-processor, a relatively slow process. Physical arrangement of processors forming this logical array is conventional, processors being mounted on printed-circuit cards. To improve responsiveness and achieve overall array control, a single additional processor may be provided, with direct connection to each of the processors in the Hypercube array. Though in principle possessing great generality, present multiple/parallel processors of this kind place heavy responsibility for efficient use of the array on programmer or user.

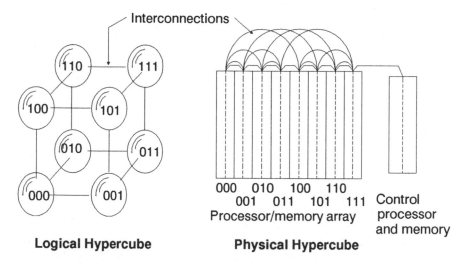

Figure 2.9. The *hypercube* connectivity concept for multiprocessors. Each processor is logically located at a binary intersection in an *n*-space containing 2^n processors.

Yet another class of parallelism is represented by *data flow processor* architecture. Usually intended for signal processing applications on real-time data, the overall signal transformation is subdivided into small parts, each being carried out in a separate processor, as the signal finds its way to the output (Figure 2.10). Some data flow machines actually place a general-purpose processor at each node and can thus perform complex tasks in each step. Like all other paralleled structures, data flow computers rely heavily on programming designed to make best utilization of their special capabilities.

2.2.3 Software architecture

Within its computing engine, every computer contains some form of instruction pointer, arithmetic/logic unit, registers, and memory. In the earliest digital computers, these were manipulated directly by binary programs loaded directly into memory. Most modern computers, however, make continual use of a number of programs, collectively referred to as the *operating system*. Their function is to simplify the use of the computer and to improve its efficiency in performing work required by its users. Because these programs are so critical to effective use of modern computers, they add a significant and growing share of the architectural elements which make up the computer system architecture. Software programs, even operating systems, do not constitute proper subsystems of the computer system. They are in a way *connective*

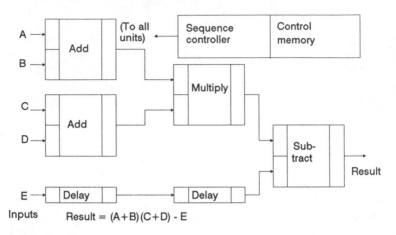

Figure 2.10. A data flow computer concept in which each primitive processor performs only one arithmetic operation. Data streams from left to right through the machine.

elements, somewhat like glue or welds in a physical structure, or the score played by members of a symphony orchestra.

Operating system programs serve: (a) to communicate between the processor and peripherals (storage, input, and output) units, and (b) to manage the computer resources when several different program tasks are concurrently being executed. The programmer developing application programs is thereby relieved of the responsibility of understanding details of how to read data from storage disks, write characters to a screen or printer, allocate memory, and other complex and frequent tasks. A part of the operating system has responsibility for loading application programs into memory, including program modifications required in light of the memory locations available for it.

The mechanisms whereby an operating system's services are invoked are central to the computer system's architecture. Where an operating system allows several programs to be executing at once (referred to as *multi-tasking*), it may switch operation periodically, from one program to another. Some operating systems provide a *priority* scheme for this assignment. The operating system is responsible for assigning available memory and other system resources to programs requesting them. An application programmer's view of the system architecture is thus very much influenced by the operating system's presence and features. In fact an operating system may be capable of *emulation*, making one hardware system resemble another which is physically quite different.

Since programs are able to make a computer system resemble another, does *architecture* lose meaning when applied to programmable systems? No, but one can envision a computer system to consist of a hardware core which is modified by successive *layers* of programming which alter the user's perception of the system as viewed from outside each successive layer (Figure 2.11).

A computer system feature of growing commercial importance is the ability to connect independent computer systems to form a *network*. Computer networks are distinguished in terms of the "tightness" of their coupling. The most tightly coupled of all are parallel multiprocessors, whose coupling is accomplished in large part through hardware. (For these the term *network*, though appropriate, is seldom used today.) In most other computer networks, coupling is implemented largely via software, though with hardware supporting high-speed intercomputer communications. In Local Area Networks (LANs), coupling between computers is effected by modifying the operating system, in particular by intercepting requests applications make to the individual computer's operating systems. For example, a request to a computer's operating system to read a file physically stored in another computer's storage is intercepted by LAN software, and a request made over the network. The data sent in response is inserted into the requesting computer's memory just as if it had arrived from the local disk storage. Data transfer may be at 1 million bits per second, or faster. The user and the application program need not be aware of network operation. If, however, an application program bypassed the operating system to read data *directly* from a local disk store, this "violation" would make it impossible to operate the program on the network.

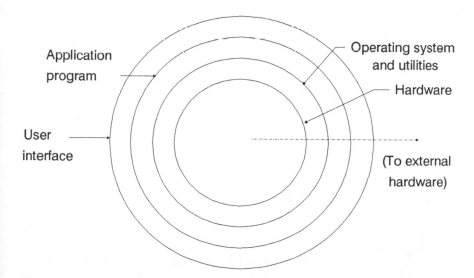

Figure 2.11. A computer system may be compared to an onion, the analogy suggesting that the system has many levels on which action occurs, each at a particular level of control.

The most common networks are less formal (i.e., "loosely coupled") ones, operated directly by user application "communication programs"; these use data transfer rates compatible with telephone-line communication (50,000 bits per second or less, more typically 9600 or 2400 bits per second). In most cases the network is not maintained continuously but only when data transfer is required. Users must initiate network requests themselves or obtain application programs which can access communications lines and achieve connection with a remote computer.

In any network situation, the system may contain several or many independent computer systems. Each can be aware of the others *only* through its communications with them. Correct network operation is critically dependent on logical and consistent definition of the signaling and data-transfer schemes connecting the networked processors. A desirable objective is that a computer network be able to securely, safely, and reliably connect itself to other networks. These difficult objectives can be attacked mainly through software architecture and design, aided by special hardware. Network architecture of all types and all levels (e.g., handling image information) is currently a most significant architectural frontier in computer systems.

2.2.4 The essentials of computer architectures

It should be evident, after our discussion of building and computer architectures, that each class of systems possesses a characteristic assortment of architectural elements. These may, for example, include: characteristic or standardized structural features, modes of operation, interfaces, appearance, and limitations.

In the case of computer architectures, the architectural features of interest to hardware integrators may relate to *electrical* interfaces. Those most important to operating system designers are the primitive instructions obeyed by the hardware, and the interface they design, between the operating system and application programs. Those merely operating a computer via a user-friendly application program should not require knowledge of the computer's design.

The internal, or instruction set processor architecture, as we have seen, is usually based, even if loosely, on the von Neumann model. In detail, it relates to the memory, general-purpose registers, and special registers, to the instruction set and other modes of operation such as interrupt handling. The external view may relate more to the forms which data entering the system would normally take in given applications and the source of that data (terminals, communications lines, sensors, etc.). Alphanumeric (text) data may, for example, be expressed using the ASCII symbol set, or the IBM-defined EBCDIC. (ASCII represents *American Standard Code for Information Interchange*, while EBCDIC means *Extended Binary Coded Decimal Interchange Code*. ASCII is based on 127 different 7-bit symbols, and EBCDIC on 256 8-bit symbols.)

For non-von Neumann computers, additional architectural elements are needed to distinguish different architectures. These features relate to the ways parallelism or concurrent processing are accomplished and how programs and data are moved into and out of and are stored in the system.

While the layperson's view of a computer has traditionally been that of a data manipulator or some sort of "synthetic brain," computer architects and operating system programmers see computers principally as data *movers* and arrangers. A significant advance in computer architecture, in general, must address movement of data, and the interaction of that data with programs in an improved manner. *Users* of applications programs, on the other hand, usually envision the computer system as a problem solver. To them, improvements must be ways to satisfy their needs more quickly or reliably, or to simplify and speed their interactions with the computer system. Technological developments which are exciting to experts in one of these communities may be of much less interest in the other.

2.3 Architectures of familiar systems

2.3.1 A simple example: The can (food tin) opener

Man-made architectures can be identified not only in monumental constructions such as pyramids but also in simple systems we use every day. The familiar utensils which we use to open food tins demonstrate a modest but remarkably varied architectural evolution. Products of different manufacturers, though not identical, bear strong resemblances and are based on common operating principles (Figure 2.12). At one end of the spectrum of products serving this function are one-piece manual openers. Their historic derivation was, no doubt, a knife-blade, which with skill (though with certain risks to its user) can be used to slice through the thin steel of a "can."

This kitchen appliance family appeared soon after the invention of food preservation by canning in the early nineteenth century. Over time it became more specialized, more efficient, safer to use, and cheaper through volume production. It evolved a levering member which could be rested against the edge of the tin, and then replaced the conventional knife blade with a short, heavy, crudely sharp piece of metal, equipped with a pointed end for the essential initial piercing action.

When the opportunity became available to electrify this by-then essential piece of kitchen apparatus, the manual design (as has often been true at the onset of automation) seemed poorly suited to automation because

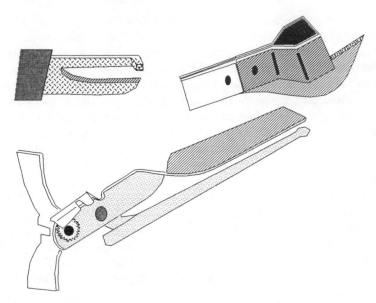

Figure 2.12. A pictorial history of the manual can-opener in the United States. Top left, a knife blade opener; top right, the most common manual opener in the 1930s; bottom, a more modern manual opener.

of the reciprocating motion it required. Though there is some electrified equipment whose motion is reciprocal (rather than rotary), mimicking the up-and-down motion used with the manual opener would have led to a formidable device, probably of questionable safety.

An architectural departure then took place in which a rotating, sharpened circular disk sliced through the top of the tin at its circular edge. New solutions were needed for other major "system functions": A lever-operated clamp, rather than the user's hand, held the tin tightly. Cutting power was provided by a motor-driven toothed wheel which pressed against the thickened raised edge of the tin, and rotated both tin and contents past the cutting wheel so that the latter did not need to be powered. Later a permanent magnet was added, positioned so that it could both capture and retain the cut-away top of the can. This configuration can now be said to represent an architecture; its principles have been repeated in almost every electricity powered can opener ever manufactured.

Another manually operated class of designs later became popular, however, and it owed its origins to the electric models. The toothed wheel of the electric opener was retained as a hand-turned key, the circular cutting disk was replaced by a simple blade (perhaps to guarantee rapid wearout and sale of replacements). Its price range could be well below that of motorized openers, while its ease of use and, perhaps most important, its safety far exceeded those of earlier manual openers.

The truly curious reader will wonder if there have been any *successful* significant variants to these basic architectures. We know of none which have taken hold in the United States. However, in the mid 1980s a manual opener was sold briefly which, rather than cutting through the *top* of a tin, cut the cylindrical side-wall, removing both the top and the raised edge or flange. Our urgent efforts to purchase one (at a price $12.95, for which one could at the time have purchased the standard electrified version) failed when the design quickly disappeared from sale within weeks. We suspect that its apparent market failure was due to the safety problem posed by the sharp edges it left on the tin.

This diversion into home appliances was intended to show that many common objects possess distinctive architectural features. Natural systems, represented by plants, animals, ecosystems, solar systems, and galaxies have architectures. Perhaps the most important of these, in the largest sense, is galactic architecture (whose characteristics follow at least in part from the physical *laws of motion*), and the *double-helix* molecules found in living cells. Many natural architectures have evolved significantly, while maintaining a common basic form. Skeletons of higher animals, even a million years in the past, reveal strong architectural relationships to today's species, while the propulsion mechanisms of ancient and modern sea life feature the same architectural elements (in terms of function, structure, and arrangement).

2.3.2 Top-down system design

Most large systems, especially ones owned or controlled by separate organizations, are now designed by what is known as a *top-down* design procedure, which involves:

- Establishing a set of overall requirements or objectives
- Dividing (decomposing) these formally, to define a set of parts or subsystems
- Purchasing or building the subsystems
- Assembling (integrating) them correctly

If this process is well planned and well implemented, subsystem design and production may be assigned to independent sources, with good assurance that subsystems will work together as intended.

This top-down system design technique was used in design of the U.S. interstate highway system, begun in the 1950s. The federal government established architectural criteria or design rules governing maximum allowable percent grade, minimum acceptable radius of curvature, lane dimensions, and other details—including even detailed specifications for information signs. Compliance of state road-building agencies with these rules and with a master plan were made conditions for states to receive federal subsidy payments, derived from taxes on gasoline, which typically covered 90 percent of highway costs. Without this fiscal leverage and an overall plan, it would have been difficult to ensure that the highways within one state connected with those in adjacent states. The architectural elements included cloverleaf intersections, medians, and breakdown "shoulders" which were not found in the earliest high-speed highways that were built in Germany and the United States in the 1930s.

2.3.3   Major classes of man-made systems

Our primary interest in this book is in *man-made systems*. These range from such items as food processors to global communication systems, or transcontinental highway systems, or orchestral music notation. All of them require design efforts, though of widely different levels and with exercise of widely varied skills.

We will acknowledge that systems which are no more than *methods*, such as rules for committee procedures, techniques for sewing clothes, or even software programs, could correctly be characterized as systems. Our emphasis in this book, however, will be largely on *physical systems*, or systems which include at least some physical elements. Systems containing large, complex, or heterogeneous combinations of elements pose more difficult problems for the designer and more interesting examples for us. Perhaps most importantly, they emphasize the multidisciplinary nature of systems architecture and design.

There are now enormously many classes of man-made systems. In dealing with nature's systems, humans take the role of observer. Thus, emphasis in natural science has been heavily on classification of animals, plants, and natural phenomena. However, for man-made systems there is relatively little interest in rigorously classifying types of systems, or even for developing *bases* for classifying them.

A particular system's class, functions, physical scale, or complexity may serve to distinguish it from other systems. While both a child's bicycle and a racing motorcycle could be said to be *bicycles*, the two classes are quite distinct. What's more, one type of *complex* system seldom serves as a useful analog or a surrogate for another. We can, for example, infer little or nothing about computer architecture by studying automobiles or human brains. Nor do toy trains reveal much about the operation of full-scale railroads. A simple system can seldom serve as a useful model for a far more complex one. However, a complex system can often be well characterized as being made up of a set of interacting, simpler constituents: subsystems.

One of the generally accepted bases for classification of systems is in terms of *portability*. Some systems are *fixed*, others *movable*, others *portable*, and still others may be described as *self-propelled*. Unless one's interests are in systems comprising *vehicles* (automobiles, aircraft, trains, spacecraft, missiles), this attribute may not be the central one; however, designers learn repeatedly that it is one thing to design a system which will complete its useful life in an air-conditioned building, but yet another to design a system with identical functions and performance but which is to operate in a moving vehicle in all weather.

Probably the largest class of man-made systems can pragmatically be described as "black boxes" which perform some function. They are basically one-piece and not particularly characterized by their physical motion or geographic extension. Those "box" systems which interact importantly with their environment often contain sensors to input from it, or transducers to output force, motion, or energy to it. In this large class are, or course, telescopes and television sets, most home appliances, business machines, and much factory production machinery. One might classify a subset of these as *converters* which take in materials or a form of energy and convert it to another. We shall have no further use for such taxonomies and will leave their definition to those with less important work to do.

A second large class of man-made systems are notable for their *physical extension*. Telephone, highway, and rail systems are important examples. These are classified as *networks*, and are discussed in the next section.

The third major class of systems of great interest to designers are *vehicles* of all sorts. These have special requirements because of their motion and independence and because they are called upon to perform well recognized functions. This class is discussed in section 2.3.5.

2.3.4 Transportation, transmission, and distribution networks

Highway systems are only one example of networks. Because they are familiar, they seem to have relatively few complexities of a *system* nature; however, as with other systems, closer looks reveal more and more subtleties. Although highways can be used by almost any ground vehicle, rail networks place requirements on rolling stock (passenger and freight cars), which must fit the track and maintain safe clearances. Additional constraints arise from the need to couple them into long trains to be propelled (and stopped) by one or more locomotive units. Rail networks, importantly, include built-in control subsystems for avoidance of collisions. Unlike automobiles, trains cannot merely pull off the road in case of failure. Unused rolling stock must be stored on tracks which are connected to the rest of the network.

Highways and railway systems are only two members of a larger class of networks by which we transport, distribute, or collect physical objects or materials. This class also includes water supply and sewage systems, heating-gas distribution networks, natural gas pipelines, and large-diameter pipelines for transporting solid materials in liquid as a slurry.

Some of these are one-way networks, for distribution (e.g., for heating gas or drinking water) or collection of sewage and storm runoff water. One-way networks typically possess the form referred to as a "tree" (Figure 2.13). Higher animals' blood-circulation systems comprise a supply system (the arteries) and a return system (the veins), connected by the capillaries and the heart/lung subsystem into a continuous loop which visits near each cell in the body.

Mass-transporting networks require some form of fixed or traveling motive power. This ranges from the engines of vehicles on a highway or railway, to the pair of pumps in a single enclosure in a mammal's heart, or to pumps distributed throughout the network, as they are in transmission pipelines. Many mass transportation systems also provide some type of storage subsystem. These include elevated tanks or standpipes in city water systems; railroad yards, garages, and highway rest areas perform similar functions. Their purpose is sometimes to ensure a continuous flow, or merely to avoid need for vehicles to remain constantly in motion. Depots, pumping stations, storage tanks, and similar localized resources on a network are often referred to as *nodes*, a term of graph theory. This term may also be applied to junctions at which routes in a network coalesce, cross, or branch.

Even simple nodes such as road crossings or branchings, require design which is distinct from that of mere network paths (*arcs*, in graph terminology), and may have special architectural features of their own. A railroad

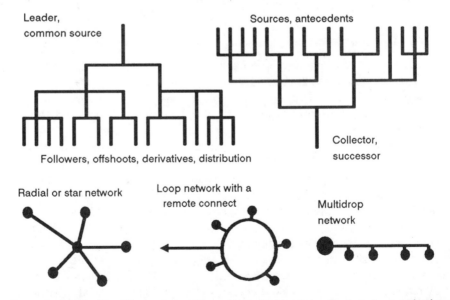

Figure 2.13. The "tree" is the shape of many networks in nature, of corporate organization charts, and of water and sewage systems. Radial, loop, and multidrop network forms (below) are common in wired communications.

"switch," which enables selection of one of two alternative paths, is designed (Figure 2.14) so that a train arriving from the unselected one of the two branches will shift the switch to its branch (rather than falling off the track). Branching strategies (such as the two-way rail switch) are often important features, or limitations, of networks. Though highway lane-switchings might appear to present more than the two-way alternatives of railway switches, a vehicle must occupy each of several lanes in some particular order, to select some particular route. *Crossings* in mass-carrying systems may represent network branches in close proximity without intended interference or interconnection, e.g., if one network path passes over another. However, in many situations this requires an overpass or some other specific structure. If such a crossing does involve physical interconnection, or the the paths must share common space as at highway intersections, some form of "traffic control" is often needed to avoid unintended collisions. In highway networks, these will take the form of stop signs or traffic lights.

Many modern networks carry *energy* in the form of electricity, or *information* as telegraph, telephone, radio, or television signals. Among these applications there are many *distribution* networks (as for electric power or cable TV) but fewer *collection* networks. Typical communication networks may be in loop, radial, or multidrop forms (Figure 2.13). Information signals are useless when many are merely added together, but there are *multiplexing* techniques whereby many low-data-rate signals may be collected and passed through a high-data-rate network.

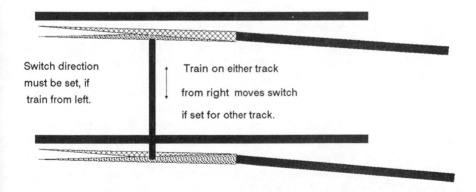

Switch direction must be set, if train from left.

Train on either track from right moves switch if set for other track.

Figure 2.14. A railroad switch. The "points" must be manually operated to select which direction a train approaching from the left will take. It will automatically change for trains approaching from the right.

Telephone networks can do this, providing two-way communication on a single transmission network in the familiar tree pattern.

Early electrical distribution networks had a single energy source at one end, with multiple energy users at the branches of a treelike network. The first such network, developed by Thomas Edison in New York City, was a direct-current (dc) system. Modern electrical power systems use sinusoidally alternating voltage/current (ac), and transform the voltage level from its highest value at the source to lower values of 110 to 220 V (*root-mean-squared* value) for application. Higher voltages, up to 750,000 V (rms) imply lower currents and permit smaller-diameter wire to be used for long-distance transmission. Large electrical distribution networks have a much different form than a simple tree. In cities they often take on a gridlike configuration mimicking the street patterns which they serve. They also usually include multiple power generators, so that any of the thousands of electricity users may receive power produced by any generator on the system.

In the United States, electrical distribution networks form what is referred to as *power grid*; the many cross-connecting lines between regional networks permit electrical utility companies with excess capacity to sell power to operators of neighboring networks which have greater peak power demand than generation capacity. Interconnection is economically very desirable. In case of major failures in parts of such a system, however, larger regions may experience power loss because their local networks are coupled together. Many complex system issues are involved in the coupling of local electrical utility networks to form the "national power grid."

Communication networks also carry time-varying electric voltages, though they operate at a few volts and at very modest power levels. Without reamplification of these signals, transmission distances would be limited to a few miles, beyond which signals would be overcome by electrical noise. Long-distance transmission is made possible by placing amplifying devices called *repeaters* at regular intervals, along transmission lines. Signal level is never allowed to drop below some minimum well above the electrical noise-power, in part the result of a natural phenomenon which is present at any temperature above absolute zero.

Because it is most economical to transmit many different information signals concurrently on a single communication network, the problem of crosstalk—interference of one such signal with another—is a major concern in information transmission systems. The signals may be separated by being carried on different wires in close proximity to one another, being displaced (shifted) to nonoverlapping frequency bands (frequency-division multiplexing), or by being assigned different series of nonoverlapping time intervals (time division multiplexing).

The most recent forms of communication network are linked by radio broadcasting. Here there is no physical network. A virtual network is created, by having different sets of users transmit and receive in nonoverlapping frequency bands, within appropriate portions of the electromagnetic spectrum. Thus AM (amplitude modulation) entertainment radio provides one-way communications using 10-kHz-wide frequency bands in the 550- to 1500-kHz portion of the frequency spectrum, while entertainment FM (frequency modulation) broadcasts use 0.2-MHz bands in the regions from 88 to 108 MHz, and television (via the U.S. NTSC standard) requires broad channels of 6-MHz width. If parties wish to communicate by radio, they must usually share an assigned part of the electromagnetic spectrum with other communicators. Broadcasting stations, however, are assigned exclusive right to transmit on a certain frequency band, transmitting from a fixed antenna location to cover a prescribed area.

Whereas with physical networks (wire, cable), absence of interference between communications can be guaranteed by using channel isolation features in the physical network, interference-free broadcast communication requires cooperative action by all users or a given channel. Cellular mobile telephone service is a relatively recent example (Figure 2.15), in which radio-frequency channel selection takes place without actions on the part of the communicators. The equipment ensures that each caller has nearly continuous access to some radiotelephone channel while passing through any geographic "cell" of the system. As with other communications networks, however, cellular radiotelephones impose limitations on the number of simultaneous communications to or from a given "cell."

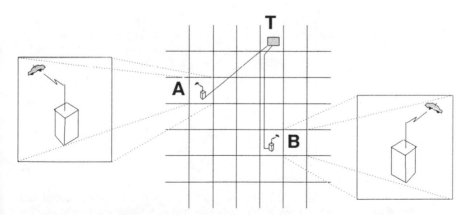

Figure 2.15. Mobile "cellular" telephone covers a wide area. Caller *A* communicates directly with a small transceiver in its vicinity. That unit uses wired telephones to communicate the cell in which respondent *B* is located. The call is automatically rerouted through the cells occupied at any time.

2.3.5 Vehicle architectures

Vehicles are in many ways the hallmarks of our modern industrial culture. They occur in myriad forms: private autos, school buses, high-speed intercity trains, ocean-going tankers, submarines or other warships, transoceanic jet aircraft and sophisticated fighter-bombers, the U.S. Space Shuttle, and earth satellites. Most share the characteristic that they are designed to operate fully while on the move. There are some (house trailers, for example) which are not intended to be utilized while in motion.

Vehicle architectures, in contrast to network architectures, are often self-contained within a single structural unit. Their subsystems may be highly interdependent. This is increasingly so in complex vehicles such as submarines, military aircraft, or spacecraft, even in less space-limited vehicles. This subsystem interdependence raises the system engineering effort required to develop, manufacture, and test the systems.

We could, somewhat arbitrarily, divide vehicle features or subsystems into certain subsets:

- Those which primarily serve to *interface the medium* in (or on) which the vehicle moves, e.g., wheels, wings, propellors, hull
- Those which *produce motive power*, e.g., engines, drive trains, axles, jet/rocket nozzles, fuel storage and supply systems, and propulsion controls
- Those which attend to *passenger comfort or other needs* and/or cargo requirements, e.g., access doors, seats, cargo spaces, windows, ventilation, cooling, oxygen supply, interior lighting, toilet and food service facilities
- Those which reveal or connect with outside objects or agencies, e.g., windows or ports, navigation systems, radios, radar, television, data communications, weather or ground-proximity warning systems
- Those which describe vehicle internal status to operators, e.g., dials, indicators or displays of subsystem status or problems
- Those which permit operator(s) to control vehicle movement or other functions, e.g., steering, control surfaces or jets, brakes, clutches and transmissions, subsystem displays and controls
- Those provided expressly for the safety of the vehicle or its passengers and cargo, e.g., fire alarm and extinguisher systems, seat belts or cargo tie-downs, weather sealing, covers, or other protection, emergency exits and ladders, emergency lighting
- Special equipment required for vehicle's functions other than mere transportation, e.g., on military vehicles: target locating and tracking systems, weapon storage spaces or connections, arming and release mechanisms, attack warning and jamming systems, etc.; on farm vehicles: power takeoffs, mowing, fruit-picking or grain-threshing mechanisms, insecticide distribution and spray nozzles, etc.

Vehicular variety is enormous, if one includes special-purpose vehicles such as fire-fighting pump-engines and aerial-ladder trucks, medical evacuation helicopters, earth-moving vehicles, and many others. Each such vehicle class has features or subsystems special to its application. Even vehicle subsystems which perform the same functions, in different classes of vehicles or in different environments, may be described by different names. The *drive train* of an automobile is analogous to the *power plant* of a sea vessel.

Most vehicles are built following well-established practices which define many salient features of their architecture. Many different personal or service vehicle designs are, for example, derived from a single starting point, the "panel truck," a large box on wheels with engine and operator controls at the front. Submersible vessels consist centrally of cylindrical chambers capable of sustaining high external pressures but may also include ballast tanks and other subsystems housed within non-pressure-tight streamlining shrouds. Similarly the streamlined noses of rockets or aircraft are often merely aerodynamic shrouds which protect, and reduce wind resistance of, the functional but aerodynamically awkward subsystems inside.

In a vehicle, what constitutes architecture depends largely on the narrowness of the vehicle's function. Large 18-wheeled tractor/semi-trailer "rigs," current backbone of U.S. interstate and local light shipping, exhibit several salient architectural elements. Perhaps most important is the required articulating joint between tractor and trailer. This is a necessity, if the lengthy vehicles must negotiate small-radius road curves or city streets laid out in horse-and-buggy days. Another architectural aspect relates to standardized or limited external dimensions, thus guaranteeing unrestricted passage on interstate highways, bridges, and tunnels. Had these dimensional standards been applied earlier to clearances of bridges and tunnels, we would less often be treated to news photos of trailers jammed helplessly against bridge supports or tunnel ceilings.

Less obvious than the visible similarities between vehicles of this class are a number of other standardized elements, such as a standardized method of attaching almost any trailer to a tractor, standard brake pressure-hose couplings, standard tire sizes, and others. Taken together, these elements can be said to make up an architecture for this class of vehicle.

In this book we will often cite vehicle architectures, particularly those of aircraft, as examples. Because of their unit character, and the typical weight and/or cost constraints, vehicle design encourages good integration of subsystems and intelligent sharing of system function between subsystems. This allows designers to exercise innovation to make the most from the least. Seldom are design skills more highly challenged than in the design of spacecraft, where weight control is essential. Though cost must be secondary to function and reliability, increasing a spacecraft's operating (on-orbit) lifetime almost proportionately reduces annual cost of ownership.

Linked or staged vehicles With the arrival of space exploration and intercontinental missilery, it has become increasingly common to *stage* transportation vehicles. The objective is performance or economy. The vehicle may collect, or shed, modules during its course. A *linking* concept has been applied to railway vehicles, since the early nineteenth century. A single locomotive and crew may collect and drop off a series of railway cars at different points along its route. (The tractor/semitrailer rig has the same feature though in a more limited sense.)

The staging objective is even more essential, in the case of extra-atmospheric vehicles. The maximum range of a rocket which has only a single stage is remarkably limited. If the fuel/oxidizer containers and engines are not ejected from the vehicle during flight, they represent large deadweight loads which must be propelled throughout. An offsetting economy, of course, is the ability to reuse fuel/oxidizer tanks or engines. The 1970-period Apollo moon landing system shed each stage successively upon expenditure of that stage's fuel (Figure 2.16). The 1980s Space Shuttle, however, returns its solid-fueled booster-stage rockets to earth by parachute, discards its large fuel/oxidizer tank unit, but retains its main engines throughout its mission.

Other common examples of vehicular staging can be found. Much transoceanic cargo now travels via *containerships*, freighters designed to accommodate large numbers of modular cargo containers. With the addition of running-gear, these containers convert to conventional eight-wheel trailers to be hauled to inland cargo destinations. Economic benefits derive from labor, time, and breakage saved by avoiding repeated packing and unpacking.

Vehicular staging need not involve transfer or shedding of portions of the vehicle or of its propulsion subsystems. Aerial refueling (of aircraft) is not in itself an architectural characteristic save for the implementation of a refueling subsystem. However as a designed-in capability it materially influences the range attainable by military aircraft. Newspapers reported, in the 1980s, that a flight of U.S. Air Force fighter aircraft enroute from the U.S. to overseas bases was refueled as many as 22 times before landing. The design feature which made this possible in the fighters was a midair-fueling receptacle equipped to couple automatically with a refueling boom suspended from the adjacent "tanker" aircraft.

Missile-equipped fighter aircraft aboard an aircraft carrier represent a three-stage vehicular concept which in World War II proved devastating to conventional navies formed around heavily gunned and armored battleships. It might be feasible for aircraft with the operational capabilities of the carrier planes to be flown from a ground base to certain targets,

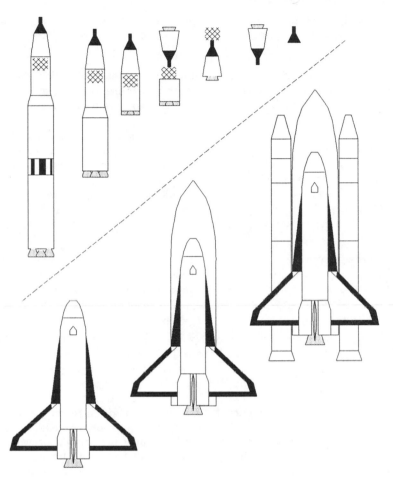

Figure 2.16. The Apollo lunar rocket system's many stagings are shown left to right at top. The space shuttle, which has much less distance to travel, has only three different flight configurations (shown *right to left* at bottom).

perhaps with aerial refueling. The carrier provides the only means by which they can, in effect, loiter near potential trouble areas and remain prepared to make an attack quickly.

EXERCISES

1. Examine in detail the exterior of some sizeable *historic building* in your area (for example, a courthouse, rail station, or theater).

(a) Record and categorize those features you believe are important architecturally, by virtue of their functions, prominence, or repetitive use.

(b) Describe any features which you perceive to detract from architectural integrity of the structure, and explain why you feel so. (The objective of this problem and the next is to develop architectural awareness of design areas in which you have no expertise.)

2. Repeat Exercise 1 for a major *modern* building.

3. From information in an encyclopedia or some other publication, define at least six important architectural elements which are, or were at one time, representative of a *tank* (an armored, treaded weapon system, a class of vehicle first employed during World War I).
(a) Discuss several salient features which appeared on early-model tanks but not, in the same forms, at least, on later ones. Learn this from reading, or if appropriate material is not available, you may speculate on the reasons why certain features present in earlier verisons have disappeared.
(b) Locate several features of military tanks which appeared *at the time of World War II or later*, and appear to have since become standard. Give your views or cite a more knowledgeable source as to the reason why these features appeared.

4. From the beginning of World War I to World War II, submarines featured a cannon prominently, on their decks, either forward or aft of the conning tower (now called the *sail*). Submarines built more recently lack this weapon. Trace literature to find *when* the cannon stopped appearing in submarine designs; then try to learn *why*.

5. Although many westernized people wear shoes nowadays, the *sandal* was and is the most common footwear in both the ancient and modern worlds. The central feature of any sandal is the sole which protects the foot against "road damage." A sandal's other features, where not purely ornamental, serve to anchor the sole from moving on the foot. The cheapest, hence most popular, sandal in the world today, apparently of Japanese origin, is termed in the U.S. a "thong" or "surfer." (Most habitues of swimming pools own at least one pair.) By experiment, and using simple sketches, show how thong architecture carries out four basic anchoring functions (against vertical, lateral, fore/aft motion, and rotation).

6. Consider a *three*-wheeled automobile as an architectural alternative to the common four-wheeler. Such a design was prototyped after World War II as the "Davis car," and used a single front wheel. Small European autos which appeared around that time from Messerschmidt (Germany) and Fiat (Italy) were not three-wheeled but four-wheelers, with a pair of narrowly spaced wheels at one end.
(a) Discuss the three-wheeled auto in comparison to the traditional four-wheeler, citing advantages and disadvantages (e.g., stability, handling, ability to deal with varied roads, understructure servicing, and tire changes). You may find it useful to create a miniature "prototype," using parts from a child's construction set, for example.
(b) Discuss the alternatives of a single front wheel versus a single *rear* wheel. (For the latter, consider a front-steered vehicle containing a single bench to seat one or two persons.) Factors such as stability during acceleration, braking, and cornering should be considered, as well as the arrangement of required internal components.

7. Cite a number of ways in which modern *materials* or *manufacturing methods* appear to have influenced fundamentally the architecture or design of your own automobile. Show this by a table with three columns. In the first column, list the architectural feature, and in the second the material or method by which it has been fundamentally influenced. Use the third column for an explanation or reference citation.

8. *John von Neumann's* most significant architectural contribution, many believe, was to suggest placing computer programs *and* data into a single, common memory. What, in your view, might have been some of the consequences if electronic computer architecture had evolved toward *separate* (functionally similar, perhaps) memories for data and programs?

9. A modern computer's *interrupt* feature allows a signal from an external (i.e., peripheral) device to cause a processor to obtain its next instructions at some memory address associated with the occurrence of the interrupt. Early computers did not possess such a mechanism, thus had to *initiate* all contacts with external devices.

For (a) interrupt-capable and (b) non-interrupt-capable computer designs: Describe in simple but explicit terms the steps the programmer of such a processor could take to communicate with a large number of input devices requiring rapid attention at unpredictable times. Consider two interrupt cases: (1) each input device has a unique interrupt signal, and (2) all input devices must interrupt the processor using a common signal.

10. Given a set of 64 numbers to be added together, each number or their sum being representable satisfactorily in a single computer word. Assume you have a multiprocessor comprising 8 processors, each of which can be independently programmed and can send "messages" (e.g., data) to *any other* processor. Further assume that each processor has local memory, but that the multiprocessor has no shared memory. During one clock cycle, each processor can either add two numbers stored in its own memory, *or* send a message containing a number to the memory of another processor. (The receiving processor *does not* devote the clock cycle to message reception.)

(a) Lay out a sequence of processor addition and message-sending steps whereby 64 numbers initially in the memory of a single processor can be added and their sum eventually returned to the same processor. (You may find that a graphic description is a good way to develop and describe steps.) Count *the numbers of cycles required*, making certain that intermediate results are sent where they need to be.

(b) Repeat (a) for a computer containing 16 processors. Then (c) comment on how a "general" solution for adding any quantity of numbers might be developed. Comment also on the general utility of this type of multiprocessor, comparing it with a single processor, in dealing with this class of problem.

11. Obtain and study descriptions of Digital Equipment Corporation's PDP-11 16-bit computer architecture, and its more recent VAX 32-bit architecture (they are related). Describe under what limitations programs prepared for PDP-11 computers could operate correctly without alteration on VAX machines. That is, what are the constraints on upward compatibility between PDP-11 and VAX?

(a) Describe in a few paragraphs the functioning of input-output architecture in the IBM Personal Computer, which you can, for example, learn from the IBM publication *IBM Personal Computer – Technical Reference*.
(b) Using a description of the 8088 microprocessor employed in the PC, discuss the ways in which the I/O architecture of the PC is defined (or constrained) *by the microprocessor* rather than by other components of the computer.

12. (a) For some complex personal computer application program familiar to you (e.g., a word processing, database management, or spreadsheet program), describe and characterize what you consider to constitute the program's principal *architectural features*.
(b) From a *software program designer's* point of view, the architectural decisions which define a program include: (1) how it uses available memory for data and program, (2) whether it keeps currently unused program routines (and possibly data) on storage disks until they are actually needed in memory, and (3) whether it can output to a printer, while its user continues other work at the console. For the program you are familiar with, what architectural characteristics of this sort do you believe you can characterize based either on use of the program or actual documentation?

13. Examine some limited-access highway in your area (i.e., a freeway, turnpike, interstate, motorway, or autobahn). List and describe briefly the *architectural features* which you perceive to distinguish it from other less important highways with the same numbers of lanes. For each such feature, give views as to its function and value.

14. Large retail stores, for example, sometimes use a *pneumatic* order-delivery tube system. (This sort of system was popular a half-century ago, nearly died out, and has regained popularity again.) Air pressure, or vacuum, moves small cylindrical containers from one part of the store to another. (A separate tube is ordinarily provided for each required route.) If there is such a system in a store in your vicinity, you should attempt to examine it.
(a) Describe system features and characteristics which should be insisted upon by intelligent buyers installing such a system.
(b) Describe features of an architecture for a hypothetical system of this sort, based on an assumption that it operates from a central air pressure pump. In short, what *rules* must a layout designer follow, in order to design a particular system?
(If you are an engineer, you may wish to include in your rules any interesting ideas for minimizing air noises from the system and preventing an open port in a circuit from degrading overall system performance.)

15. Describe the important *architectural elements* of a high-quality modern *residential plumbing supply and drain system*. Give a detailed description of all types of *elements* used, including distribution pipes, interface devices, and appliances. (Given standard elements, a plumbing *design* should require only showing locations of piping and of interface elements, and locations and types of fixtures—sinks, toilets, water heaters, etc.)

16. A single-masted sailing vessel (e.g., a sloop) will include some design details and refinements defined by its designer plus others included because of architecture and design traditions. Define those elements essential to

propulsion, control, and seaworthiness of such a vessel (with, or without a keel, your choice, but not catamaran). Ignoring refinements directed toward maximizing speed or capacity.

17. Electrical appliance manufacturers have for at least four decades attempted to market a special type of kitchen appliance. This employs a powerful electric motor built into the countertop which can operate any one of a number of replaceable applicance heads: a mixer/beater, blender, ice grinder, fruit juicer, food processor, or knife sharpener. Discuss, in some detail, the architectural and design problems which must be satisfied by the designer of such a product. Provide your own views to explain the modest popularity achieved by any of these products.

18. Consider architectural issues associated with a class of vehicle based on use of revolutionary "antigravity modules." (Physical realizability of the module is not your problem!) Each module is 12 in in diameter and 18 in high, weighs 100 lb when inactive, and is able to produce a *constant upward force* of 800 lb (gross), while consuming a special fuel at a rate of 300 lb/h. The module develops force 1/10 second after full fuel flow is established, and ceases generating force within a few milliseconds after reduction of fuel supply to below 300 lb/h. The "anti-gravity" force produced is independent of *orientation* of the module; a module can be viewed as having negative weight of 700 lb.
(a) Devise an architectural concept for a class of vehicles which would use numbers of these modules. What would be some essential elements of their configurations? (You might find it useful to begin from a concept similar to an airship, but without huge gas bags.)
(b) What control elements and features would be needed on vehicles using this module, and how could it be controlled, in each of the three axes and rotations, while moving through air or space?
(c) Describe six or more vehicles, having different functions, which could be evolved from a common architecture.

BIBLIOGRAPHY

H. Aspray and A. Burks, *Papers of John von Neumann on Computing and Computer Theory*, MIT Press, Cambridge, MA, 1987. Works of pioneers such as "Johnnie" (as he was known to colleagues) reveal concepts still unexploited.

Basil Collier, *The Airship - a History*, G. P. Putnam, New York, 1974. Few important systems of our age can be viewed from cradle to grave as can the airship. Though its resurrection is repeatedly suggested, its many practical limitations are obvious to the persistent reader.

Michael Collins, *Liftoff*, Grove Press, New York, 1988. This user's eye view from a former astronaut covers Apollo through early space shuttle experiences. As we have since come to learn, the seemingly flawless manned lunar trips had many small glitches.

Margaret Cooper, *The Inventions of Leonardo da Vinci*, Macmillan, New York, 1965. The master's own drawings illustrate beautifully many concepts which were far ahead of their time and available technology.

E. M. Cortright (ed.), *Apollo Expeditions to the Moon*, National Aeronautics and Space Administration, 1975 (U.S. Government Printing Office, Washington). Many of the Apollo system elements are depicted and described in this official record of achievement.

V. K. Garg and R. V. Dukkipati, *Dynamics of Railway Vehicle Systems*, Academic Press, New York, 1984. Modern high-speed rail transportation requires careful analysis of train motion on nonideal roadbeds.

Herman H. Goldstine, *The Computer from Pascal to von Neumann*, Princeton University Press, Princeton, NJ, 1980. Dr. Goldstine directed computer research for Dr. von Neumann at the Institute for Advanced Studies, Princeton, N.J.

P. E. Green (ed.), *Computer Network Architectures and Protocols*, Plenum Press, New York 1982. Not for the casual reader, but a well done treatment of data networking.

M. Harb, *Modern Telephony*, Prentice-Hall, Englewood Cliffs, NJ, 1989. A short technical treatment of many aspects of telephone systems. Requires a modest familiarity with electric circuit principles.

M. T. Heath (ed.), *Conference on Hypercube Multiprocessors*, Society for Industrial and Applied Mathematics, Philadelphia, PA, 1986 and 1987 (1st and 2nd conferences). Each structural form of multiprocessor, such as the Hypercube, has attracted its collection of enthusiasts.

Jane's publishes annual or semiannual updates of a number of books describing modern vehicular systems. *All the World's Aircraft, All the World's Fighting Ships, Armour and Artillery*, and *Weapon Systems* are titles of likely interest. Jane's, New York.

P. M. Kogge, *The Architecture of Pipelined Computers*, Hemisphere Publishers, New York, 1981. Pipelining has been used since the early supercomputers, and is now finding its way into microprocessors.

C. Lazou, *Supercomputers and Their Use*, Oxford University Press, New York, 1986. A readable overview of the machines and their applications up to about 1985.

J. J. Norwich (ed.), *Great Architecture of the World*, Random House/American Heritage Publishing Co., New York, 1975. Though not exhaustive treatments, chapters on each important phase of building and monumental architecture portray the unique aspects of each style.

J. L. Potter (ed.), *The Massively Parallel Processor*, MIT Press, Cambridge, MA, 1985. This term has been applied principally to a planar array of processors which was developed for NASA.

Peter Thompkins, Secrets of the Great Pyramid, Harper and Row, New York, 1971. The implications of "secrets" presumably enhance the popularity of Thompkins' books. He does collect a great many interesting facts and a wealth of drawings and photos.

Major computer manufacturers publish extensive literature which may be purchased or examined at their local offices. Contact International Business Machines at 112 East Post Road, White Plains, N.Y. Digital Equipment Corporation publishes many paperback books describing its computer products and applications. Contact them at 146 Main Street, Maynard, MA, 01754. Intel Corporation publishes detailed descriptions of microprocessors, support chips and circuit boards; contact Intel Literature Department, 3065 Bowers Avenue, Santa Clara, CA, 95051. (The three are from the mainframe, minicomputer, and microcomputer industry, respectively.)

Elements of System Architecture

We have discussed computer architecture at some length because it provides a wealth of examples and lessons. A second reason is that computers are increasingly used as critical subsystems of modern systems of various kinds from home appliances to spacecraft. Though we shall return to discussion of computers from time to time later in the book, at this point we must enlarge the scope of our discussion to architectures for a wider variety of systems.

The reader must be warned that many relatively simple collections of parts or of methodical steps are, at least popularly, referred to as systems. Those systems demanding the most painstaking attention to architectural definition are those of most interest here, since they are usually large, expensive, or complex, and may involve a range of technologies and disciplines dealing with:

- Structures which must support static and dynamic loads without failure and without impeding system operation
- Mechanisms subject to friction and vibration which are required to operate reliably
- Electronic controls and displays which must operate reliably and meet the needs of users
- Control and data-manipulating software which must be closely integrated with the system
- Propulsive machinery which may operate at high speeds and temperatures for extended intervals
- System operating environments; deep underwater, airless space, corrosive liquids or vapors; tropical, desert or polar climates; one-shot, intermittent or continuous operation
- Maintenance ranging from unmaintained deep-space systems to systems requiring significant rebuilding after each use

Modern military and commercial aircraft and spacecraft provide many examples. Though an individual system architect seldom deals with such variety, good approaches to architecture and design have much in common, independent of the kind of system.

3.1 Addressing system requirements

No system should be developed unless there is (or is believed to be) a need for such a system. System requirements may emerge formally from a user organization, but even here they may represent the unresolved views of a group of individuals. Even at best, users are generally not certain just what they really need. This is especially true if (a) they have had no experience with similar systems, and (b) no such systems are currently available. Thus:

- Requirements will in most cases be fragmentary, at least in some respects.
- Requirements will usually address high-level system needs, but key requirements may be omitted from attention.
- Systems requirements are often overstated, beyond actual needs, because special groups of users pose their own requirements independently without overall control within the using organization.
- Fine-detail requirements are often included among true top-level ones.
- Seldom are requirements prioritized; users are always hopeful that they will achieve every need or wish.
- It is not uncommon to find sets of inconsistent requirements for any large or complex system.
- Some requirements may be given whether really needed or not, and some may be given merely because it has become traditional to ask for them. This is often true where there are documented standards, since naming a standards document is often simpler than deciding what is truly required.

For systems designed to order for each customer, presumably an ideal starting point for a system design would be a thorough, well-thought-out formal requirement statement—developed by a knowledgeable user organization which understands the state of relevant technology. For speculatively developed systems, one should ideally have results of a broad and comprehensive survey of knowledgeable and cooperative potential system buyers. In practice, however, the starting basis provided to the system designer is usually far from these presumed ideals.

The novice system designer may be worried if a requirement does not address all design-decision topics. The well-qualified system designer views overrequirement as an evil and welcomes those which permit the maximum leeway in the design. If the requirements appear to be delinquent either through insufficient or overrequirement, the experienced designer will take steps to understand the needs of the user(s) and establish priorities.

3.1.1 Identifying and resolving incomplete or ambiguous requirements

Careful and unbiased study of requirements data by expert system designers, perhaps with the assistance of detail design specialists, will be required to identify potential problems in documented requirements. Or, if there are no documented requirements, experienced analysts must work with the users to develop them.

If requirements have been amassed via a market survey, they will assuredly reflect the biases introduced in the design of the survey questionnaire. If the results of the survey show unusual or unexpected biases, it may be necessary to modify and repeat portions of the survey with the same survey respondees, or test the original survey on an independent group of potential users if they exist. A reality in market surveying is that those questioned respond from within the contexts of their personal experience—which is, of course, why they are asked. This experience often breeds conservatism. Hence, a proposal to create and market a system notably different from existing systems performing given functions will usually not be enthusiastically supported by current system users—unless they find serious limitations in present systems or unless a proposed system promises truly remarkable benefits.

The words used in a survey or a requirement may mean different things to developers or to users, since they are usually involved in quite different activities. Ambiguities, both major and minor, often need to be resolved. The best way to do this is face to face. If a system is to be developed for a single customer, ambiguities in the users' requirements should be resolved formally, and documented to the satisfaction of both parties, before a contract is executed.

If systems are to be developed speculatively for a number of potential purchasers who are perceived after surveys to have conflicting views as to their requirements, it may be necessary to distinguish these by defining market segments which would require different systems. If wide disparities are recognized, certain parts of the potential market should be ignored so that development resources can be better concentrated on filling the needs of the more favorable market segments. Small firms attempting to introduce and market products in fields dominated by large producers often choose portions of the market not well served by their larger competition.

Those system requirements which address certain system details are often the most difficult to deal with, particularly if they constrain overall system design and thus lead to undesirable system characteristics. For example, a requirement that a particular, relatively unproven technology be used for some small system part might make that part incompatible with connecting or adjacent parts of conventional types.

Incomplete requirements may omit mention of system characteristics or features known to be critical to system appearance or operation. The system designers might on one hand assume that the user had no concern about those features. On the other hand, the omission may have been an error. Informal discussion of likely alternatives with the user can often establish whether the omission was intentional or an oversight. Certainly the designer will not benefit from insisting that every loose parameter be firmly tied down prior to design!

In resolving requirements inadequacies, systems designers should emphasize those inconsistencies or omissions which will influence top-level (i.e., architectural) design. They may decide to leave unresolved certain details which can be better understood during the actual design effort. There should be no attempt to fill in all the details, particularly if needed specialized experience may not be immediately available. Leaving significant "holes" in requirements is always a calculated risk for the developer. Their importance may not be correctly assessed during the top-level design analysis.

Requirements documents for large or complex systems are often difficult to revise in toto. Unless needed changes are fundamental and very numerous, it is easiest to revise such documents by describing the changes in approved amendments. Where requirements need massive overhauls, experience has, time after time, proven it best to delay system acquisition while requirements are reworked. For good results, a revised or amended requirements document must form central input to the system definition steps to follow.

3.1.2 Architectural design: Concept, structural arrangement, partitioning, modularity

Objectives It is now almost universally accepted in system design circles that the best system designs are carried out in top-down fashion. Designers begin with an overall concept of the system, then work out further details in an orderly and mainly hierarchical manner. In this way the more specialized details can be dealt with by the specialists in relevant fields.

Desirable as it may be, this ideal is seldom attained in toto. One reason is that a user requirement may be specific about a desire for certain detailed features. Another might be that an objective of a new system design is to incorporate certain new technology. Still another is that the system may be expected to incorporate particular existing systems, subsystems, or system components.

The major exception to true top-down system design occurs when a truly revolutionary objective is required. This must carry with it the availability of adequate development time and resources, and a certain amount of good luck. The first designs of nuclear weapons were largely of this character, since there were major uncertainties at each level of design. The recently developed "stealth" aircraft designs had some of these same uncertainties, since they were driven by a primary objective never before achieved (nor sought).

At the outset, the experienced system architect will often separate the known requirements into two subsets: a set of objectives and another of design constraints. Figure 3.1 depicts these in a Venn diagram-like representation. An outer boundary represents the objectives, by the "universe" of systems which might achieve those objectives. But application of each constraint (inner circles) reduces the possible number of system solutions.

Taking those requirements which form the set of objectives, the system architect will often find it desirable to add complementary objectives not addressed via user requirements. These may also be highly significant objectives such as minimizing system development and construction cost. These are important to the system developer, especially if the system is being developed speculatively. These additional objectives may form a set

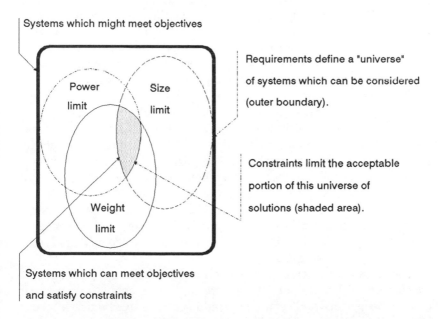

Systems which might meet objectives

Requirements define a "universe" of systems which can be considered (outer boundary).

Constraints limit the acceptable portion of this universe of solutions (shaded area).

Power limit
Size limit
Weight limit

Systems which can meet objectives and satisfy constraints

Figure 3.1. Both system objectives and constraints under which the system is designed tend to limit the available design space, as described here using a Venn diagram. Although *objectives* could be met by systems within the outlined design space, weight, size, and power *constraints* limit the scope of acceptable designs more narrowly.

of rules of practice, in the developing organization. By one of these corporate principles, system designs might be expected insofar as possible to use only proven system components. Or, the architecture must be extendable to meet known or predicted future needs. When the architects or designers address later how proposed system solutions meet user objectives and constraints, some of the added objectives may prove incompatible with essential user requirements and may accordingly be scrapped. If they cannot be, the development itself may need to be terminated. For example, holding to firm limits on system development or production costs is meaningless if the resulting system will be noncompetitive or will fail to meet basic user requirements.

First steps: A system concept The main theme of a system architecture is often referred to as the system concept. If one were developing a new urban transportation system, a variety of alternative vehicular concepts should emerge: perhaps an underground rail system, light surface rail system (trolley cars), trolleybuses, omnibus, small omnibuses ("jitneys"), or perhaps schemes as unusual as moving sidewalks. For many system developments, there may be only a single realistic concept; the choice is essentially automatic. In most cases, however, a range of alternatives can be identified and analysis is needed.

When one of several different concept alternatives must be selected, it is necessary to evaluate the important implications of each alternative against the set of system objectives. In this work, the good news is that one seldom needs to fully define each alternative, to locate one or a few best solutions. Thus, when selection between major alternatives is required, one may proceed far enough along the path we describe below to be able to make, for example, cost, schedule and performance estimates. Later, one should retrace these steps more carefully, to reach a more detailed definition or to select from among several qualified alternatives.

In evaluating the set of major alternatives, firm requirements must be the primary guide. Assuming that the requirements definition activity has ranked or prioritized the requirements, each concept alternative should be weighed against that ranking or priority list. Selection means may include mathematical analysis, simulations, the opinions of experts based on previous experiences, and, where available, opinions of prospective system users. (Methods for assessing alternatives are principal tools in a complete systems-engineer kit.)

Circumstances may dictate early selection of a single system concept. Where cost and schedule permit, system designers should carry along more than one concept and certainly should continue to evaluate alternatives for challenging aspects or subsystems. Selection between a gasoline- or diesel-powered vehicle concept might, for example, be deferred after most other aspects were determined.

System structuring and partitioning Almost any system concept implies some sort of structure. The structure may be a physical one, such as the sections of roadway, overpasses and interchanges which form a limited-access highway system, or the airframe (and its contents) for a military aircraft. Some systems may be better expressed as functional rather than physical structures, for example, the processes of a particular chemical plant or computer application.

At the point of envisioning the structure of the system, one should not always presume traditional system partitioning. If the system is of a familiar kind, certain structural or functional subdivisions may be obvious. If a system must meet new or special requirements, it is unwise to jump to traditional assumptions as to the most favorable partitioning.

Most system structures can be most easily understood by diagrammatic constructions similar, for example, to the diagrams of computer systems in the previous chapter. The structural diagram for a transportation system would of course describe, though perhaps crudely at this point, its routes, depots and storage, and maintenance and operations facilities.

A major objective of diagramming the system's structure is to begin the process of partitioning which will ultimately serve to define the subsystems or component parts which must be purchased or developed, and the interfaces between these parts (Figure 3.2). Interfaces must be defined

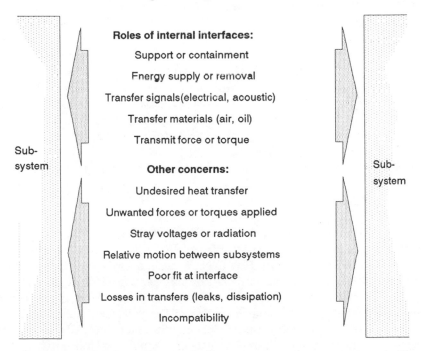

Figure 3.2. Interfaces intended for certain functions (top) frequently have incidental functions (bottom) which also require design attention.

so that subsystems or parts developed (or purchased) independently can be expected to operate correctly when eventually connected as a system.

The terms *subsystem* and *component* or *part* are not precisely defined but depend very much on custom and usage in the systems field. A costly and complex microwave oscillator in a radar system may be called a component part, for it contains no parts which could be used elsewhere. A much cheaper ac/dc power supply in the same system might be referred to as a subsystem because it contains many standard electronic parts.

In a large or complex system, actual system partitioning may extend to a hierarchy with many levels of detail. A top-level system representation, however, should seldom show more than two or three levels of subsystems. This avoids confusing those who will use the diagram. What's more, at this level of detail there should seldom be need for revision during system development.

Using a familiar example, a major subsystem of an automobile, clearly, is the engine. This is in turn partitioned into an engine block, cylinder head, carburetor, distributor, etc. The block, in turn, contains pistons, valves, crankshaft and other components often designed and manufactured not by the engine-builder but by specialty firms.

Partitioning of a computer system will reveal its need for a memory, ALU, instruction execution unit or control, display subsystem, etc. Each of these will, in turn, be composed of circuit boards, integrated circuits, and other components.

An initial partitioning of a system, for example von Neumann's five-part stored-program computer of Figure 2.5 will seldom describe all the subsystems which will emerge in system development. It should, however, identify a list of the system functions to be carried out by each major subsystem and by the combination. This caveat is not idealistic: Functions overlooked in the early system partitioning may later be found to be possible only by costly revision of several subsystems and the extensive redefinition of system interfaces.

A first level of partitioning may serve only as a vehicle to describe and examine the system concepts. This partitioning may, however, reveal the most important system interfaces: i.e., the internal interfaces between major parts of the system itself and external interfaces between the system and the outside world including system users. More detailed partitioning will expose more interfaces. We will examine interfaces at length in a following section. However, even in a first-level dissection of a system, it is useful to characterize not only the functions of the subsystems or components but also those of interface connections. If this is not done early, the problems and priorities of the interface design will not become clear early enough. Limiting numbers and complexity of system interfaces can have significant benefits in simplifying and unifying the subsystem designs.

The following process may be useful in arriving at a suitable system partitioning, assuming one begins with a system concept and the important requirements:

- List required and desired features and functions of the system. List also the "outside-world" systems, users, or other entities which must influence or be influenced by the system.
- If any of the functions must occur in some predetermined sequence or otherwise have order or priority, arrange them in a suitable order or sequence.
- Group functions which are of the same type, i.e., which might be carried out by the same, or similar means.
- If certain system functions are conventional or can be performed only by some specific type of subsystems, identify these and group them.
- Characterize each functional grouping as potentially defining a subsystem. Diagram it, along with the required interconnections or relationships with other subsystems.
- Identify the necessary or desirable couplings between outside systems or agencies and these tentative subsystems, and add these to the diagram. (Usually, system inputs and outputs can be identified as communicating with particular subsystems.)
- Examine the results, looking for rearrangements which will reduce duplication and simplify the necessary interfaces. To simplify an interface between two subsystems might mean making the connection less time-critical, or more universal, or selecting it from available standardized interfaces. It could also mean reducing the quantity or variety of information, material, or energy which must move across the subsystem interfaces.

After a tentative first-level system layout and partitioning is accomplished, it is usually valuable to examine briefly the next level or levels of subsystem partitioning. This may reveal unforeseen opportunities for beneficial top-level rearrangement. (Looking well ahead of current state of understanding is a hallmark of good designers.)

As with other engineering activities, system layout and partitioning is often an iterative process. One learns more with each reexamination of the problem. After two or three intelligent and determined attacks, however, it is best to move on. And, as partitioning extends to greater detail, there will usually be less need to revise top-level system concepts, partitioning, or interfaces.

In any top-down process like this, a significant fraction of design time should be used in dealing with the higher levels of design decision making, and as little as possible in pursuing fine details which have known and acceptable solutions. The best argument for this is that one cannot "fix" an inadequate system concept by struggling at detail levels. Since a great many details must be pursued in arriving at a system design, however, the numbers of designers involved will expand as work moves into more detailed stages. Specialists should be brought in to deal with specialized subsystems or objectives. Their assistance may also be needed for dealing with top-level issues of a specialized nature.

Software required for system operation may be considered as a subsystem or group of subsystems associated intimately with a processor or group of processors, even where the overall system is not strictly an information system. Though perhaps in the future, system designers will consider software a proper subsystem, this is not valid today except for software-only systems which will operate satisfactorily on available computers. In early system development it is important to use services of both hardware- and software-competent designers, working together. Software cannot compensate for inadequate hardware, while programmable hardware without software is useless.

Even after carrying out a complete system architecture/design exercise, a designer may be uncertain as to where the architecture left off and where design began. There is no hard-and-fast rule. If one were designing a family of software-compatible computers, the architecture of the family would constitute hardware and software features and functions common to members of the family. If, on the other hand, one were designing a one-of-a-kind system, the architecture could be described as comprising:

1. A basic layout or means of operation which identify the system as a member of a familiar class (e.g., a von Neumann computer architecture, or a factory based on assembly-line operation)
2. The types of interfaces the system presents to the outside world (e.g., a point-of-sale information system using bar-code reader inputs, or a tracked, or wheeled combat vehicle)
3. Familiar or repeated elements within the system (for example a conventional rear-wheel drive automobile using a diesel engine, or a balloon-framed house)

Once familiar elements (with perhaps a few nonfamiliar ones) are defined as being essential parts of a system's architecture, there will be improved understanding and communication between all its designers.

Repetitive and modular elements in architectural design Repeated elements in the form of structural and decorative components are familiar in large constructions, from the uniform stones from which Cheop's great pyramid was built, to the steel girders in modern skyscrapers and the memory chips in present-day computers. Some of these repeated elements are fully modular. For example 8- by 16-in concrete blocks can be used to build a masonry wall of almost any desired size in multiples of 8 in long and 8 in high. Or, 1-megabit memory chips make it possible to construct a computer memory containing any multiple of 1024^2 words, having any desired number of bits per word.

Several economic benefits may derive from repeated elements. There is often economy of scale in their production and purchase. Economy is also achieved by reducing the numbers of spare parts types required for system maintenance. These important considerations drive system designers to use—if feasible—the same model of computer repeatedly in systems requiring multiple computers, or employ a standard-size printed-circuit electronics board in systems or subsystem.

The term modular is another which cannot—and perhaps should not—be precisely defined. The modularity of bricks is useful only with the addition of a plastic medium (mortar) which fills in and conforms to irregularities. Mortar is also essential to strength, for it distributes forces over the bricks' surface and makes a wall much stronger than any mere pile of bricks. The pre-Columbian Inca culture in Peru used nonmodular stones of a range of sizes and shapes, with each stone hand-fitted to its neighbors so precisely that this construction has withstood earth tremors far better than modern masonry. In the same vein, the modularity of modern wooden residences always entails sawing modular cross-section lumber to arbitrary lengths.

Most major systems (the automobile, for example) offer ready opportunities for repeated or modular elements. The designers of the 3.0-L Mercedes diesel engine extended the block of the existing 2.4-L engine to accommodate a fifth cylinder. It is more convenient to wipe water from a windshield with two identical wiper blades than it is to design a drive mechanism which can sweep a broad windshield using a single blade. Tires, wheels, bearings, spark plugs, and many other items are used in multiples in automotive vehicles.

Repeated elements also prove to be a convenient mechanism to extend system lifetime or utility. Mammals can process liquid waste satisfactorily with only a single functioning kidney, carry on near-normal activities with a single eye, or hear adequately with one functioning ear. (The well-known time traveler, Dr. Who, star of British television, also has two hearts.) Many earth satellites contain multiple subsystems. Each may take its turn performing some essential function, until its eventual failure and replacement by a stand-by duplicate. Some multiprocessor computers are designed so that on failure of one processor, its work can be shared by others.

Repeated or modular system elements almost always offer economic benefits if the elements have well-proven quality. Even if such an element must be developed, it may be worth the trouble if there are no other good and readily available alternatives. However, this is seldom affordable except for quantity system production. Few low-quantity systems, except for certain high-priority military ones, can justify added costs of development and production for new technology, new subsystems, or major components such as special integrated circuit chips.

Microprocessors, per unit processing power, are much less expensive than mainframe computers or supercomputers. Because of this, they have long attracted the designers of low-budget high-speed computers. As yet, however, no broadly effective architecture has been developed which will allow arrays of microprocessors to operate effectively on a wide range of large-scale computing problems. Where the character of a pariuclar problem allows it to be partitioned into many parallel, near-independent computations, such microprocessor arrays can be highly effective.

3.2 Interface identification and definition

A widely held view in systems engineering is that successful system design depends heavily on good interface design. Even when subsystems are able to function correctly, systems may fail due to poor interface selection or design.

Nonengineers may envision as interfaces only the most obvious couplings and connections. However, in a complex system, interfaces perform a wide variety of functions and are subject to unavoidable influences ranging from mechanical shock to temperature extremes and external radiation. Figure 3.3 gives some idea of the richness of interface phenomena. Each bearing in a mechanical system is an interface, and subject to wear. Every plug and receptacle in electronics is an interface. The operating system of a computer serves principally as an interface

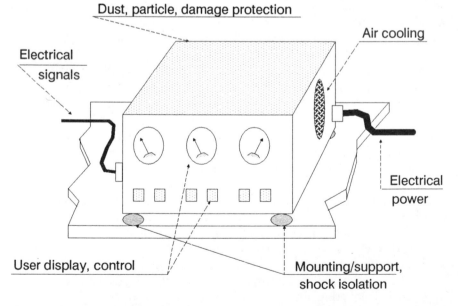

Figure 3.3. This sketch illustrates some of the many intentional interfaces and incidental ones for a small subsystem.

between the application programs and the computer hardware. Controls and displays by which a user operates a system are also interfaces. The system designer, called upon to define each interface specifically and accurately, realizes their complexity and importance.

The criticality of interfaces in a large system is easily understood when it is recognized that they are boundaries between parts of the system designed or defined by different individuals, sometimes by nature itself. A subsystem designer's knowledge of some subsystem connected to an interface of a subsystem he or she designs may be fragmentary and incomplete. Since the system designer must be held responsible for compatibility, well-worked-out interface definitions are worth the large effort they require. The expert system designer expends the necessary effort to select or design interfaces which minimize risk due to a time or knowledge hiatus between the subsystem designers.

Interface definition (including selection or design) should be principally part of architecture rather than design detail. It must be the key to partitioning which allows complex parts of a complex system to be developed independently by different groups. There must be high confidence that the system will operate correctly if its parts operate correctly and adhere to the system interface definition. Inadequate or tardy attention to interface definition will bring on problems at system integration time or, even worse, after the system has been fielded.

3.2.1 The many facets of an interface

An interface may represent a transfer portal or medium; for example, interfaces between computer systems and their users. What is transferred across the interface may, as in that example, be merely signals. In other instances it may be energy, or materials such as fuel or water.

Some interfaces are not specifically intended to transmit material or some form of energy but merely allow one part of a system to physically support another. The interfaces between the tires of an automobile and the roads on which it travels, or the bearings which allow the auto's crankshaft to rotate almost frictionless within the engine have a supporting role.

While most of the interfaces with which the system designer must deal are created for particular purposes, unplanned and incidental, though nontrivial interfaces between system parts may produce undesirable effects. An example is the heat generated by an auto engine, which (if the firewall is not thermally insulated) will cause the car's interior to be uncomfortably warm in summer. While this can easily be overcome by considering the firewall as a thermal interface and insulating it to reduce heat flow, other unintended system interfaces are less easy to deal with.

For example, an aircraft or vessel often has on board a variety of radio transmitting and receiving radio equipment, operating in nearby or overlapping frequency ranges. This can impose serious system problems, if a transmitter radiates power from its antenna directly into the receiving antenna of another subsystem. If the receiver is not protected in some way, its input amplifiers could be damaged unless protected in some other way from the high signal levels found near a transmitting antenna. If there is only a single transmitter and receiver, the problem is handled merely by not transmitting while receiving or expecting reception.

Even if so protected, unless properly designed electrical filters are installed in the transmitter, receiver, or both, the transmitting subsystem could interfere with reception if both subsystems operated on nearby frequencies. A system-level control scheme might be required to prevent unintentional operation of both the transmitter and receiver on the same frequency, if that were a real possibility.

Not only do interfaces come in wide variety, for the most part they involve the understanding of specialized areas of technology: heat transfer, tribology (science of friction), electromagnetic theory, or electric circuit operation, to name a few. Where interfaces use electric signals of low voltage and power to transfer signals across an interface, additional technology such as spectrum analysis or digital circuit design may be important. Although the need for or presence of certain interfaces may be recognized by a system architect, specialists must usually be involved in detailed solution of interface problems.

Defining, designing, or selecting interfaces   We refer to the definition or design of interfaces only for those interfaces intended to transfer signals, energy or matter or to support parts of the system. Interfaces which are incidental but deleterious to system operation will be avoided (e.g., by placement of system parts to minimize undesired interaction) or controlled by treating them in some manner.

Like a system, a typical interface can be defined at different levels of detail. At the highest level, a particular interface might be defined (for example) as a data path, a connecting pipe, an alphanumeric display, or a roller bearing. This may be an adequate architectural description, assuming that the subsystems facing one another (across the interface) are compatible (i.e., one is capable of reading in, and the other of accepting, the same information, flow, torque, etc.).

This should not imply that designing the interface will be a trivial matter, for there may exist incompatibilities between the subsystems joined by the interface. In some instances, incompatibility may be extreme, with one subsystem having a capacity, flow rate, or other limits substantially different from those of the other.

The two subsystems may be fully compatible: a pump and a hydraulic motor operating at the same pressure and flow, or a modem producing

9600 bits of information per second, and a computer able to transmit or receive at the same rate. In this case, the interface may be relatively simple, since it requires no conversion or manifolding. The interface may be as simple as a plug/receptacle coupling, or a short electrical cable with a plug on each end.

But is that so simple? A glance through an electronics parts catalog reveals a wide and competing variety of plugs, receptacles, and cables, almost any one capable of handling information signals. Likewise hydraulic fittings are available in a wide variety of sizes, attachment methods, materials, pressure ratings, and fluid-type limitations. Experience is needed here not so much to ensure defining a "perfect" interface but to avoid a poor choice.

Even the simplest of interfaces, such as an electrical cable and plug, requires attention in its design or selection. Whenever possible, most system designers select proven designs which meet all applicable user requirements and industry standards. An unnecessary deviation from familiar, proven, interface designs often imposes extra design and testing. Discounted bargain-basement components are out of keeping with quality system design unless they meet all pertinent standards.

There can be great complexity even in simple-appearing interfaces. Communications interfaces such as the nine-pin plug and receptacle commonly used to transmit computer signals serially (1 bit at a time) over two wires involve these parameters:

- A *physical* interface involving physical placement and diameter of the conducting pins and of mating housings, locking screws and screw holes
- An *electrical* interface involving current-carrying capacity of each pin
- A *maximum allowable voltage* between pins
- A *minimum leakage resistance* between adjacent pins in high humidities
- A *maximum electrical capacitance* between pins (which could cause signals to interact undesirably)
- *Mechanical requirements* such as an expected lifetime measured in connect/disconnect operations
- *Resistance to deterioration* from certain corrosive vapors for some period of time
- *Weatherproof qualities*
- *Thermal interface criteria* such as maximum safe operating temperature

Here, we have dealt with only a single group of interface parameters, namely, physical ones which many novice designers may take for granted. In an actual modem-computer interconnection, voltage levels representing binary signals, the interpretation of voltage changes as representing binary information, and eventually the interpretation of the information in relationship to the system operation are equally important parts in a complete interface definition.

It is not surprising that qualified system designers will adopt a recognized standard where one exists. It is a welcome alternative to facing the myriad of issues which must be resolved—especially when interfacing subsystems are to be acquired from outside sources. Even when "standard" interface definitions are invoked, careful evaluation of subsystem compatibility must be made. Many standards are too general or incomplete, or may include alternatives incompatible with a particular system's needs. As a simple example, the computer-modem connection (Electronic Industries Association, RS232 interface) leaves the data rate and message format to be defined by the system designer.

Where subsystems are not fully compatible, their minimum essential interface may be much more complex. What amounts to a separate subsystem may be designed to accommodate subsystem differences at their interface. Sometimes a given subsystem must interface a number of alternative subsystems. This is true of the interface between an automobile carburetor, and then air-fuel intake point for each cylinder. The interface device in this instance is known as an intake manifold. It divides the air-fuel mixture from the carburetor as evenly as possible into separate flows to each cylinder. Its design is nontrivial, since distances from the carburetor to the cylinders vary. In some racing auto engines, this manifold is designed so that each connecting pipe is of the same length as the others, giving the manifold an appearance reminiscent of a bunch of bananas.

As another example, certain electro-optical cameras (notably ones operating in the infrared part of the spectrum) produce a two-dimensional image formed of horizontal lines, each line representing output from a separate horizontally scanned detector. A conventional TV raster display produces an image by back-and-forth, up-and-down scanning of the screen by a single electron beam, whose current is modulated to produce changes in intensity. The "interface" allowing the camera image to be displayed on a conventional display tube must convert a composite signal consisting of a number of parallel single-line signals into a raster signal. This can, for example, be accomplished by storing the camera image in a memory capable of storing a value for each picture element, then reading out the stored data values in proper order, converting them to a beam-control signal and sending this to the raster display.

Another classical complex interface is that between stages of a ballistic rocket booster. The U.S. Titan IV booster, for example, has an upper stage engine much smaller in diameter than the payloads it boosts. In addition to providing a mechanicam interface, control signals to the first stages must pass through these interfaces, and the interfaces must support the separation of the stages during flight.

There are usually system advantages to be gained from making interfaces simple and few in number, i.e., composed of only one or a few physical subsystems. The microprocessor, usually containing the ALU, registers, and control in a single integrated circuit, is an excellent example of this principle. It achieves its speed from the short distances electric signals need to travel, its reliability from the small number of plugs or solder connections needed, and its economy from the almost fully automated production.

However, combining many different system functions in a single subsystem may only appear to reduce interface requirements, i.e., by "hiding" them. Good system design arranges subsystems and allocates functions between them in a way which will result in fewer and simpler interfaces. It does not combine unrelated system functions into a "black box" subsystem and pass along undefined the associated integration responsibility.

3.2.2 Internal interfaces in systems

System interfaces can be characterized as internal—within the system itself, and external—between the system and the outside world and users. The careful system designer will have concern not only for the internal interfaces which *must* be managed (those, for example, *joining* subsystems developed by others) but should also show concern for important interfaces *within* subsystems. As is the case with computer buses, internal interfaces may also appear, in whole or part, at an *external* interface to the subsystem. If subsystem internal interface management is slipshod or error-prone, this is a measure of the quality of the subsystem. Ultimately it will probably produce bad results in the larger system.

For example, if a hydraulic subsystem has inadequately designed internal seals or its systems connections cannot withstand system hydraulic pressures, a short-time pressure peak elsewhere in the system could produce leakage or rupture.

Defining internal (subsystem and below) interfaces When inside a system component or subsystem, many interfaces may be informal, for example, conductors connecting a printed circuit board to a panel-mounted switch, or a length of flexible plastic tubing interconnecting instruments on an aircraft control panel. It cannot be expected that every single electrical connection, each hydraulic passage, and all those areas through which cooling air is admitted to a subsystem will be the subjects of intense study and detailed interface definition and documentation. However, in most of these areas, design rules can be used to provide a reasonable basis for successful systems. Such rules, described later in more detail, are usually invoked as part of design, to ensure uniform treatment of many informal interfaces. One such rule might require that any group of two or more electrical wires passing from one point to another across a space longer than, say, 2 in must be grouped using a special string or plastic "tie." Like most design rules, it relates to parts which are built or assembled ad hoc for the system. Clearly some form of inspection is required to verify design and construction practices conform to the design rules.

Principally because of the advance of technology, design and manufacturing rules applicable to an older technology (e.g., string-bound bundles of wires) may be inapplicable for new materials or technology (e.g., specially formed plastic ties). Since these considerations apply mostly in the details of subsystems and components, the system designer may have little control over them. The key step is to verify, through required inspections, that subsystem detail designers and manufacturers follow design rules whose adequacy can be verified.

Special needs of a system may impose economic or technical need for important internal interfaces of nonstandard types. In an information system, for example, local communication between certain subsystems may suggest a multiwire parallel bus connecting the subsystems. There may be no appropriate standard in this situation. The system's designers therefore must create an interface to meet their needs. It is naturally desirable that designers study how others have attacked similar problems. Many examples of internal computer buses have been described in professional or trade publications.

In establishing an ad hoc interface definition, it is difficult to envision the breadth of its future applications, in the system or outside it. If designed merely to meet current requirements, it is unlikely to be used elsewhere. In this case it may even prove marginal merely to meet future requirements of the system for which designed. However, if it is designed with broader application in mind, more design effort is needed. The small quantities required will also increase production costs. Interface definition strategy is thus in large part a cost issue.

This enigma may sometimes be intelligently resolved by modifying some existing interface device or standard. For example, an interface specified at one speed, pressure, transmission distance, voltage, etc., might be boosted in performance, though seldom by large factors. In "pushing" proven interface designs, one may be moving into regions of operation originally provided as a margin of safety. The modified interface may thus lack the robustness expected from the interface which served as a model. *Scaling up* is a term in general use to describe more strenuous conditions than those for which something was originally used. It can bring on a variety of problems: Commercial equipment must meet substantially more stringent reliability requirements than scaled down home-appliance counterparts. Digital signalling methods applicable to twisted-wires telephone cables are not appropriate to fiber optic cables. However, modest scaleup of interface capabilities may prove satisfactory where the interface environment is carefully controlled.

3.2.3 Interfacing to external systems

Where a system under development must support a complex interface other systems which may not yet exist, it is even more important to validate system behavior for all aspects of the interface definition. This is nontrivial. The problem is even more challenging if the external interface definition was an ad hoc design, incomplete and possibly subject to individual interpretation or intentional deviations from standard practices. If this is the case, the only same approach is to meet with designers of the other system(s) to establish a common understanding for the interface definitions. In very special cases, when even the identity or characteristics of the external system is unclear, or it is the product of a competitor, it may be necessary to provide some flexibility (e.g., adjustable parameters or ready programmability) into your interface design to allow for a range of external-system behaviors. The term foreign interface has been used to describe this sort of uncertainty.

A common interfacing difficulty is merely a consequence of the independence of two interacting systems. The system with which ours is supposed to communicate may be "turned off" and therefore uncommunicative. Good design on our part requires that this condition not cause our system to fail; if practical, it should report such problems to its operators. If our system is to operate together with others as part of an even larger system, it should be designed with interfaces which allow it to be started or stopped or its status to be monitored from a remote control point. A system not originally designed for such a "slaved" operating mode may require significant modifications, if this is later required.

A conventional information system interface often includes a communication protocol, usually implemented in software, between one information system and another. A simple example is shown in Figure 3.4. Only a few such protocols have been standardized. Many actually implemented for use between a manufacturer's own products are neither standard nor available in competing products. Most communication protocols are initiated via a "handshaking" process similar to the informal "Hello" " . . . Hello" of telephone callers. Once both computers understand the states of one another, one can make a request to be satisfied by the other. Computers using well-designed communication protocols should not become confused if the transmission line has failed or if the system on the other end has failed or has software errors. In improperly designed or implemented versions of well designed protocols, systems may, for example, wait endlessly for responses which will never arrive.

Caller sends		Purpose		Listener sends
SOH		Are you ready?		
		Acknowledgment - yes	←———	ACK
SOT	———→	Start of transmission		
Data	———→	(Send message or file one character at a time.)		
....	———→	May not include SOH,SOT,ACK,NAK		
....	———→	XON, XOFF characters.		
		Command to pause	←———	XOFF
		Continue sending data	←———	XON
Data	———→	Data continues		
Data	———→	....		
EOT	———→	End of transmission		

Special conditions

If nothing from caller for n seconds   ←——— NAK

NAK   ———→   If nothing from listener, after XOFF, for n seconds.

Figure 3.4. A simple *protocol* for transmitting a message to a device which, from time to time, must pause to process and store characters. The special control characters (*SOT, ACK, NAK, XON, XOFF and EOT*) never appear in message data. In the absence of error there should be no confusion at either end. In this simple example, no provision is made for detecting or correcting an incorrect reception.

The principal interface consideration will usually be compatibility with some external system. If the term *compatibility* is encountered in advertising, requirements or specifications, this could be a clue that similar though incompatible interfaces also exist. This is often the case in both digital and analog information systems. Television receivers intended for use in the United States are useless for operation in Europe where signal standards are varied and different from our own. Seasoned travelers in many foreign lands are also well aware of differences in electrical power systems: 220 V rather than 110, 25 or 50 cycles per second rather than 60, or perhaps direct rather than alternating, current at the wall outlet. Fortunately, under these circumstances, the power receptacles are usually incompatible with U.S. electrical plugs. These same power source problems can plague large-scale systems if not anticipated.

Where a system requires a dependable source of fuel, this can also be considered an important interface. Fuel octane levels are not standardized, and fuel quality varies around the world. Even in the U.S. military, Air Force aircraft are designed to operate best on a jet fuel termed JP-4, and Navy aircraft on JP-5, while all commercial aircraft use a slightly different blend termed JET-A. In a pinch, any of these will serve; however, JP-4 is superior for low-temperature liquidity and favored for high-altitude long-duration flights, while JP-5 has a slightly higher flash point and is safer on a hangar deck.

A major source of incompatibility between systems or subsystems designed and manufactured in different parts of the world is their adherence to national standards. There are few international standards save for weights and measures, certain aspects of radio and telephone communications, and items sold in international commerce for which incompatibility would ruin business. Perhaps our most embarrassing example is continued reliance on "English units" of measure by the United States, England and a few other nations, when most of the world has moved to the metric system. This is particularly difficult in mechanical interfaces, for which there are no directly compatible inch and millimeter bolt diameters, bolt head diameters, or screw-thread sizes.

In system designs where it is evident that a problem will exist in interfacing to external systems, because of adherence to different standards, it may be necessary to add conversion to the interface functions. That is, the interface may need to convert to compatibility with other standards. This adds to complexity and cost, though not nearly so much if planned from the start rather than as a late-design "patch."

3.2.4 Interfacing to system operators and other users

System interfaces to users include all means by which humans are involved with system *operation*. This may involve system output by direct visual observation of the system, indicating or recording meters, alphanumeric or graphic displays, generation of natural or synthesized sounds, system movement, or printout. System input by users may involve manual or pedal controls, a keyboard, pointing devices (such as a joystick, mouse, light pen or stylus), voice commands, even head or body movements (e.g., for physically disabled users).

For each important class of system there usually evolve characteristic input and output means. For example, those of the automobile have changed little in 50 years. Those of computer systems, for the first 2 decades of their evolution, were punched cards for input and printed forms for output. However, computers have since been evolved to use a wide variety of inputs and to produce almost any desired form of output.

A generation ago, serious concerns were raised about the strength, effort, speed, or coordination of movement required to operate certain systems, particularly in military applications. The specialty termed *human factors engineering* was developed rapidly to respond to these needs. This field's practitioners base their work on physiological and psychological tests and comparisons to determine the most appropriate control, display, seating arrangement, etc., for a system operator. Some of the most important aspects of human factors engineering have since been accepted into good engineering practices, and the techniques have taken their place in comprehensive engineering handbooks.

While normal system *operation* has become highly automated in most major systems, certain system *maintenance*—particularly that which is infrequently needed—may be physically or mentally challenging. A high level of genius should not reasonably be expected in system maintainers. Maintenance operations requiring superhuman strength to disconnect, lift, or attach system parts should be avoided where reasonable, and if unavoidable these operations must be aided by special tools which are always available to maintainers. (Consider changing a tire on an "18-wheeler," for example.) Since maintenance tasks are often assigned to individuals having little training in overall system concepts or technology and very limited training in maintenance procedures, system maintenance has now become a frontier of need for better interfaces.

Architecture of user interfaces External electrical or mechanical interfaces can usually be adequately defined in terms of technical measurements, drawings, and materials lists, the same media used in describing the system's internal parts. For human interfaces, the architect or designer can describe the technical (system) side of the interface in this way. The human side of the interface, however, requires verbal and/or graphic description of operating sequences, and should in addition define special capabilities or training required by the human user.

For competent design of the technical aspects of human interfaces, one must consider factors including:

- The number and type of outputs which the system must present to the operator via meters, recorders, displays and status indicating lamps
- Priorities among this information and how priority will be represented, e.g. by flashing displays, color, audio tone, or size or intensity of indications
- Schemes for accessing and presenting secondary indications, interpretation, time-sequence records, and other information needed to interpret system operation but which must be requested by the operator or provided automatically in particular system operating states
- Operating controls required, including hand levers, pedals, knobs, toggles, keyboards, and graphic pointing devices

Humans do possess great physical and mental adaptability. Perhaps because of this, system designers have often neglected to consider human limitations in their designs. Humans can be overstressed by time-urgent requirements or an overabundance of information. In that condition, they may respond in less timely or less accurate ways than required for correct system operation. Humans under stress, when properly trained, will concentrate both visually and mentally on the most important set of tasks within their ability, disregarding the less important information. The system design and the physical layout of system displays can recognize this by placing critical information in a central display area and critical control functions readily at hand. Time-critical control sequences should involve only a small number of steps. Good design also dictates that *accidental* triggering of emergency responses (such as ejecting from an aircraft) be unlikely to occur.

User interfaces often require design of safety features. In rail vehicles, it is traditional to arrange the throttle such that, unless pressure is continuously applied by the operator's hand or foot, the vehicle will be automatically braked. This proves sufficiently tiring that operators may defeat such mechanisms by makeshift holding devices. The result is that the system is then even *less* safe than without the automatic braking device. Technology is now available by which an operator's state of alertness can be monitored without effort on the part of the operator. Systems subject to catastrophic events can be provided with automatic shutdown sequences which, if desirable, can be overridden by specific operator action.

3.2.5 Software interfaces

The basic interface between computer hardware and software is the computer's instruction set. This is fixed insofar as design of a system incorporating computers is concerned. Other critical software interfaces include the software tools (e.g., compilers, debuggers) and utility programs employed in developing and operating the system.

For most *data processing* applications, this software is also obtained from the computer manufacturer, not designed by the application designer. However, in manufacturing systems or military systems including built-in computers, almost all software may be designed specifically for the application. Immature or limited-capability operating systems or other executive software, interfacing the hardware to the application software, can pose the most serious limitations and risks to system performance or even to correct operation. Because the operating system or executive is the deepest of the layers of software which isolate users from the hardware, their errors are the most critical and often most difficult to identify. Corrections to them, by the operating system (or executive software) developer, may inadvertently cause other outer layers of software to become incorrect. Because it is *too easy* to change (in comparison to hardware), software change has frequenlty been used carelessly for changes late in the design cycle. Results have often been unsatisfactory.

An important software interfacing means, in systems containing organized stored data, is a *data-base-management system*, or DBMS. This provides procedures for building and maintaining organized data, as well as access programs whereby users or other programs can request and receive subsets of data based on formatted queries. Like a full-featured operating system, a full-capability DBMS program is too complex to justify its unique development for most applications. A proven, mature DBMS should be selected during system design, based on system data management requirements and the system's overall architecture. The quantity and variety of data stored and the frequency and variety of access must be taken into account in selection. A DBMS which is highly adept at locating a special item of data may be less effective for selecting, arranging, and outputting a large mass of data, for example.

Another key software interface feature is the programming language in which applications will be described for computation. For some application areas, well-designed *application-specific* languages exist. For example, ATLAS is the language commonly used for programming of *automatic test equipment*. Where such languages exist and are used, programming is usually carried out by application-area specialists rather than professional computer programmers. For many applications there is no applicable special language, and a *general purpose programming language* must be used.

Where a system development organization's programming personnel are most familiar with a particular language, there will be pressures to use it. (This will often extend to applications for which the language is a poor choice, as for example, using Fortran in a typical business data processing application.) The system client may in some instances specify a language. The Department of Defense is attempting at present to enforce almost universal use of the language Ada. (Ada is described in the standards document ANSI MIL-STD-1815A-1983, Jan. 1983, published by U.S. Department of Defense, Ada Joint Program Office, Washington, DC 20301.)

A system requirement may call for a particular programming language, or conceivably for a particular computer architecture. (Such requirements are often founded on a user's resources for system operation and maintenance.) This will often limit choices to ones where compatible hardware and software already exist. Developing a new compiler and perhaps operating system programs will be more expensive and time-consuming, than selecting a computer for which the software is already available. Where the developers have little or no experience with a selected computer architecture or programming language, program risks have been found to increase very significantly.

Major system-unique software interfaces Some systems may require special software interfaces to permit computer applications to support particular subsystems. These may consist of device drivers implemented in software which can manipulate the hardware of displays, indicator lamp banks, recorders or plotters, or of digital communication subsystems. Good software architecture invariably isolates applications software from the software which manipulates subsystem hardware such as a printer. Subsystem changes such as installing a new printer should require changing only a software "printer driver."

If a software-driven system has many modes and is expected to evolve functionally, software architecture should include central and utility software which "manages" the interfaces between applications and the hardware subsystems. Subsystem driver software can be controlled through an *executive program*, which may itself operate within the overview of, and obtain services from, an *operating system* program. (See Figure 3.5.)

This arrangement also provides a suitable framework for execution of diagnostic test software, user training, recording of system activities, and other features desirable in modern systems. A human user may be further isolated from the operating system, when not executing an application, by a *shell* program used to simplify the operating system's user interface.

3.2.6 Interface standards

Wherever applicable, system architects and designers should select a standard interface from among standard definitions developed and published by such organizations as the American Standards Association, National Bureau of Standards, Electronic Industries Association, Institute of Electrical and Electronics Engineers, Society of Automotive Engineers, and the Society of Heating, Ventilating and Air Conditioning Engineers. If the system is developed for a U.S. government agency such as the U.S. Navy, or NASA, contract documents usually specify particular military standards or other governmental standards to which the system is required to adhere.

Incompleteness and ambiguity in standards Most standards describe interfaces *or interfacial criteria*. Even standards of length or mass are employed to enforce compatibility between original equipment and spare parts, or between products acquired from different manufacturers. Standards are developed by human beings. As with all products of human endeavor, they are subject to a degree of incompleteness or ambiguity. Sometimes these reflect incomplete maturity of technologies to which the

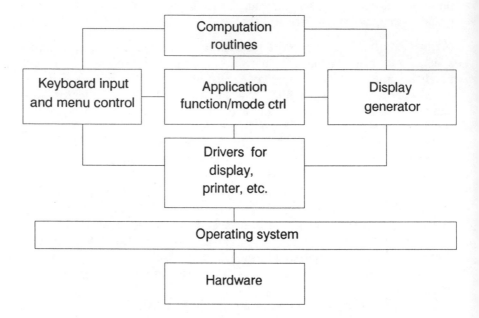

Figure 3.5. A block diagram of a computer system running a program able to interact powerfully with user, display, and printer. Most of the blocks represent utility software. Many of them may be parts of the application program.

standards relate. They may represent differences of viewpoint between standards committee members or their firms.

If a published standard is mute on certain aspects of an interface, designers often assume they have carte blanche in those areas. This can lead to incompatibilities between systems or subsystems supposedly equipped with standard interfaces. The only sure way to verify compatibility between existing products from independent sources is by experiment using the equipment. When the equipments are both under development by separate designers, an interface definition team should be assigned, as early as possible in system design, to resolve the uncertainties.

The right amount of standardization If application of standards is desirable, it might be assumed that *more* standards would be even more so. Not so, from the designer's point of view, since standardization represents constraint on design. Imposing more rigorous standards than needed results in higher system design costs, greater manufacturing costs, and longer developments. Imposing too-rigid standards will make it difficult or impossible to meet all system requirements.

Consider military standards for nuclear radiation hardened electronics. The objective is that systems in high-altitude strategic bombers or military satellites should have reduced immunity to distant detonation of a nuclear weapon. Applying the standards to every military system would be wasteful. Electronics which can satisfy rigid nuclear-radiation tolerance ("rad-hard") requirements cost more and usually deliver less performance than those with greater radiation susceptibility.

Understandardization is as great an evil. This implies avoiding use of accepted standards—taking an *ad hoc* route. The big penalty is in ignoring the large design, production, and operational experience embodied in mature standards. A frequent example of this in digitally communicating systems, is design of ad hoc signaling *protocols* for transfer of (digital) information. Standard protocols may be rejected as "too complicated." The ad hoc protocols substituted for them often contain design errors and limitations not understood by their designers.

Evolution of ad hoc interface designs If there exists no suitable standard interface to meet a particular system need, an ad hoc interface is the only alternative. Typically, too little effort is spent on the design. Designers should consider likely changes of subsystem technology, or their own system's evolution, over the life of the system. A well-designed interface will have some extensibility. For example, a digital system may initially require 20 different attached subsystems to be addressed, using parallel address conductors on a binary digital "bus." Five conductors can do the job, since they can address 32 (i.e., 2^5) binary addresses. However, if the system grew to include more than 32 subsystems, this interface would limit system growth. Making room for generous expansion usually costs little in the original design.

An interface design may be recognized as an ad hoc standard, after repeated use. Recognition as an ad hoc standard requires that the design be available to, and used by competing firms. If it has been patented, by law its originator can bar others from using it for a 17-year period or charge a royalty for its use.

Companies which are industry leaders face difficult decisions. Their interface designs, especially the external interfaces of their systems, are often patentable. If they wish to interface to certain systems or subsystems built by other firms, those firms must be allowed to employ identical interfaces. To avoid legal complexities, the interface definition is usually placed in the *public domain*. (This means anyone can use it.) This, in turn, may create opportunities for less creative competitors to cash in on the industry leader's work. The alternative, limiting access to proprietary interface definitions, may also constrain the market for the industry leader's products.

Extending existing interface standards An accepted standard is often used as a basis for defining a new interface. However, even modest extension of existing standards can in some instances prove difficult and costly.

Existing interfaces can sometimes be modified and extended and remain compatible with the original. Zylog Corporation's *Z80* microprocessor chip is fully compatible with the earlier Intel 8080 chip in that programs which operate on the 8080 will operate identically on the Z80. However, the converse is not true if the Z80 software is designed to take full advantage of the features of the Z80 chip. This sort of interface is sometimes referred to as having *upward compatibility*.

In extending existing standards to meet special needs, it is more common to have no such compatibility. A consequent danger is that some system designed for the modified interface will be connected to one meeting the original definition. The results are sometimes unpredictable and always unpredicted. If upward compatibility is not required, it is wise to design the modified interface to be so highly incompatible with the standard, that misuse is discouraged.

3.3 Other architectural considerations

3.3.1 System integrity

The term system integrity, as used here, refers to systems which are organized and designed to meet standards of quality and consistency, perhaps with a long useful life. A seven-course dinner served in a formal dining room, with fine vintage wines in hand-cut crystal goblets would be expected to employ fine linen rather than paper napkins. Plastic tableware would be jarringly inconsistent. Likewise a modern information system using high-speed computers but deriving inputs from a Morse telegraph key also seems a likely loser.

These silly examples only emphasize our point. A complex system may be constructed mainly of modern, proven subsystems and components. If only *one* of its critical components is marginal in performance or reliability, the system could fail to meet its users' requirements or expectations. The annals of the U.S. government—as recorded in the public press—provide many examples of systems with severely limiting components or subsystems: in 1988, the electronic warfare system of the B-1 bomber failed to match the aircraft's other capabilities; in 1987 the M-X or "Peacekeeper" intercontinental missile's inertial measuring unit failed to meet reliability standards; 1986's most notable technical disaster was the space shuttle *Challenger*, whose destruction was brought about by failure of a $10 rubber seal in a poorly designed rocket joint, on an unusually cold Florida morning.

The presence of unproven technology or unsatisfactory design elements in key locations in a system may be most obvious during architectural and early design phases if it appears at levels of detail examined during those top-level design phases. Once subsystem design is assigned to independent designers or suppliers, the information required to ensure integrity in the overall design may be overlooked. Certain nagging problems, if believed to be minor, may be ignored in the hope that they will "go away" or that a solution will be found later. No subsystem designer or subcontractor wants its product identified as a weak link. Personal corporate egos often tend to suppress potentially bad news. Suppression of problem *awareness* makes it even less likely that solutions will be found since little or no effort will be spent. Hence, a major responsibility of the system designer is to identify real, suspected, or potential problem areas *as early as possible*, requiring adequate efforts be made to overcome them and demanding that alternatives be studied vigorously.

There is no design formula or process which can guarantee a system design and its real-life embodiment which have uniform, high integrity. The personal and corporate integrity of designers and their organization(s), openness of communication between all those involved, and willingness of developer and client management to cope with "bad news" are essential ingredients. In addition, easy availability of recognized experts in critical areas of system design will make feasible the analysis or experiment required to scope problems and set about overcoming, or avoiding them.

3.3.2 Interoperability: Interfacing with other systems

A system architect faced with design of a system which is to interoperate flawlessly with systems developed by others, elsewhere and at other times, has special problems. These are not insuperable, provided that the systems to which one must interface all have well-designed, correctly implemented interfaces and that their *design and user documentation* is adequate and well organized.

It is not uncommon to find complex interfaces, designed to meet a particular standard, which were allowed to deviate from that standard. Even if such an interface has been actually *tested against* other systems, or simulations, which have been verified to satisfy the standard, certain deviations may not have been detected. A radio transceiver which has not been demonstrated to communicate with others operating with the same frequency and signal standards might very well be useless. (As a boy living on a farm, the author once built a high-frequency two-way radio from *Popular Science Monthly* diagrams. He never heard a voice on the receiver, never knew if the unit worked.)

Interoperability problems are very common in commercial and entertainment systems, especially where interfaces are provided in anticipation of future capabilities. When time comes to put such an interface to use, it may be found inoperable without major modification of the system. Stereo interfaces were, for example, built into FM radio receivers prior to the time FM broadcasts were available and before consumers had stereo amplifiers or multiple loudspeakers. These may have influenced sales but were otherwise mainly impractical.

Demonstrating the correct operation of complex external interfaces depends greatly on the availability of external system(s) with which the system is expected to interoperate. Prudent actions by the system designer will take different forms:

- Where the external system is a *proven design and in use*, there may exist a thorough interface specification. An inquiry should be made directly, through your system client, or perhaps to another owner of the external system which is a prospective customer for your system. Unless the external system is considered a proprietary product or no interface definition was prepared, documentation should be available, perhaps for a fee. A potential risk is that the system's developer will cite standards with which the system was intended to comply, while in reality it does not.

- If *prototypes* of the external system exist, and one can be obtained, it could be used to test the interface. This will often require that *inputs* to that system be generated or perhaps simulated. Correct or representative operation of the external system may require a special environment as in computer-controlled antilock brakes for a vehicle. It may be necessary to piece together simulations, in cases where some aspects of the actual operating environment or system configuration are impractical to set up.

- If a system with which one's system must interface is *under design*, its designers should be requested (or required, if one's contractual situation permits it) to participate in interface definition. This should usually be welcomed by the other system developer, because of the useful and perhaps complementary points of view brought by the sets of designers. However, if the interfacing system's developer builds systems competitive to one's own, cooperation may require persuasion by the client who will pay for these systems. The situation is particularly complex if many firms have similar systems under design; interfaces may be considered highly proprietary and used as vehicle to capture the market. Whenever access to actual versions of the system is infeasible, care in simulation and modeling is even more important.

- If there is *no access* to the interface definition or to a copy of the system itself for use in experimental testing, it may still be possible to actually measure the system interface on some version of the system, using appropriate measuring techniques. If the interface is subject to toleran-ces in its parameters (accuracy of some frequency, diameter of bolt circles, precision of fit, etc.), where feasible, several or all available examples of the system should be examined. Though this is not guaran-teed to reveal true *design tolerances*, experts will be able to infer their likely values from the data and their own experience.

Manufacturers in competitive fields may use interface specification tolerances as ways to maintain their competitive position. A major manufacturer of computer punched-card equipment and of cards might, for example, issue a formal specification calling for more precise dimen-sional accuracy in cards than that actually required by its machinery or met by its own cards. The intended result would be an increase in competitors' costs of card manufacture. Though eventually this ruse should be discovered by any competitor willing to invest in extensive tests, in the short term it might be competitively effective.

3.3.3 Proving new architectures

It is seldom possible, or even desirable, to develop a new system architec-ture "on paper." Though an architecture need not be a system design, it is unwise to propose use of some new architecture without demonstra-tions of its capabilities by some test process. Credibility of architectures in paper form is usually poor. With his artist's skills, Leonardo da Vinci described numerous revolutionary architectural innovations, for systems as diverse as grinding mills and parachutes. In few cases did they achieve realization, though admittedly Leonardo was far ahead of his time.

What is convincing by way of demonstration usually depends on the nature of the system and its departure from recognized architectures. In some cases, extensive computer analysis and modeling will suffice, while in others experimental proof may be demanded. An interesting example was a U.S. Department of Defense proposal for an aircraft employing severely forward-swept wings. This had not been seriously proposed earlier. A forward-swept wing designed and built traditionally would twist under increased bending loads and *increase* "wash-in" (i.e., increase the angle of attack of wingtip to the airstream). This in turn would increase lift at the tip, which would further increase bending, until a wing of normal strength and stiffness would break. The experimental X-29 (Figure 3.6) used fiber composite material in its wings, with the fibers oriented specially. Under increased loads, the wingtips "wash-out," that is, they twist so as to *reduce* the angle of attack. As a result, the traditional breakage problem does not occur.

It may satisfy some technologists to prove theoretically that a principle is sound, but decision makers may take more convincing. In this case, an experimental aircraft embodying the novel wing design was built and

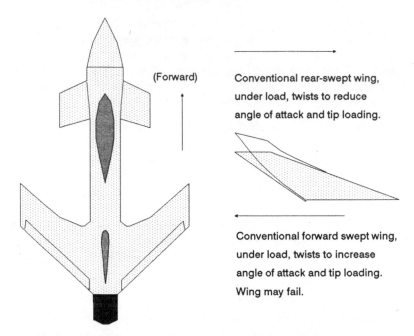

Figure 3.6. A *forward-swept* aircraft wing, if designed in the traditional way will twist under maneuvering loads so as to increase its angle of attack and loading until it breaks. This structurally unstable design is replaced in the Grumman Corporation's X-29 aircraft by a "tailored" composite-fiber design, which twists oppositely to make a forward-swept wing structurally feasible.

flown successfully. Although radical departure from traditional design in one area (the wing) could promote other departures in design, this was not considered essential to the feasibility demonstration.

When a new architecture relies on unproven design methods or technology (the atomic bomb was another example!), it is usually essential to actually demonstrate feasibility and often *utility*, in competition with existing system designs. It is seldom necessary to build and test a full-scale or full-featured prototype. In some instances a carefully orchestrated computer simulation, combined with partial models demonstrating features or construction methods, may demonstrate successfully that the new architecture can deliver what it promises.

3.4 System architecture: Summary

3.4.1 The nature of system architectures

In the art of painting, the visual *appearance* of the product is the major user concern. A painter might argue, however, that arrangement, line, texture, color, etc., are also aspects of user concerns, as indeed they are.

It is characteristic of system architectures to contain many different features and facets of importance to system users.

In commercial building architecture, for example, a pleasing visual appearance of the exterior is certainly appreciated, but there are also many other requirements to be met: arrangements for moving people or things into, out of, and through the building; providing for comfort and other personal needs of building inhabitants; communications, internal and with the outside world; structural, electrical, and cooling provisions for installing and operating machinery for the building's users; and provisions for securing people and facilities against fire, theft, flood, windstorm, etc. The building's architect applies certain design elements repetitively throughout the structure to address each of these needs within space, cost, and other constraints.

In building architecture and in other complex designs, a great deal of creative compromise must be made, more than is typical of traditional engineering design. In a sense the system/building is optimized to meet the needs of all its users. To refer to architectural design activities as processes of optimization would, however, be misleading. System users frequently have divergent views about what is optimum. Traditional math-based optimization theory, in the presence of partially conflicting objectives, leads to mundane least-common-denominator system solutions. Outstanding system architects have the foresight to recognize opportunity and at times to ignore users' viewpoints, which are after all based on past experiences with *different* systems. While the architect respects sound, well-founded traditions, even these may be questioned for applicability to a novel situation.

There is no clear definition of the boundary separating system *architecture* and system *design*. One reason for this is that architectural decisions made without testing through actual system design may later prove unworkable. If architecture and design activities are treated as a continuum, it should be feasible to do modest revision to an architecture, based on design results, without junking it. An architecture which exists only in theory is an academic curiosity, and often more misleading than useful. This was certainly true of many early flying machines, especially those in which designers tried to imitate the wing structures and motions of birds.

A living architecture often continues to evolve with changes in technology or applications. One hallmark of a well-laid-out architecture is *how little* change is needed, despite great changes in the environment where it is used or the nature of its technology, components, and subsystems. Thus it is desirable to define an architecture in terms which are fairly abstract though unambiguous. Because it was defined in an abstract way without rigid definition of the electronic devices to be used, the von Neumann computer architecture of the 1940s has survived from vacuum tubes to semiconductor integrated circuits, from Williams tube storage to semiconductor memory, and from punched-card backup storage to laser disks.

System designers who apply architecture developed by others, without being aware of it, often believe their new product or system to be unique. While this may serve the designer's ego, it seldom serves well the need of the organization to evolve new products from its own earlier products or their markets. The most successful system developers usually create designs evolved from proven architectural approaches rather than seeking change for the sake of change.

3.4.2 Assessing the architectural merits of systems

It is of great value, in an organization which must carry out system design or merely acquisition decisions, to be "able to tell a good one from a bad one." Not all proposed new system configurations are successful. Some may turn out disastrously, as was true for certain nuclear power reactor designs. Selecting an unproven approach has risks. But when a system designer meets unprecedented system needs or requirements, there may be no proven choices which are suitable.

The author believes that an expertly architected system stands out from the ordinary, in certain characteristic ways. These are often readily discerned, even by diligent students of system architecture:

- It evidences an overall unity—its parts do not compete but complement one another, and are similar in quality, durability, and utility.
- It has no parts which appear to be afterthoughts. Likewise there is little waste in its operation, no duplication of parts except that required to fulfill functional, performance, or reliability objectives.
- It exhibits balance, order, and symmetry from many points of view: e.g., internally (through its structure and organization), externally (through its appearance and its ease of access and use), logically (through design relationships), and functionally (through economy of design, meeting the objectives without waste).
- It has not only a sound top-level scheme but its quality holds up in detail as well—close examination of its parts reveals the same qualities and soundness as does the system as a whole.

In practice, far more mediocre systems are designed than outstanding ones. Probably the majority of systems do *not* follow a set of coherent architectural definitions and standards. Design or user bureaucracies in many parts, each with separate objectives, characteristically produce pedestrian or Byzantine system solutions. Design efforts guided by an outstanding designer or by closely cooperating designers, embodying architectural knowledge and awareness, are examples to which other systems are compared.

No system is perfect. In most cases it is impossible to achieve every one of the users' wishes or even the designer's objectives. There are simply too many practical impediments (schedule and cost being principal ones). But if objectives are well defined and properly articulated by the architect(s) and clearly appreciated by designers at all levels, results will be superior to those obtained without such leadership and organization.

3.4.3 Innovative architecture

Architecture typically reflects the distillation of designs which have been attempted and have proven useful or otherwise desirable. It may appear that the term *innovation in architecture* is a *non sequitur*. This is not the case, though architectural innovation may change the form of a system solution so greatly as to make it a distinctly different object, not even identified with predecessors. For example, simple addition of a gasoline engine and drive chain and sprockets to a bicycle had the effect of changing the vehicle to a motorcycle. The bicycle continued to evolve on its own while a new branch of vehicle evolution was begun.

Over several generations of designs a successful system architecture will become refined and consolidated. Those designs which prove successful may survive in easily identifiable form for millennia. What we call the *Venetian blind* was no nineteenth-century invention. Examples were found in ancient Rome. As with architectures of plants and animals, system architectures may if desired be represented by evolutionary tree structures (Figure 3.7). Peculiarly, rate of architectural evolution of a given system is often high for brief periods but static for far longer ones. Evolution of Egyptian pyramids and Gothic cathedrals each took only a few hundred years. Evolution of the modern skyscraper occurred over less than a century following design of the Eiffel tower. In each instance, environmental changes have caused reevaluation of the utility of the architecture.

Rationales for ceremonial royal tombs in Egypt changed drastically, following the building of the huge Giza pyramids. Because the pyramids failed to protect the pharaohs' earthly remains from robbers, and because the Egyptian capital had been moved south, to Luxor, pyramids were no longer in fashion. The new generations of royal tombs were secreted underground. This new arcane architecture proved as unsuccessful as the pyramid in preserving royal relics.

Gothic cathedral innovation took the principal form of opening side walls of large churches, to let in more light, through ever larger stained glass windows. The roof was also raised, ever closer to heaven. However, no new materials or technology was added to aid this structural evolution. The result was self-destruction of some of the later, and taller, cathedrals. This demonstrated, dramatically, practical limits to certain cathedral

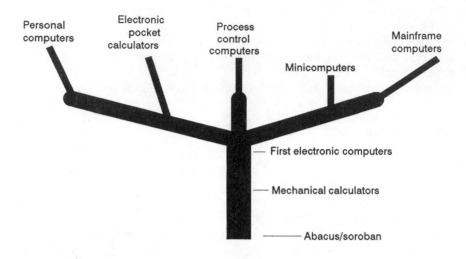

Figure 3.7. A computer "family tree," tracing digital technology from the abacus to supercomputers.

dimensions. These limits have been carefully observed in later designs making use of the same materials and construction technologies.

Though evolution of the skyscraper has certainly not reached its technological limits, the rationale for high occupation density in U.S. business districts has fallen to improved communications and to new management styles and employee lifestyles.

Architectural innovators are seldom aware, as they do it, that they are creating something of permanent or lasting value. Our observations, for more than 40 years, of the evolution of electronics and of computers suggest that innovators seldom recognize immediately what turn out later to be the most important innovations. Those who remain unaware of the history of past attempts to refine and innovate systems in their areas of interest are, however, less likely to come up with important advances.

EXERCISES

1. (a) Develop personal requirements for a system which would consolidate your personal computer and peripheral equipment, study table or desk, and associated storage needs into a single new system. Imagine this as a system comprising one or more coordinated pieces of furniture which satisfy various requirements for convenience, protection, comfort, lighting, storage access, power supply, and other features. In this part of the exercise, do not attempt

to visualize a design, i.e., do not concern yourself with anticipating conflicts between requirement solution elements. Merely list a dozen or more requirements, in the order in which they come to mind.

(b) Organize the list of part (a), collecting requirements in functional groups, and ranking them according to importance, within each group.

2. [We recommend that you work out parts (a) and (b) of this exercise prior to reading part (c).]

(a) Characterize the principal architectural features of your public telephone system as you, a user, understand them.

(b) Define a set of user requirements which, if public telephone systems had not yet been created, might have led to this system. The requirements should not define the communication medium (wire), selection mechanism (dial or push buttons), or any other hardware details.

(c) Given the requirements you used to answer (b) above, in what ways (if any) would a radio communication system, with transmitter and receiver at each residence or place of business, have failed to meet them?

3. Clients, consciously or unconsciously, sometimes adjust their requirements to suit characteristics of some available solution. List separately, in descending rank order, your personal top six requirements for a personal motor vehicle. Alongside each requirement, describe briefly the choices of motor vehicles you believe you would have available, based on currently manufactured vehicles, if that requirement and those above it in the list were applied. (Have you limited yourself unreasonably?)

4. State, in one sentence, what you believe to be the system concept (the function of the device or system, and the means used to achieve it) for each of the following familiar items: (a) a conventional manual typewriter; (b) a man's electric shaver; (c) a paper staple (not the stapler!); (d) a ball-point pen; (e) an elevator; and (f) a rotary lawn mower. (A test you can use to rate your description is that the function and nature of the item it describes should be understandable, even if it had not been invented, but technology to build it was known.)

5. (a) Using an automobile as an example of a system, partition the system functionally, that is, according to functions and subfunctions, down to a second level, or hierarchy, of decomposition. Be careful to keep function definitions as generic as possible.

(b) For each functional item you have identified on your two levels of detail, describe the physical subsystems of an auto which perform the function.

(c) Repeat (a) so that, while it remains a functional decomposition, each of its elements corresponds to specific physical parts or subsystems in the car. (You may need to redefine functions to avoid the need to include a given physical part in more than one functional subsystem.)

6. Use the seven-step partitioning method described on p. 61 to describe a stereo entertainment center whose features and layout you are familiar with. For the first step, create a list of features and functions which represents the real system. If you reach any point where the process appears to fail or become ambiguous in this application, attempt to explain why.

7. For a specific personal computer model, define all interfaces with which a user might be concerned. Describe in your own words the function of each interface.

8. Examine the building in which you live. Identify all visible construction elements which appear to be modular in dimension, given dimensions and shape of each module. (Module dimensions are usually characterized by module-to-module spacing after installation.) Include items which are modular in decorative patterns, and identify those modular parts for which special versions are provided, for use at corners or edges.

9. On a conventional automobile, what parts are intentionally used in quantities of more than one, in a modular sense, and why is each used this way? (Disregard wheels and tires. A four-wheel layout is not merely a motorcyle designer's way to distribute vehicle weight over many tires!)

10. (a) Name the different types or examples of external interfaces appearing on a single-family suburban home with attached garage. Describe special functions, characteristics, or requirements placed on each. (Note: You should consider even windowless exterior walls as interfaces.)
(b) Some interfaces are passive or unchanging, while others are changeable or have control features. Identify interfaces in this second category and the nature of the mechanism and its function. (Hint: Doors can be opened to admit people or animals.)

11. Assume that some great experiment requires you to remain, alone, in a small, completely enclosed windowless container, for an extended "mission" of several months. You have available as the sole interface area on the exterior of the container a 3 in diameter opening, which is connected to "outside" through a 3 in diameter curved conduit 100 ft in length. Your container lacks an internal water supply, waste reservoir, electricity source or any other utilities, but your container has capacity to store enough food and other supplies (except water) to last 20 percent of the mission length. What kind of nonremovable interface would you design to fill the 3-in diameter opening? (If it will make you feel better, imagine that the container is buried 100 ft deep in the ground!)

12. Open the hood of your automobile. Identify (a) the interfaces which connect the engine with other subsystems, and (b) the function of each of those interfaces (what it does). (c) Give views on the critical aspects of each interface (for example, " – must withstand alcohol or ethylene glycol at temperatures above 212°F at internal pressure slightly above atmospheric").

13. Describe the interfacing means whereby the steering wheel and column (and optionally, power steering mechanism) in your auto interfaces with the arm connecting the two steering "knuckles" attached to the front-wheel assemblies. What functions does this interface perform? What other criteria must it satisfy?

14. What provisions does your television set or video cassette recorder make for interface to other systems? (Suggestion: Use the owner's manual as a source of information.) Name the external subsystems to which it can be interfaced, and list any criteria (qualitative or quantitative) defining the interface,

whether you understand them fully or not. What selectable interface alternatives, if any, are included?

15. (a) Describe the operation of the present interfaces available, for services other than local phone calls, on your local telephone system. For each service, describe how it is accessed.
(b) What is the rationale for assignment of dialing digits to these services?
(c) If, as a system architect for phone systems, you wanted to introduce 100 complete new services, what arrangements might the customer use to select any one of these services?

16. List at least 10 commonplace user-operated interfaces, found in homes or workplaces, which are provided with a safety feature intended to prevent injury to a user. For each interface, describe the nature, function, and limitations of the safety feature.

17. Based on several interactive computer programs with which you are familiar, describe the sequences employed in user (keyboard or mouse) interfaces to control or select different types of software features. (One such interface function might consist of pressing a key corresponding to the first letter of an action word, to select that action from a list of alternatives.) For each such interface function, describe: the number and type of primitive step(s) taken to initiate, carry out, and terminate the function, and the function performed (selecting between alternatives, drawing a line, ending a pause, etc.).

18. Obtain a copy of the Electronics Industry Association's RS-232C (or subsequent version) serial interface standard at your library or from the Association at 2001 I Street N.W., Washington, DC 20006. Read it, then summarize (a) what aspects of serial communication with your personal computer it covers, (b) what aspects it indicates it does not cover, and (c) which, of the aspects covered, have manufacturer- or user-determined choices of parameter or function.

19. Examine your automobile with the critical eye of a system designer. What interfaces (including, but not limited to, operator controls, door and window actuators, fastening devices, seals, spacers and spacings between parts) can you find which appear to lack the integrity (e.g., quality, uniformity, lifetime expectation, or user friendliness) represented by the rest of the vehicle design? In what way(s) is each such substandard interface inferior to others found on this or other vehicles?

20. Describe at least six examples of interoperability required between systems found in homes or workplaces, and for each how interoperability is assured. It should be interesting to search out examples in which systems of different sources, types, or ages can all be interfaced. (One example, in the United States, is the 2- or 3-pin 110-V electrical outlet, whose voltage, maximum current capacity, and pin dimensions are standardized.)

21. As an initial part of the design of a new underground urban transit railway, you are assigned to plan the overall system architecture. Because of limited underground space available for tunnels, it is decided to select a slightly narrower than standard track width, with smaller minimum radius of curves.

Trains will consist of cars of an existing design but will be equipped with narrow trucks to fit the nonstandard tracks.

Describe a model or computer analysis method whereby the design of tunnels, tunnel forking or joining rail segments, and curved tunnel sections, laid out on drawings, could be checked for maintenance of a physical clearance of at least 12 in between any car and a tunnel wall. Cars are tubular, i.e., of uniform cross section throughout their length. The car cross section is roughly oval but not definable as a simple mathematical function. Trucks (i.e., wheel assemblies) are mounted at standard distances from the ends of the cars. Tunnel walls have a circular cross section whose radius is a function of the number of tracks in the tunnel. (You should sketch some typical situations which require clearance checking, to get a feel for the problem.)

BIBLIOGRAPHY

American National Standards Institute (ANSI, New York) is the major U.S. *interface-standards* organization. (Standards developed by other professional or industry organizations may later be accepted as ANSI standards.) *American Society for Testing and Materials* (ASTM) deals with standards for materials used in systems of many kinds, from metals or alloys through paints and glues. Some professional and trade societies are also active in standards activities. In the computer field, for example, *Electronic Industries Association* (Washington, D.C.) or *Institute of Electrical and Electronics Engineers* are sources of important standards. *Annual indexes* to ANSI or ASTM standards can be found in many libraries.

K. R. Boff, L. Kaufman, and J. P. Thomas (Eds.), *Handbook of Perception and Human Performance*, Wiley, New York, 1986 (two volumes). A major reference on the human side of human-machine interfaces.

E. Ozkarahan, *Database Machines and Database Management*, Prentice-Hall, Englewood Cliffs, NJ, 1986. A wide-ranging treatment of the handling of collections of data, including discussion of unorthodox architectures.

C. G. Ramsey and H. R. Sleeper, *Architectural Graphic Standards*, Wiley, New York, 1988 (8th edn.). Illustrates building techniques graphically. Shows modular systems of many kinds.

D. V. Steward, *Systems Analysis and Management*, Petrocelli Books, New York, 1981. Sections on system structure and partitioning are excellent - a personal view, with good examples.

F. M. White, *Heat Transfer*, Addison-Wesley, Reading MA, 1984. A mathematical treatment, as necessitated by the subject, but containing many practical examples.

Design of Systems

4.1 Introduction

A system design represents the application of system architecture to a specific design requirement. It is the essential step if an architecture is actually to be put to use. In schools of (building) architecture, student and faculty efforts often result in elaborate scale models, which reveal even to lay persons the salient physical features of the buildings. However, these models necessarily exclude all of the actual *function*, and most of the design details. Models may be supplemented by physical realization of critical details. . . a modular section of a curtain wall, samples showing cross sections through the proposed structure, representative floor surfacing material, and hardware items. Though this is still far from the finished product, it is able to convey the important facets of both the architecture and the design details. In fact, the real system/building often does not display architectural concepts as well as a model can.

In evolving complex systems such as vehicles or machines, a combination of scale models, mathematical analysis, and computer modeling of system functioning, plus limited experimentation may be used to demonstrate feasibility of system functions. This activity may constitute most of the architectural portion of system design activity.

Expensive systems to be built in quantity must usually be demonstrated convincingly, at full scale, with most system functions operating. Architecture and conceptual design activities may be followed up by design and construction of a *prototype system*. If only one or a few systems were required, a prototype system may not be afforded. In this case key subsystems, and especially ones which are multiply used, may be separately prototyped to verify feasibility.

Prototypes are seldom complete in all details. For example, most prototype military aircraft (such as the X-29 forward-swept wing fighter, for example, as illustrated in the previous chapter) lack many sensor, weapon, and self-defense subsystems, though these constitute a major part of the cost and of the weight of an operational aircraft. The prototype may serve mainly to convince users of the performance and handling capabilities of the airframe, engines, and flight control equipment. Missing subsystems can for the most part be evaluated separately on existing, non-prototype aircraft platforms. Prototypes may include noncritical components taken from existing designs. In the final design, these will be replaced by new components.

The *engineering design*, which follows architecture and optional prototyping activity, must deal with many additional system aspects. These include manufacturing, assembly, testing, training, compatibility with other systems, etc. Designers must deal with many of these issues. Accordingly much of the rest of this chapter deals with design in the presence of real-world concerns.

4.2 Prototypes and simulations

Why design a prototype system? A "paper design" of an expensive system is seldom an acceptable basis on which to justify its investment. Development of an engineering-prototype system, or perhaps several duplicates, may be very cost-effective when the production system is to be produced in quantity. It is particularly so if there are unknowns in the application or the applicable technology, calling for convincing demonstration. For one-of-a-kind systems, critical but hitherto unproven subsystems will often be prototyped as a form of insurance against failure.

An engineering-prototype development will typically cost less than one-half that of full-scale engineering development. System features or subsystems not essential to resolving uncertainties may simply be omitted from the prototype. Some of the prototype may be built up out of parts from existing systems. Available parts or subsystems which would not meet final durability specifications may be used, so long as the prototype tests can be concluded. Software, too, may be scavenged from suitable sources, the pieces strapped together adequately to carry out limited demonstrations, without assurance of correct operation outside those bounds. Some parts may be hand-crafted out of materials lacking in the durability needed in the production system.

A prototype may even take the form of a scaled-down working model as a means to save costs and reduce certain kinds of risks. Figure 4.1 is a photo of the NASA/Rockwell HIMAT (HIghly Maneuverable Aircraft Technology) scale-down aircraft, directed toward reasonable-cost studies of high maneuverability aircraft technology. It did not impose on test pilots the extreme accelerations of which it was capable.

Even when a prototype cannot perform all critical tasks which have uncertainties, it may still be valid as a departure from which to extrapolate to the production design. However, any prototype must be convincing

Figure 4.1. *HIMAT*, a subscale remotely piloted aircraft whose flight characteristics emulate those of a highly maneuverably military fighter. Extensive instrumentation allows it to be used to examine maneuverability characteristics outside the capabilities of a wind tunnel, and at G-force levels beyond the limits of a human pilot. The unpainted aircraft here reveals the graphite/epoxy composite structure. It was designed and manufactured for NASA, by Rockwell International's Los Angeles Division.

in what it is supposed to demonstrate; a prototype racing automobile limited to 25 mi/h is hardly convincing! A prototype may not employ technologies and operating principles identical to those in the production design. If critical features of the design were faked or crudely modeled in the prototype, it could deliberately mislead and would thus be unethical. The late *Preston Tucker*, who shortly after World War II proposed what was at the time a revolutionary automobile, hand-built 50 prototypes. Tucker was praised as a visionary by many, branded a charlatan by traditionalist competitors. To his credit, most of the 40-year-old prototypes were still operable in 1987, when his story became the subject of a popular motion picture.

A prototype automobile which could not be driven might faithfully represent the styling, but it is hardly a reasonable basis for investment in a new auto firm. Prototype system designs may be highly convincing, especially when they closely resemble in appearance the proposed system. A client, after paying for a prototype design, may raise questions as to the necessity for the expense of a new design for production. The designer, explaining the prototype's many limitations, may appear self-serving. It is thus important to explain clearly, before prototype design and demonstration, how the prototype design will help resolve system issues, and how it can differ from the production system.

Commonly, only a single prototype is built. Complete or partial *copies* may be arranged for if there is significant risk that a prototype may be damaged or destroyed during its testing and demonstrations. In development of systems which will be produced in large quantity such as automobiles, *successive prototypes* may carry the design forward incrementally or serve different demonstration objectives. Both an *engineering prototype*, and later, a *manufacturing prototype* may be developed, if a system is to be manufactured in large quantities.

System simulations may be employed where it is not feasible to demonstrate certain essential system features or capabilities via an affordable prototype. Some simulations stand alone. Others may complement partial-system prototypes or demonstrate system features still under study. Simulations are also useful for studying the characteristics of large systems or large *numbers* of systems, as with communications systems or weapon systems. Even though one cannot reasonably prototype a global network of optical-fiber communication, demonstrations on a limited network plus modeling of larger ones represents a workable compromise.

4.2.1 Technology demonstration prototypes

A prototype may be intended to serve as a showcase for new technology or as proof that an untested technology has practical applications. In such cases, it may be so early that the forms which systems applying the technology should take are unclear. While the designers of aircraft jet engine in England and Germany had waiting applications in military aircraft as World War II approached, the designers of the first internal combustion engines had no ready market among coachmakers. If there is no valid user requirement for a new technology, it may mean merely that users are unaware of what is possible.

If a new technology is well suited for application in a known and important class of systems, then the technology demonstration prototype should where possible be a familiar and well-regarded system. A new concept in aircraft wing variable-cross-section was demonstrated by the U.S. Air Force by installing a wing using the new technology on one side of an F-111 fighter-bomber, with the original wing on the other side. Since the technology was not expected to make drastic change in flying qualities, at a cost little more than one-half that of two wings, both the compatibility and benefits of the new technology were demonstrated.

There are many other ways to reduce cost of technology demonstration prototypes. *Parts* of proven existing systems, which may be readily available at quantity-production cost or as spare parts for the system, can be incorporated to reduce prototype cost. Even obsolete systems rescued from the junkyard have provided a useful basis for technology demonstration prototypes.

Many prototype systems allow for cost-saving by omitting subsystems unessential to demonstration. Systems which in final form would contain more than one of a certain component or subsystem may be demonstrated using minimum population of these items. Although combining several new technologies in a single prototype system may increase risk, the cost *per technology* can be significantly reduced. This is particularly valid if the technologies are not required to operate in concert. Demonstrations *combining* untried technologies are often frustrated by inability to make all the untried elements work at the same time.

Technology demonstration prototypes usually do not emphasize "-ility" issues such as reliability, maintainability, availability, or ease of use. But these ownership factors continue to grow in importance even where technology advances are most important. If "ility" factors such as *reliability, availability,* and *maintainability* (discussed in Chapter 5) must be dealt with, prototype costs can be expected to *at least* double, because of the extra design and analysis required and the emphasis on long-term utility.

The term *breadboard prototype* is applied, especially in electronic systems, to prototypes which bear little physical resemblance to the production system. Well into the 1950s, prototype electronics assemblies were often assembled on actual boards sold for use in slicing loaves of bread. This prototype, though no longer on an actual breadboard, is usually packaged casually and built quickly, and is often limited to testing in a laboratory environment. It often represents only a partial-system demonstration. A similar low-cost scheme for solid-propellant rocket motor tests may employ a motor casing made of heavy metal. This can be repeatedly filled with propellant and test-fired, its nozzle pointed upward, to determine characteristics of propellants, nozzle designs, etc.

The term *brassboard prototype* is applied to prototypes which are closer in size, weight physical appearance, and operational details to the final system but may differ significantly in constructional detail. This class of prototype is usually designed for successful operation in the real system's operating environment. Its function is often the demonstration that critical performance, size, weight, or power requirements can truly be met. This does not mean that in every respect it will resemble the final system. It may, for example, be limited in its number of hours of operation before it is no longer safe or reliable. Or, it may be deficient only in certain ways which do not detract from its utility as a demonstration system.

4.2.2 Prototype limitations

Prototypes are usually put together on tightly limited budgets, most particularly when they are speculative offerings of a firm which is attempting to locate enough customers to warrant a system's development and production. Prototype testing can be very expensive not only because the system environment must be conquered or simulated but also because extensive technical support may be required merely to keep a prototype operating, to repair accidental damage, and to vary and record many system parameters.

If multiple prototypes are demanded by destructive prototype testing—such as for missiles which will be test-fired—these early units are likely to vary more, unit to unit, than would final production units. This often means that tests of prototype copies require some interpretation in view of slightly different characteristics for each prototype system.

Reliability and dependability of prototypes are often poor because these factors can be afforded too little attention during design. Prototype systems also may be far from optimum in ease of use if, during tests, they are to be operated only by design engineers and development technicians.

Prototypes are frequently equipped with adjustable or replaceable elements to allow for parameter adjustment and optimization. This "fine-tuning" may not be provided in a fielded system for a variety of reasons: adjustment may not be possible, in the operating environment; time, essential test equipment, or personnel may not be available during field operation; or costs of providing adjustments may not justify them. Too often, provision of adjustability may increase likelihood of *mal*adjustment by well-intentioned but inept personnel.

System developers or their clients should not as a general rule expect development of prototypes to reduce overall *cost* of system developments. It should, however, be designed to reduce the risk that systems will fail to meet user requirements. Reducing full-scale engineering-design costs by incorporating hastily assembled prototyped designs *in the production system* is usually unwise. The cost and inconvenience of maintaining and enhancing these systems will, over their life cycles, offset initial savings of time or funds.

4.2.3 Systems engineering simulations

Though not usually classed as prototypes, system engineering *simulations* may be a substitute for them in some cases. Their usual role is to resolve technical uncertainties, prior to initiating engineering design. Simulations offer additional potential for showing potential clients how the system can be expected to operate in their system environment and how their operators would run the system. Simulations may also represent certain problem situations which might damage or destroy prototypes and injure or kill personnel. Engineering simulations may also make possible quick experimentation of the results of system parameter changes of sorts infeasible with a prototype. However, in other respects simulation may be inferior to building an operating prototype.

Engineering simulations assume a myriad of forms including:

- Full-scale prototypes lacking certain features or characteristics required in the final system, for example experimental aircraft which cannot take off from the ground but must be launched by a large aircraft, as were the supersonic X-1 and X-15 aircraft
- Subscale models, for example wind-tunnel models of aircraft, model-basin models of ship hulls, or scaled-down radar antennas operating at frequencies scaled-up from those used in production systems
- Unmanned subscale versions of manned systems, remotely operated through radio telemetry or programmed to follow a specific sequence
- "Hardware-in-the-loop" simulations, in which prototypes of certain critical subsystems are combined with computer simulations of other subsystems in a simulated operating environment
- Manned flight simulators used to adjust and optimize handling characteristics of the production aircraft
- Environmental simulations, wherein a prototype system or subsystem is exposed to simulated environments similar to those expected in operational use. The environment may be climatic, as in refrigerated test facilities simulating polar environments, or using simulators of stressful electromagnetic or nuclear-radiation environments
- Scenario simulations, in which computer models of many systems (including ones under study, as well as others) are simulated in an interactive mode to study interoperability, or a system's competitive performance

Engineering simulation differs fundamentally from *training simulation*. In the former, the objective is understanding and modeling the technical characteristics of the system. Training simulators (Figure 4.2) model a completed system, often in a simulated environment, so that system *operators can be more safely or less inexpensively trained*. While the typical engineering simulator may include many adjustable system parameters or a variety of software routines for modeling a variety of systems over their useful lifetime, training simulators usually model a particular system. Training simulators often include elaborate means to test and score the performance of system operators before, during, and after a course of training.

Simulators are essential to designing any systems for which realistic, full-scale tests are impractical. In the initial evaluation of missile seeker designs, for example, it may be too costly to fire statistically significant numbers of instrumented prototype missiles. Each seeker would be destroyed after only a single test and causes of failures might never be

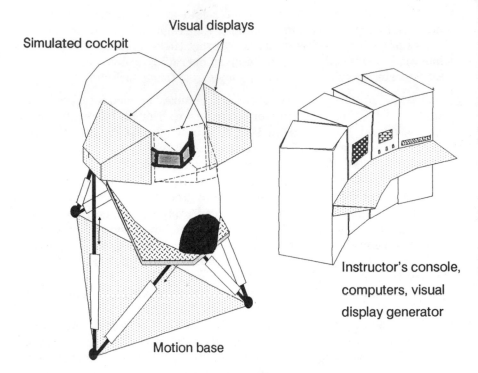

Visual displays

Simulated cockpit

Instructor's console,
computers, visual
display generator

Motion base

Figure 4.2. A flight simulator cockpit may be mounted atop a hydraulically positioned platform which can simulate motion. Pilots see computer-generated scenes on large displays. Instructors set up simulations and imitate flight controllers.

learned. A hardware-in-the-loop simulation usually places prototype subsystem hardware in a simulated system environment, for example testing only the missile seeker, with simulated missile and simulated operational environment. The environment is usually produced by some combination of computer and physical device functions. In a missile seeker test, an actual infrared (heat-seeking) seeker could attempt to "track" a heat emitter located at a fixed distance, in the laboratory. The seeker is mounted in mechanical gymbals which, computer-controlled, simulate the relative angular displacement of the target produced by the relative motions of a real missile and its aircraft target.

Many tests for nuclear weapon effects on military systems are simulated in whole or in part since testing using above-ground nuclear blasts is no longer permitted. In this case, *blast effects* may be studied using conventional explosives and scaled-down models of the systems being tested. For testing nuclear radiation effects, a variety of simulations are required, including controlled underground nuclear explosions, high-energy-particle accelerators, X-ray generators, and voltage-impulse sources.

Simulated operation is often used to determine useful life of system elements, and modes of failure in long-term operation. Automobile doors, brake linings, or power tools may all be tested to failure by "exercisers."

Engineering simulation techniques, hardware, and software used by system contractors or major system-acquiring clients may evolve over many years and through design of many systems. This equipment is among the most expensive of development laboratory aparatus. For example, a jet engine test facility which was constructed in the early 1980s, able to test large engines with representative airflow around them, cost over one-half billion dollars at the time.

4.3 Engineering design processes

4.3.1 Introduction

A system developer's *engineering design process* would include all of the following, if every aspect of design and system construction were to be performed by the system developer:

- Concept study and definition (affirmation of requirements)
- Partitioning (subsystem definition)
- Modeling of critical system and subsystem features through subscale models, computer simulation, or fabrication and testing of hardware and software elements which are critical or previously untested in a system
- Specification of subsystem hardware and (if applicable) software requirements including interface definitions, in sufficient detail to ensure the desired operation
- Detail design of hardware parts, including necessary housings or mounting hardware and ancillary parts such as switches, valves, control devices, power, or fuel supplies
- Development of initial or prototype versions of software programs
- Fabrication of one or several full or partial sets of hardware parts, depending on whether destructive testing is required
- Initial testing of software on simulated systems
- Assembly of hardware subsystems, and that testing which can be carried out at component or subsystem level (The term *component* usually refers to a part acquired from a specialty manufacturer which is not disassembled by the user or system developer. When no longer serviceable it may be be discarded, replaced, or "traded in" for a new or rebuilt component.)
- Assembly and partial testing of ensembles of subsystems which may be referred to as "major"

- Assembly of the complete system followed by testing which includes both operation in simulated system environments and diagnostics testing to discover failed or failing parts

These engineering design activities are carried out using methods which are characteristic of the type of system under design. Electronic systems, for example, are designed with extensive use of computer-aided design (CAD) programs used to design integrated circuits and printed-circuit boards. These design tools also support simulated "testing" to verify if performance will meet requirements. Mechanical parts and assemblies are now designed using CAD support. Computer aided design, however, uses two-dimensional computer displays. For three-dimensional systems or parts of complex form, physical models are often needed to study appearance, fit, or clearances and as a basis for design of production tools.

In developing almost any kind of system hardware, the first one or more sets of hardware are usually produced by methods less automated than will be used in final production. The full-featured and expensive production tooling (hardware and software) is generally not produced until the design is mature or a production contract is in sight. Electronic circuits may be built up on generic circuit cards using discrete wire conductors rather than the etched copper sheets of "printed circuits." Sheet metal shapes such as aircraft or automobile bodies may be hand-formed on handmade wooden molds. Noncritical subsystems such as power supplies or fuel supplies may be replaced, for initial tests, by laboratory power supplies or fuel piped from a "tank farm."

The activities listed at the beginning of this section might characterize a system developer which also develops and fabricates hardware. Many system developers develop and fabricate *only* elements which unify the system from subsystems developed and produced by others. The system developer must develop adequate specifications for use by subsystem contractors. It should take central responsibility for developing the system-unique software. The organization must have expertise in system testing, diagnostics, and evaluation, though they will often carry out these activities with the assistance of subcontractor personnel.

In the United States, aircraft developers are representative system firms which subcontract a large proportion of subsystem design and fabrication (Figure 4.3). An aircraft firm usually designs at least the airframe, fuel, electrical and hydraulic subsystems, and critical aspects of flight control. Other subsystems may be developed and manufactured by specialty firms, but installed and tested by the system developer. The engines are developed and manufactured by one or more of a small number of specialty firms. Commercial jet transports may be designed to operate with alternative U.S. or foreign-built engines. When the cost of engines is added to those of other purchased subsystems, the total cost of purchased parts and subsystems may exceed 50 percent of the aircraft's total

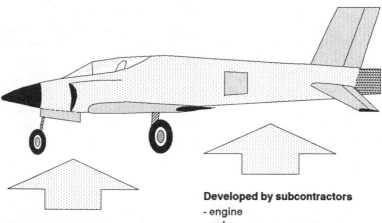

Developed by system builder
- airframe
- structural interfaces for subsystems
- flight controls
- landing gear mechanisms
- electrical and hydraulic systems
- electrical distribution and control
- interior and exterior lighting system
- stores management system
- central control software
- fuel system and controls

Developed by subcontractors
- engine
- radar
- navigation system(s)
- flight control computers/software
- cockpit displays & instruments
- radios
- wheels, tires, brakes
- antennas
- hydraulic valves and actuators
- radar warning receiver
- jammers
- ejection seat
- hardware and fittings

Figure 4.3. A representative *physical decomposition* of a modern fighter aircraft into subsytems and parts developed by the system contractor and those obtained from subcontractors.

cost. Military *shipbuilders*, whose development and manufacturing activities may be limited principally to the hull, account for an even lower portion of design and manufacturing costs.

By contrast, major auto manufacturers have positioned themselves to design and manufacture a large fraction of parts and subsystems, including power trains, bodies, and trim, and most of the luxury and comfort subsystems. This approach represents *vertical integration* of the business, i.e., doing a large fraction of the manufacturing, from raw materials to finished systems. Although vertical integration is found in a few large computer firms, electronics-industry firms are mainly assemblers. Both auto and computer manufacturers in the United States make extensive use of the products of specialty firms which produce huge volumes of automotive or electronic parts for both new-production and the spare-parts markets.

Architectural design of new systems is usually assigned to a small part of the system developer's organization, which may be titled "Conceptual Design," "Architecture," or perhaps even "The Skunk Works." Engineering design is characteristically distributed widely over the organization and external subcontractors. The system contractor may also distribute schedule and cost risks by writing contracts which require subcontractors to invest in tooling or provide penalties for late performance. This results in subcontractors sharing part of the financial risk of poor market acceptance or inability to meet system goals.

It is always difficult to truly *unify* system design when most major and minor parts are designed and built by independent organizations. Most successful system developing organizations, therefore, maintain design and (at least low-volume) fabrication capability for critical subsystems needed in their products. This is especially true of parts or subsystems essential to system integrity and performance, which represent the system firm's competitive edge. Since most modern technical systems make use of programmed computers, most system firms now require competent in-house software developers.

4.3.2 Partitioning and interface definition for integration and testing

Earlier, we discussed briefly *system partitioning* as an architectural activity. The major objective of this activitiy is to define a set of subsystems which will reduce complex interdependencies between subsystems, thence to define subsystem interfaces whose definitions exist or can easily be defined.

These objectives remain part of engineering design. Now, interfaces and subsystem descriptions, often incompletely described in the architectural description of the system, must be fully detailed. Where architecture is to serve as a basis for a collection of systems, nonstandard interfaces should be detailed as part of the formal architectural description.

In defining detailed subsystem specifications, one's work must be guided by anticipation of the tasks of integration and testing which will be carried out later. To be certain that integration and testing are dealt with adequately by the system engineers defining subsystems and interfaces, some of these same personnel are often assigned to manage or carry out these later tasks.

While final stages of testing usually require the system to be fully integrated both internally and with respect to its operating environment, important advantages occur if most tests can be carried out on smaller ensembles of subsystems and parts (Figure 4.4). A phrase describing this process, often applied especially with regard to software, is "build a little, test a little." With only a small portion of the system under test, isolation of *local* faults should be simplified, though some full-system faults may not evidence themselves. A second advantage is that several tests can be carried out in parallel, on different portions of the system, if test equipment and personnel are available.

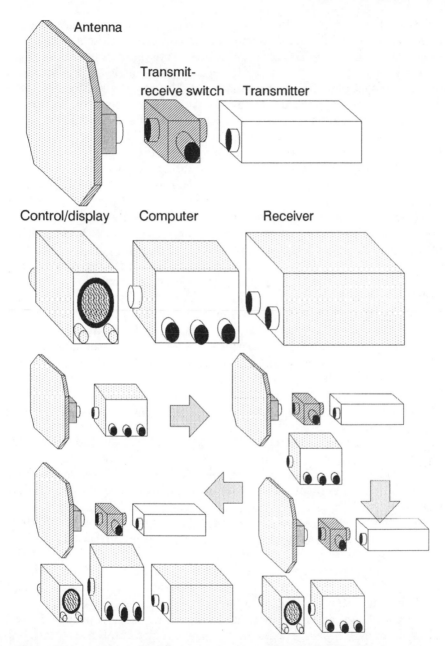

Figure 4.4. A hypothetical but reasonable sequence for integration and testing of the six major components of an aircraft radar (top). In the first phase the computer is integrated with the computer steered array. The transmitter and transmit-receive switch are added, permitting transmission tests with an external receiver. The display, whose output is computer-produced, is tested with simulated output from the computer, and finally the receiver is integrated. A variety of test instruments will be required.

From an integration and testing viewpoint, an optimal partitioning might be slightly different than one based solely on operational considerations. Imagine that some signal-generating subsystem were required as a source by only one subsystem. It might best be made an integral part of that subsystem. Then certain tests could be completed by subsystem developers, rather than being delayed until full-up system integration. It would also eliminate a need to design an interface between the generator and the subsystems it drives, since both are now in one subsystem. If, however, those signals are to be employed as inputs to *several* subsystems, designing a separate signal-generating subsystem could allow separate and parallel testing of a signal generator, with each dependent subsystem.

Consider another example: a system having unified on-line monitoring and diagnostics-testing in many or all subsystems. With nonunified monitoring, during engineering design, each subsystem design organization would need to determine measurement points, viewing ports, and a strategy for suitable testing. Overall, this could result in many new system interfaces, special test equipment and the like. If an integrated on-line monitoring concept were designed in from the start, optimal system partitioning and hence interfaces would probably change. Though this would call for additional coordination with subcontractors prior to and during design, it could reduce system integration costs and risks.

In architectural description of many kinds of systems, as, for example, the von Neumann computer architecture (Figure 2.5), little or no attention may be devoted to *physical* design and arrangement of subsystems. In engineering design, however, this is important. Take the case of an automobile engine compartment. Arrangement of engine and drive train subsystems will greatly influence the ease of assembly and maintenance. Figure 4.5 illustrates some of the variety found in production automobile drive-train arrangements, each having some benefits and other limitations.

Physical rearrangement of subsystems may suggest improved partitionings. An automobile engine's coolant pump ("water pump") is not required to be integral with the engine. In a typical engine design, a part of the coolant pump housing is integral with the engine block. This reduces difficulty in locating the drive pulley and at the same time eliminates need for additional interfaces between the pump and the engine block.

Rearrangement and repartitioning required during engineering design of a system may represent changes to, or expansions of, the architectural and interface definitions. Most often, however, they occur at levels of detail lower than those which would be termed architecture. A competent architect usually attempts to limit those rules and constraints which are defined as a system's architecture, lest these same rules and constraints be later found incompatible with needs to develop a new and improved

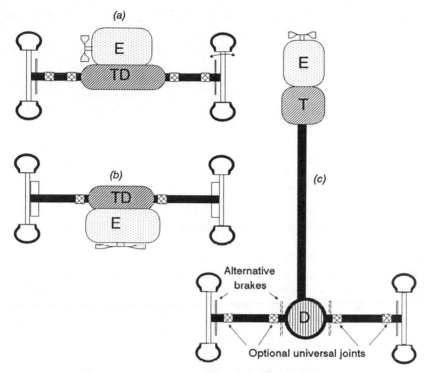

Figure 4.5. Three popular automobile drive-train configurations. Top left, that most commonly used for front-wheel drive, with two universal joints per axle. At lower left, the familiar Volkswagen "beetle" rear-engine configuration, and at right, the traditional front-engine, rear-drive configuration, usually employed with a rigid rear axle. Each configuration has inherent influences on handling and weight balance. In each case, E is the engine, T the transmission, and D the differential. A clutch, if installed, is located between engine and transmission.

system with the same architecture. A system designer, on the other hand, will often claim even a system's *least* significant features as important parts of the design.

4.3.3 Technology dependencies

Some systems are developed using only proven technologies whose principles are well known and understood, and only parts which are immediately available for purchase. Because system development firms operate in a competitive environment, however, many new systems incorporate technological advances not used in prior systems. Committing to a particular technology is referred to as *technology dependency*.

Technology dependencies usually imply *technology risk*. Less proven technologies, or technologies that are critical to successful system operation impose greater risk. These risks may include uncertainty of delivery dates for system components. Initial specifications may not be fully satisfied when the first components are delivered. And, if risky components are required in quantity, delays may occur while their production reaches suitable rates.

It is only natural for specialists in a technology relied upon for a system to be somewhat optimistic about schedules and other problems. System designers, who may be unfamiliar with some specialized or newly emerging technology, are often unable to perceive accompanying difficulties. They may, themselves, become wild-eyed enthusiasts and throw caution to the winds. Or they may be hopeless pessimists, assuming gloom and doom from the outset. If allowed only these two limiting behaviors, the latter is usually safer. Unfortunately, it is sometimes a fact that there *is* *no* substitute for the new technology, given the system's requirements. In this case, the risks *must* be assumed, but prudent steps should be taken to ensure success. Here are some things they can do:

If the risky parts are being developed in the system organization, expanded or parallel efforts can be put into place. If the parts or subsystems using the technology are being developed elsewhere, however, system designers lack control over allocatable resources. The problem may be even worse if technology development *is* within the company, but in a part of the organization where the priorities and sense of urgency are different.

Direct financial and personnel involvement of the system organization in the technology effort are desirable. If work is supported in significant part by funds of the system developing organization, its result can usually be kept exclusive to that organization for a period of time. Financial support also provides opportunity to monitor progress. Temporarily placing a few selected technical personnel within the technology organization (for example, to assist in device testing) builds a natural route for technology transfer and should minimize transition time. If no suitable personnel are available, a less satisfactory option is to call upon an independent outside consultant having good credentials in the technology. The consultant's subdued participation in reviews and visits, followed by discussions with system development personnel, should provide a clearer view of the prospects for success.

Even the most promising of technologies may fail to meet objectives, or the work on it may fall so far behind in schedule as to discourage its application for time-urgent system developments. Wherever success or failure of a system hinges on 100 percent success of challenging technology goals, a backup solution (better yet, several alternatives) is essential. Where use of some proven alternative requires easing of system specifications, system engineers should make a detailed study—in advance—of how results of that shortfall could be minimized. It is essential to develop one's fallback plans long before the situation can become critical, for any technology not fully available at the time commitment is made to a system client.

4.3.4 System integration and testing

System integration is analogous to the final examination of a course in system architecture and design. If preliminary work has been done thoroughly and well, integration may be a routine exercise. If not, it can be a jumble of slipped schedules, with wave upon wave of costly design iteration.

As what may appear an almost trivial example, consider a situation in which too low an estimate was made of system power supply current requirements for some subsystem. All critical parts have been developed and models delivered, assembly is underway, and at the last moment someone discovers that the power supply ordered "won't hack it." A likely next discovery is that no available power supply of adequate current rating will fit in the allocated space. A system size-reduction exercise may be feasible, but with very uncertain added cost and time. The only remaining option is redesign of the subsystem housing, and possibly its entire internal layout.

Even so minor a lapse as an interconnecting hose or cable too short in length or too large in diameter will add cost and delay. When multiplied by many subsystems in a complete system, profits can evaporate. Prospective customers may not wait. *Successful integration, therefore, means correct initial design.*

Mechanical parts, assemblies, and interfaces provide a significant share of integration difficulties. Depending on the nature of the system, a variety of techniques can be used to ensure near-straightforward integration. . . though no one is perfect. Techniques as simple as using wood or foam plastic blocks, cut to represent system parts, or lengths of plastic tubing to represent hoses or cables may be invaluable during mechanical design of almost any system. Even the most sophisticated three-dimensional graphic computer design aids may not reveal some problems involved in a mechnical assembly, while simple models could solve them quickly. Skill in any area of engineering design involves not only knowing the physics and mathematics which are applicable, and how to apply them, but having in hand a wide variety of applicable design methods and tools.

Simulation of unavailable system elements System integrators must often face the problem of integrating systems for which there are available only quantities of one, or a few parts or subsystems which occur repeatedly in the system. Some parts or subsystems may be unavailable, and external systems are frequently not available for testing. In these cases, skilled system integrators find ways to improvise well enough to expose a large proportion of the integration problems.

One may, for example, be integrating a system which in operation must operate 32 display terminals at remote locations. None of the remote locations may yet exist, and only one or a few terminals may be available for the system integration and test. What to do? A first ploy may be to connect the available terminal(s) to each terminal connection point in turn, verifying that in *each* position, operation of a terminal will be correct.

Simulating the computer "load" imposed by a full complement of terminals will be more difficult. A knowledgeable integrator might locate some channel in the system through which all the terminal input and output passes, and connect a computer at that point. A relatively simple program, executing on that computer, could be used to generate an approximation to the activity of 32 terminals. In any event, this setup could, in skilled hands, attempt to characterize the bottleneck and how it depended upon the number of online terminal users, their activity rate, and the distribution of activity types.

This type of integration procedure, even if it must be to some extent *ad hoc*, should be planned far enough ahead that its implementation does not postpone the system delivery schedule. Testing techniques involving substitution of a computer to simulate high rates of system activity are actually superior to "the real thing" involving human operators. It is difficult to locate 32 human terminal users who would pound away steadily in representative fashion for as long as a test required. For the computer, it's easy. However, a study of a human operator using a terminal to perform typical system activity should be carried out as a basis for the multiterminal computer simulation.

Integration and developmental testing: Procedures Developmental testing should be included in the discussion of system integration because in practice the two are closely interwoven. Each integration operation such as the connection of two subsystems should usually be followed by tests to verify that the correct interfaces are established. In integrating even so mundane a system as a simple residential plumbing system, supply and drain piping subsystems are first installed, and each separately pressure-tested. Then the plumbing fixtures (sinks, toilets, etc.) are connected, and each part of the complete system is tested.

In most cases, *components or subsystems manufactured elsewhere* should have been subjected to pre- and/or postdelivery tests. Unless contractually required, one cannot ever assume that 100 percent of components have been tested. In the case of small and low-cost units such as electrical switches or hydraulic fittings, it is possible that *no* testing is performed by the manufacturer after initial development of the item. In some cases, small items may not even be 100 percent *inspected* to verify their visual correctness or basic functioning.

The contractual conditions under which suppliers provide parts or subsystems can influence their practices with regard to testing prior to delivery. For example, if no warranty were required, a supplier might conceivably include parts known to be defective. With a warranty requiring mere replacement of out-of-spec parts, some suppliers may include enough "extras" to compensate for units found to be bad. If quality standards or limitations of the integrator's test ability require more stringent testing by the supplier, this must be negotiated in the contract. It is always prudent for a system integrator to require that suppliers of components or subsystems provide recommended *test procedures*, by which the purchaser will be able to verify that units meet specifications *after* receipt, perhaps even after extended storage. Shipping damage is fairly common. Long periods of storage are not uncommon in large system development and are the normal expectation for nonexpendable repair parts. Many parts or subsystems, especially those containing organic materials with required physical properties such as elasticity, will deteriorate to some degree in storage and may benefit from storage in oxygen-free or reduced-temperature locations.

A *schedule of integration and testing* should be developed for any system requiring careful or orderly assembly of its components and subsystems. Typically, this should describe in detail (1) steps of assembly and interface connection, (2) developmental tests which are to follow certain integration steps, (3) setups required for tests, and conditions under which the system (or parts of it) are to be tested, (4) definition of test success or failure, and (5) the "rework" required after predictable types of failure. Certain tests, for example, voltage or pressure measurements, will be of simple types and can be referenced by appropriate test standards. Assembly and test descriptions should be accompanied by photographs or schematic drawings, illustrating proper practices and warning against errors known to be likely.

Steps in the integration and testing schedule should be clearly identified by a sequence number and the associated step(s). Any minor hardware such as nuts, bolts, washers, and rivets required for assembly must be uniquely identified; even these simple parts may be critical to correct assembly and reliable operation.

In carrying out the sequence of integration and test steps, progress made against the schedule may be reported using formatted "checkoff sheets," computer forms, or direct computer entry. Completed assembly and test steps should document completion date and time, describe detailed results (including both anticipated and unexpected results), integrators' comments or observations, and full identification of the participants. Failed tests or assembly difficulties must be carefully detailed during system integration. In complex systems, it is not uncommon for the integration work to halt, and specialists to be called in, until a problem is fully understood and solved. Where parts or subsystems must be returned to distant sources for rework, a report showing the reason for failure or other rejection should accompany each unit or each group of units which failed for a given reason.

Clearly, good documentation is central to system integration. A problem may be met with again while integrating a duplicate system, or perhaps during operation of the completed system. If the initial integration activity records its experiences (both bad *and* good) in orderly form, this can serve as a valuable source for many following activities including the preparation of maintenance manuals.

Scheduling integration and developmental tests The principal objective in system developmental testing is to verify that the system *satisfies its specifications*. When this is not true, the principal objective of designers becomes the *localization* of problems. The smaller the group of subsystems to which a particular fault can be attributed, in general the easier will be its location. This emphasizes the importance (and, in many cases, essentiality) of first testing individual subsystems and then, interconnected minor groups of subsystems, prior to attempting all-up system operation. It is particularly important in integrating and testing the first, or only, system of its kind. Careful recording of results should limit the scope of the search for causes of failure and thus minimize the delay before integration can continue.

An *order of integration and testing* for system integration should be established well in advance. It should depend on the factors discussed in the three following subsections.

Subsystem interconnections In assembling those systems which process communication signals or digital data, subsytems usually form a chain (which may also be called a train or cascade) whereby the output of each is directed to the input of a following subsystem. Such cascades of system elements might be built up and tested starting from *either* end. If assembled starting at the *input* end, one requires as a testing means a source representative of the external source of signal, data, etc., in the system's operating environment; one may also require sensors, detectors, or other instruments which are able to measure the signal at the output of each subsystem when appended to the chain.

Alternatively, this arrangement might be inverted, integrating and testing starting from the system *output*. Here only one output measuring or assessing means may be required as the chain is built up. But, a new input format will be required as each "stage" is added. With many kinds of systems, a build-up starting from the input is preferable, since in this case one need not speculate on what forms correct or incorrect inputs to each stage will take.

Many kinds of systems incorporate such trains of subsystems: liquid-handling systems, petroleum refineries or other chemical plants, product assembly lines, product distribution systems, washing or cleaning systems. In vehicular systems, a motive source (engine or motor) often rotates a shaft to which is attached in succession a transmission, differential, auxiliary shafts, universal joints, wheels, propellors, etc. Here, integration is usually begun from one end or the other, or sometimes from both. An engine will be assembled, then tested on a *dynamometer* (an electric generator which can dissipate engine power through an electrical load) prior to connecting it to the other power-train components. For dynamometer tests, even essential engine accessories such as a generator, coolant pump, or fuel pump may not be attached, but their drain on motive power must be considered when evaluating adequacy of engine power for the complete vehicle.

Subsystem trains forming *closed loops* are very common. In electronic systems the result is termed *feedback*. Control theory includes analysis of feedback, though mainly for *linear* systems operating continuously or periodically. Many feedback phenomena in real systems involve nonlinear or intermittent operation, and require computer analysis.

Product assembly lines may involve looping, i.e. to repeat certain operations if errors are found. Output material from one stage of a chemical plant may sometimes return as raw material for input to an *earlier* stage in the process. *Speed governors* on engines represent a form of feedback, as do the *control rods* in a nuclear reactor. In most large systems, feedback is used to *stabilize* system operation against uncontrollable changes in environment or in certain system parameters. However, merely by reversing the sense of feedback, in a linear system, instability or continued oscillations will often be produced. Unless it is intended that the system exhibit oscillatory output behavior, use of feedback must be designed carefully.

In some systems which eventually use feedback for maintaining stability, initial testing may be carried out "open-loop," so that behavior of the subsystem elements forming each feedback loop can be verified to be correct prior to "closing the loop." However, in many situations, it may be dangerous to operate a system open-loop, as in the case of nuclear reactors, whose feedback mechanism causes control rods to be inserted into the reactor when neutron production exceeds the desired level. In such cases, the control mechanisms are themselves thoroughly tested, as well as emergency controls by which operators can shut down the system in case of trouble. Then the loop is tested, usually being only gradually brought up to full operating conditions. It is most desirable to have initially developed a *control-theoretical model* of system performance. This should ensure that the time-constants involved in the process are understood, so that feedback mechanisms can be designed to safely control time-varying operation of the system.

Physical arrangement of the system   Integration and testing order may be dictated by physical arrangement of the system. Systems incorporating a structural framework supporting other subsystems must be integrated with the framework as the starting point. Where some subsystems are *enclosed within* others, or in operating position have impaired accessibility because of adjacent subsystems, the best order of integration may not always be the most obvious one. In certain difficult examples or poorly laid out designs, the required order of assembly may indeed be unusual. The author recalls a 1964 *Chevrolet Corvair* automobile he once owned which had an air-cooled, rear-mounted engine. An electric blower near the engine heated the car's interior in cold weather using air drawn past the engine. To remove the blower motor, the maintenance manual specified as a first step the removal of both engine and transaxle!

Many systems are built on structural frameworks. Examples include buildings, land vehicles, aircraft, earth satellites, and sea vessels. In large systems of this kind, groups of subsystems may be integrated separately, later to be installed as "major subsystems."

Even where a system possesses no physical framework into which its subsystems are installed, the subsystems may be arranged to minimize drive-shaft or plumbing hose lengths, align mechanical shafts, provide access space for maintenance, or provide for other physical needs. For initial integration and testing, subsystems may be arranged less compactly than in the final system, so as to provide simpler access for special tests and modifications not anticipated in normal system operation.

Subsystem availability Where systems require the special development of some parts or subsystems, while others may be available "off the shelf," subsystem availability seldom coincides with a sequence of integration which would be optimal in regard to integration and testing efforts. To save time, and to make best use of skilled personnel, integration must be initiated in the best order possible under the circumstances. If a subsystem which is included within a train of subsystems will be delivered late, the train might for instance be integrated and tested from both ends, including all but the late-arriving part. This is usually more costly, since additional test facilities are required and tests are partially redundant. Risk is increased, for if the late-arriving subsystem does not "fit" the location intended for it (physically, or functionally), additional delay will take place while the subsystem is modified or some of the related subsystems are replaced by more suitable ones. If subsystem delivery is much later than planned, it is often infeasible to rearrange integration and test procedures cleverly enough to avoid delay in system delivery and resultant increase in development costs.

In summary: Good layout of integration and developmental test plans and schedules is a high art requiring excellent understanding of system operation at many levels of detail. It is made especially challenging, because planning should be complete before system elements and test facilities are ready. Management of integration and testing activities requires orderly processes, clear recording of results, immediate alerting of delays and other problems, vigorous and skillful prosecution of system "exceptions," and often massive rescheduling when certain types of problems turn up.

4.3.5 Operational testing

Developmental testing is always the realm of the system developer, usually tied tightly to system integration and meeting system specifications. The completed system may be subject to further *operational testing*, performed by or on behalf of the system *user*. The function of this portion of system testing is to verify that the system properly fulfills the *requirements* of the user.

Unlike developmental testing, operational testing is normally carried out in the actual system operating environment or in a simulated equivalent. The detailed technical specifications of the system and its parts may be of little direct concern to its users. For example, a chemical processing plant may be required by the user to process liquid chemicals at a certain rate in gallons per hour and/or to produce outputs having a certain chemical purity. A particular operational test might consist principally of a period of continuous operation during which system output was monitored carefully. The system may be defined as passing the test when the quantity and quality of output meet the user's stated requirements.

Failure of a system to fully satisfy operational (test) requirements may be easily correctable at little time and cost, or may require major surgery, or may, for practical purposes, be impractical. The experienced system developing organization protects itself, when possible, against capricious or arbitrary user requirements which could result in a user's refusing to purchase a system developed explicitly for that user. This protection is best accomplished contractually. Where some objective parameter is especially important to the user, the contract may be written so as to impose a profit penalty dependent on the downside departure of the parameter from the desired value. The contract may also award more profit if the system exceeds the stated requirement (while meeting other requirements, of course).

If the user organization has paid for the cost of development of the system, or if the system was developed speculatively by the system development organization, results of operational testing may determine whether, or in what quantities, the system will be purchased. It is fair to say that seldom if ever can the developer of a special-purpose system be unconcerned about the results of operational testing. For this reason, developmental testing usually includes test sequences closely resembling operation in a typical user environment. For example, after verifying correct operation and measuring the instruction execution timing of a new computer, a computer developing firm will input a "mix" of instructions or a set of programs representative of expected user exercise of the computer. It will generally be compared with the firm's previous models and with those competitive computer systems whose characteristics can be obtained or measured.

In U.S. Department of Defense terminology, the term *testing* is frequently paired with *evaluation* (as in the functions Developmental Testing and Evaluation, or Operational Testing and Evaluation). In this environment, where separate organizations are involved in acquiring, developing, testing, and using defense systems, it is important to distinguish testing from evaluation. Whereas testing consists of planning and carrying out the tests themselves, the evaluation activities consist of examining the results of all applicable tests, using computer-based models, with respect to users' operational requirements. Because complex systems seldom exceed all their user's wished-for capabilities, and because it is sometimes impossible to perform truly representative testing, evaluators must map test results into "acceptable" and "unacceptable" domains, to assess whether a system may be considered acceptable.

4.3.6 Manufacturing engineering

Until just a few years ago, many industrial firms' development activities began with an "engineering development" phase, in which manufacturing considerations were typically given little or no attention. The product of this phase was expected to be a fully operable "engineering model" of the product or system, plus a typewritten description termed *engineering specifications* and a set of *engineering drawings*. Typically, none of this material addressed the way in which the system and its parts were to be manufactured.

This first phase was classically followed by a "manufacturing engineering" phase, always carried out by different design personnel. Designs produced in engineering development were often considered to be little more than prototype descriptions. Parts or complete assemblies might be extensively redesigned during manufacturing engineering, to take better advantage of special fabrication and assembly equipment in the factory, to minimize wasted raw materials and to reduce touch labor content required for production. (The term *touch labor* refers to human fabrication or assembly—workers performing operations on parts or assemblies, rather than merely monitoring mass-production machinery.) The principal products of manufacturing engineering were part-manufacturing drawings, materials specifications, process specifications and assembly drawings. Special tools and assembly jigs required for manufacture were also part of this design. Process of manufacture and sequence of assembly was worked out to be compatible with the raw materials, facilities and personnel available. Figure 4.6 illustrates how the design of a particular mechanical part might differ in engineering-model and production versions.

The final step, production itself, was often assigned to still another group of personnel and management, in the actual manufacturing plant. They would frequently make practical revisions to the documents passed to them by Manufacturing Engineering, as they set up and applied the real processes used in fabrication and assembly.

Although this multistep process did allow for several steps of design refinement, it was time-consuming and wasteful because of duplicated efforts. Advances in materials processing and other manufacturing technologies often made feasible designs better in both performance and cost than those which could be hand-made in the engineering "model shop." It was recognized, beginning with a few firms in the 1960s, that something drastic needed to be done to (1) reduce duplication and other waste, (2) simplify transferring design to the manufacturing operation, (3) improve initial product designs beyond what could be produced using hand-crafted models. As consequences, (4) lower product manufacturing costs and (5) shorter development/manufacturing cycles were anticipated, and have been realized.

The needed methodologies arrived principally through advances in computer systems and their application to product design and to manufacturing automation. Today, in most modern product development organizations, engineering designs are created using computer programs which reflect both the potential and limitations of the raw materials, processes, and manufacturing equipment. Manufacturing engineering as a separate

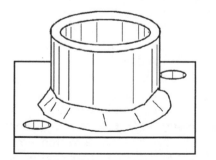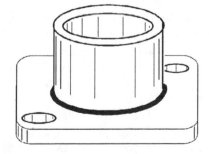

Figure 4.6. The engineering prototype part at left is made from a plate welded to a tube, roughly finished and with excess material. (It will serve its purpose in the prototype, however.) The production part at right is cast to near-final dimensions.

engineering activity is fast disappearing, either no longer needed, or else much attenuated because design engineers can (and are required to) "design for manufacture" from their earliest designs.

There still remain in the product development lexicon independent acronyms: CAD (*computer-aided design*) and CAM (*computer-aided manufacturing*). Increasingly these are merged as *CAD/CAM*. . . a single unified and integrated set of design tools is needed rather than separate design and manufacturing-design tools. At the same time, CAD/CAM design tools are providing additional facilities for both engineering and manufacturing: For example, they may support sophisticated design *simulation* for use in design verification. Modern CAD/CAM also is now capable of almost fully automated part/wiring layout of electronic subsystems. Progress in developing programs for designing electronic integrated circuits and printed circuit boards has been greater than in most other complex design problems. Computer display technology is near ideal for these applications, because designs are largely two-dimensional. Design programs allow an engineer or technician to lay out the pattern of each layer (conductor, insulator, semiconductor) in a separate color on a large color display. Designs can be built up by replication of the design of a device or group of devices. A design, or its constituent parts, can be easily stored in a computer "library," for later reuse.

System development organizations frequently transfer manufacturing to small but efficient "job shops," and often assign design of parts requiring specialized design knowledge to outside contractors. CAD/CAM systems also provide the means to communicate design requirements to subsystem designers and to return design data to the system firm. For the design of customized integrated semiconductor circuits, it is now possible for a system design organization to specify the design of a complex circuit chip containing tens of thousands of individual circuits, then send the specification to a "semiconductor foundry" for manufacture (Figure 4.7). This requires that the specification be prepared by system designers in a computer format with which the semiconductors manufacturer's tooling is compatible. At present this usually requires that system designers employ the *same* CAD/CAM software used by the semiconductor manufacturer. There is no common design language, even in this computer-based field. The U.S. Department of Defense has promoted an integrated circuit design definition standard termed VHDL (VHSIC Hardware Development Language). VHSIC (Very High Speed Integrated Circuits) was a ten-year-long advanced integrated circuit initiative funded by DoD, beginning about 1978. A formal *design language* is similar to a programming language. However, its function is to describe the parts, their relationships, and design requirements or other criteria for a complex product or system. Achieving great generality in a design language is clearly difficult, and probably unrewarding.

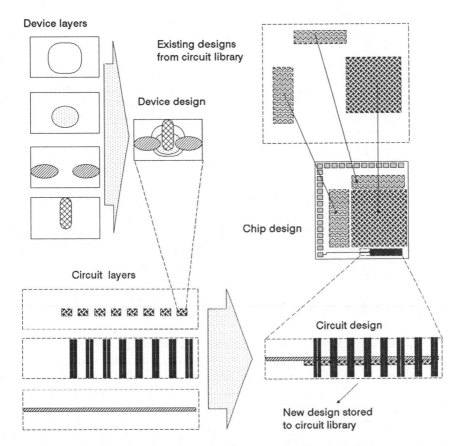

Device layers

Existing designs
from circuit library

Device design

Chip design

Circuit layers

Circuit design

New design stored
to circuit library

Figure 4.7. Computer-aided design for integrated circuits usually begins with design of a single device (upper left), which requires design of four or more etching masks ("layers").

In product areas where the expected functions and performance of products change slowly and in predictable ways over many years, a large fraction of engineering activity is devoted to manufacturing-related engineering. This is true in automotive design, home appliances, and even aircraft design. The major effort will deal with how the next-generation product will be fabricated, the materials from which it will be made, and how it will be assembled. Even in design of semiconductor integrated circuits (considered the highest of "high-tech"), from the onset of design emphasis is on layout for production.

It has repeatedly been demonstrated that a manager, if given responsibility for both short-term and long-term problems, usually focuses greatest attention on short-term ones. The principal danger in too closely integrating our research, engineering, and manufacturing functions is that revolutionary new products which depart from traditional functionality, materials and/or manufacturing methods may not emerge from a highly unified R,D&M organization which is driven strongly by production-management considerations.

4.3.7 System design aids

The most important system design activity takes place at higher levels than those details to which available CAD/CAM programs can be applied. System architects and designers have traditionally described system concepts using sketches or block diagrams. They must also create or use requirements, functional definitions, and interface definitions in prose form. The most meaningful design descriptions combine prose with graphic diagrams. Understanding system concepts or designs, so described, requires the system designer or evaluator to have a good semantic understanding of both the class of system and the application (the *system context*). There is no mathematical formalism which is both rich enough and simple enough to serve as a "universal system description language."

Present computer database systems have very limited ability to deal with system contexts in the myriad ways in which humans can. A search-phrase such as "conventional von Neumann architecture" might evoke from a computer database a diagram or description such as those shown in Chapter 2, while a designer may have in mind a variant which only vaguely resembles our description. Innovative designers frequently *combine* concepts in ways difficult to characterize well in abbreviated form, or ways which may appear in conflict (e.g., *"..a von Neumann computer system with 2-level virtual memory"*). Only at far more detailed levels of design, such as in circuit layout or mechanical structure design, can our limited-context CAD/CAM tools be brought into play.

At higher levels of design, what computers presently offer is a means for *organizing* and storing ideas and system concepts in symbolic (prose) form. A computer can also store graphics, but unless they are separately coded with keyword data, it cannot interpret them or relate them to word descriptions. Even computer graphics which contain worded titles and legends generally cannot be computer-searched to find key words in those titles or legends. The arrangement and rearrangement of masses of design data therefore still remains largely a task for the designers. If this data is well structured and ordered, and provided with brief identifying descriptions, computer data management works well. Where large volumes of standards or requirements must be dealt with, a computer can be a powerful tool by which to manage a large part of the design information.

Many *design practices* are exercised by designers of a particular class of system. Some of these are explicit and theory-based while many others are approximate or experience-based. Those most valued are often considered to be proprietary by the firms or designers which apply them. Because of these factors, it is unreasonable to expect that good top-level design aids will become commercially available for many classes of systems. More likely, there will emerge computer-network based *system design environments* for management of system design data. The users will add to these their own special and proprietary design practices. Similar capabilities are already becoming available, principally for software design, as *software engineering environments*. The acronym CASE (*computer-aided systems (or software) engineering*) may be applied to computer programs useful in software engineering, or those useful in systems engineering. Since software engineering is less diverse than systems engineering, many "CASE tools" may be of no value for designing *hardware* systems.

Significant progress in computer design-aids can be anticipated if and when it will become practical for computers to manipulate *natural language* (e.g., English) descriptions of requirements, system organizations, testing procedures, and the like. Today computers can test system descriptions for completeness and consistency, but only if descriptions are prepared using highly structured system description languages whose phrases resemble computer program instructions. System designers find it difficult and constraining to express complex design concepts in such form, especially in the early phases of design. Computer manipulation of natural language is thus a desirable, perhaps even necessary, step toward computer aid to early system design.

Using available computer-based design aids There are many computer programs available today which are of great help to system designers. An important key to utility, available using personal computers or the somewhat more capable *computer workstations*, is *direct* interaction between designer and computer. Even such simple and readily available programs as word processors, database managers, spreadsheets and graphics drawing programs have great value, if the designer (1) has ready and immediate access to them and (2) learns to use them effectively.

Verbal system descriptions captured in machine-readable form can be quickly abstracted or restructured using a *word processor*, especially if the program has an outlining feature. Data in numerical-computation *spreadsheets* (such as *Lotus 123*, not the first but the most popular such program) can be expressed in either table or graph form and incorporated directly into text by a modern word processor. Graphics programs are available at levels of sophistication ranging from simple freehand drawing to sophisticated drafting programs which provide support in dimensioning and can show synthetic three-dimensional views of an object. Some graphics software, for example, can be used to test if a subsystem could be physically moved in and out of a system assembly, without interfering with other parts.

A significant problem, both presently and in the future, is to obtain all needed computer facilities using a relatively small number of software programs. This problem is more basic than it might appear. One solution is a collection of relatively simple, independent high-quality programs, each capable of providing a certain kind of computer support. This is to some degree the situation today, though there exist many design functions for which no ready-made computer support is available. The difficulty of this approach is a lack of compatibility between data files created by different programs. The result is that many activities must be done piecewise, with data moved in simple alphanumeric form, between programs.

A totally different approach, which *a priori* might be considered ideal, would be a single "integrated" program able to deal equally well with information in a variety of forms and relate it flawlessly. Current attempts at such "integrated" programs suffer from limited capabilities and a wide variation in quality of software dealing with different phases of design.

Both approaches can be appreciated via a future scenario wherein computers would provide a wide range of support services to system designers. If a designer were required to use 20 to 100 different programs developed by as many different clever software design teams, a major problem would be to keep in mind: (1) the capabilities and limitations of each program; (2) what system design data is available from earlier design steps, and in what forms it can be used; and (3) the current state of the design. Those who now routinely use 5 to 10 complex software "packages," or several programming languages, will appreciate how seriously these computer "aids" can intrude upon the design process. A major increase in the numbers of independent programs to be learned and used is a grim prospect. Where many designers are working on the same design, this approach falls apart entirely, for it lacks an essential element of good computer-aided design: comprehensive "design library" support.

The other solution, single integrated programs, deals better with integration and library problems but usually suffers with certain capabilities which are far inferior to those available on a "stand-alone" basis. Design of a superior word processor or spreadsheet is a specialized skill, as demonstrated by wide variations in product capabilities and ease of use. If a single software firm of compact size (best for producing a highly integrated product) attempts a comprehensive product including word processing, advanced graphics, database management, library management, and other capabilities, *quality* of solutions for each function may range from excellent to poor but can be expected to be mediocre on average. This is borne out by present-day "integrated" software products, which are frequently based on a single program of exceptional quality plus others far inferior to it.

Both personal computer (PC) consoles and computer/workstations may operate as self-contained computers, or they may communicate with a more powerful remote computer. Workstation displays usually contain four to eight times as much information than is displayable by standard personal computers. Either can derive inputs from a keyboard or pointer ("mouse"). Workstations' built-in computers are typically several times more powerful than those of personal computers, and their maximum memory capacity much larger. PCs appear to be closing this technical gap, while workstation low-end prices are falling. More important than the console display or other hardware, however, is the software available for a given system. Personal computer software is available in large variety, especially for performing generic functions. Workstation software has typically been designed to support particular engineering design activities such as printed-circuit (electronics) board layout, mechanical parts design, or electronic integrated circuit design and layout.

Future design systems The economic benefits achievable with computer-based design and manufacturing systems are potentially so great that their development can be expected to continue, and even accelerate, in the foreseeable future. Some of the largest industrial firms will continue to evolve key parts for proprietary design systems; smaller firms operating in specialized fields will do the same.

Certain future directions are pointed to by leading-edge activities. One of these is extension of large-area multicolor graphics displays, controlled by a variety of "mice," "light pens," or new positioning devices. Good design systems will have immediate access to *operational simulations* or *exercisers*. For example, while designing automotive suspensions, it is valuable to be able to quickly simulate and display the dynamic response of the suspension to a simulated road surface. In electrical or electronic designs, it is valuable to be able to "measure" response rates, linearity, sensitivity to noise, or other important circuit characteristics. The future value of simulated *three-dimensional* graphic views of systems is less clear. Though it can allow designers additional degrees of freedom in examining design parameters, only in certain design areas does it appear to add significantly to the ability of the designer to conceptualize. Electronic circuit layout, by reason of the manufacturing technology, involves *layered* two-dimensional structures. Mechanical and building designers should benefit more from advanced 3-D capabilities.

Automated validation of a design, versus specifications or other description, should emerge as a future capability. The objective here will be to determine both necessary and sufficient conditions, i.e., (1) if all stated requirements have been addressed by a given design, and (2) if the design is more constrained than it is required to be. Though this kind of capability is at present not even feasible for software, the most logical of design media, work is being carried out in that area which, if successful, should promulgate into other design fields.

An important need for future computer support of systems design (and for other engineering and programming activities) is definition of standard interfaces for computer-aided design programs. This could, for example, consist of standardized data formats for design data. It would allow design files, drawings, and other data, developed using one design tool, to be manipulated immediately using other tools. Thus it should eventually be possible to:

- Formulate a design
- Check the design for completeness or adherence to requirements
- Examine it for special manufacturing implications
- Convert it to fit a particular manufacturing environment

. . . in each case using a program optimized for the task.

Future system-design software should be capable of varied levels of interaction with its users, ranging from highly automated modes to fully interactive modes, allowing the user to create a design, explore design-parameter spaces without need for interaction, exercise designs through computer-based simulations, adjust design parameters, and repeat steps.

The intellectual exercises involved in system *architecture*, where one's objective is often to find innovative and useful combinations of system elements, can be supported by computers only to a limited degree. A useful computer aid for architectural activity would be one which facilitates verbal, schematic, or pictorial description of a new architecture or feature, permits prose, graphic, and (where needed) mathematical descriptions to be rearranged and combined swiftly, and provides means for "testing" the concept, for example, comparing it with existing concepts. In fact, we already have some of these facilities, though the mechanics required to access them often impede the innovation processes. For this type of intellectual activity, the easiest to use computer aids, suited to broad intellectual domains, will probably prove the most valuable.

EXERCISES

1. Prototypes may sometimes fall far short of being full-scale, full-featured models of the systems they represent. The criterion of their design is usually that they be convincing demonstrations of features of system design which are *not adequately demonstrable via drawings or explanation*. What should be the features and character of a prototype which would be suitable to convince *you* of the credibility of each of the following systems:

(a) An automobile whose new-design brakes and linings would operate for 300,000 mi without replacement

(b) A projection television display unit which was claimed to be able to project a clear picture onto a specially designed 30- by 40-ft stadium screen in daylight

(c) An automobile incorporating a fundamentally new engine fuel system, capable of carrying six adult passengers at 60 mi/h while using only 1 gal of gasoline every 250 mi

(d) An airplane whose radar reflection characteristics (radar cross section) were said to be less than 1/1000 as great as those of conventional aircraft designs

(e) A bomb, twice the size and weight of a typical present-day fission-triggered fusion weapon, but capable of 1000 times as much energy release

2. Let us suppose that one wanted to develop a *simulator* for a new automobile design with improved suspension, brakes, and other handling-related features, to demonstrate to a prospective investor by a realistic simulation, that it was "almost impossible to get into trouble, even if driving it hard." The simulator is to consist of a driver's seat, controls, and displays which represent views of the instruments and simulated views through front, side, and rear windows. The entire simulator is placed on a base that permits it to move through limited travel which roughly simulates motion sensations in the vehicle.

(a) Describe the *road situations you would want to simulate*, for the most effective demonstrations, including presence and movement of simulated vehicles on the road.

(b) If the simulation was to be *nonresponsive* to controls (no human in the loop), diagram and describe a system which could implement the simulation.

(c) Repeat (b) for the more difficult case in which the simulator would be responsive to the controls in a driver's seat.

3. Imagine that each of the following simulations is to use computer display only, with displays showing action for benefit of observers or operators but would not attempt to simulate faithfully the operator environment. For each case: (1) Describe what you think the simulation might be designed to demonstrate. (2) Describe the sort of simulation you believe would be suitable. (3) Describe in what ways other participants in the simulated scenario would interact with the simulator if they were not merely to be computer-simulated. (4) Describe what observers of the simulation should be able to observe. (5) Where applicable, describe the operating sequences which should be demonstrable. (6) What features of the real system and operating environment do you believe it most important to simulate with high fidelity? (7) What features are of little or no importance?

(a) Computer simulation of a new air traffic control system which automates runway assignment and ensures that aircraft enter the landing pattern with adequate separation (tasks now performed by humans).

(b) Computer simulation of a "totally automated football linesman" system which operates from permanent locations in the stands, tracks the football's position by means of a miniature transmitter in the ball, and signals the referee—via a radio tone signal—when the ball has been moved 10 yd for a first down, or by some other distance for a penalty.

(c) Computer simulation of the remote operation of a deep-sea research submersible capable of unmanned operation into the deepest parts of the oceans, and containing robot arms, storage bins, and cameras.

(d) Computer simulation of a high office building subjected to a severe earthquake.

(e) Computer simulation of an aerial "dogfight" between two military fighter aircraft, with a variety of simulated weapons.

4. Imagine that your firm produces custom-designed home entertainment systems. Systems engineering in this relatively simple case involves: (1) selection of externally designed components and cabinets to hold selected equipment, (2) developing equipment arrangement and interconnection layout, (3) specifying and acquiring interconnection cables, and (4) preparing directions for installation and testing by technicians. Develop a brief description *in the form of a checklist* covering work items you or your employees may carry out, for each of these seven phases of the work: (a) requirements development, (b) system design, (c) subsystem and parts acquisition, (d) assembly, (e) integration, (f) testing, and (g) installation.

5. Alarm or voice communications systems often serve all floors of a high-rise building, from a central control or switching unit usually located on or near the ground floor. Each handset, sensor, or alarm unit is connected to the central unit by a pair of wires. Consider the following *alternative methods of distributing* the pairs between devices on a floor and the central unit.

(1) One wire pair is run *direct* from the central unit, to each location where a phone or alarm device is to be installed, with wire pairs threaded through partitions or run on wall surfaces as required.

(2) Wire pairs run from the central unit through *many parallel vertical conduits* spaced throughout the floor plan of the building, so there is *at least one vertical conduit passing though each room*. Each device is connected directly to a wire pair in a conduit passing through the room.

(3) All wire pairs run from the central unit through a *single vertical conduit*. A fraction of pairs are terminated in interconnection boxes on each floor, from which direct connections are made from the interconnection box to the devices in the rooms.

Compare, using commensurate measures, the *cost and risks* of dealing with each of these arrangements during (a) initial installation; (b) addition of one device, on a single floor; and (c) repairing a problem due to a broken wire somewhere in the wire between the central unit and a device.

6. The problem of layout of tables at a restaurant is comparable to that of physical arrangement of parts in a system. Both may often be effectively solved by graphical methods. Sketch an intimate, windowless square dining area 25 by 25 ft. Provide an *entry area* where two or three couples can stand while awaiting seating, and a 2- by 6-ft table for supplies, such as china and flatware, to be used by waiters. Tables for 2 to 4 diners are 3 ft square, and round tables for 8 to 10 diners are 6 ft in diameter. A diner requires 2-ft clearance between the table edge and the back of his or her chair for easy access to or from the table. Waiters and busboys use carts which are 18 in wide and 3 ft long to bring certain food to the tables, and they carry trays 24 in in diameter. The

restaurant owner wishes to leave two round tables in position, under usual circumstances. Square tables can set two when put together for large groups, three when set against a wall, or four when separated from wall and other tables.

Devise a *process for designing table arrangements* in this environment. Then, design a table arrangement which can handle a near-maximum of diners assuming that 50 percent of *diners* arrive as couples, 10 percent as singles, 10 percent as groups of three, 25 percent as groups of four, and 5 percent in larger groups.

7. Consider a simple stereo system consisting of a record player, frequency compensating preamplifier, power amplifier, speaker selection switch, and two pairs of speakers. In this simple system, it would probably be most effective to connect all together before testing.

What *specific* procedures would you specify to installation personnel for: system testing, recognition of problems, and correction of problems? Assume service personnel carry only a simple device which, when its range is correctly set, can detect if signals of reasonable strength are present at the output to which it is connected.

8. Examine your automobile's underhood region and devise an *order of integration* for the following: engine/transmission, radiator, radiator hoses, air-conditioning compressor, battery mount and battery, and steering column. Assume any *unlisted subsystems* are installed at the most appropriate times during the main sequence. Assume this is for routine assembly of production vehicles, not for initial testing of the design. Give reasons for your choice of assembly order.

9. Using a private automobile as your system example, describe a set of *tests* which the automobile manufacturer might carry out, as: (a) developmental tests, to verify the vehicle's performance, and (b) operational tests, to verify appropriateness for its expected use. For this problem, consider only tests which can be carried out using a *completed* vehicle. Separate each set into a subset dealing with testing for *limitations*, and another dealing with *suitability* for vehicle operations.

10. If you are familiar with the use of a computer-based line-drawing or CAD program, describe how it could be used as a design aid for the restaurant-table problem (Exercise 6).

11. If you are generally familiar with use of a computer spreadsheet such as LOTUS Corp's *1-2-3*, describe how it could be used to assign 12 aircraft to scheduled weekday flights involving routes between a "hub" city and 12 cities surrounding it at various distances requiring from 0.5 to 2 h flying time. Allow no less than 1-h terminal time after arrival and before a following departure. Equipment should be available for all flights, i.e., you may need to keep aircraft in reserve at certain locations, but total equipment requirements should be kept near a minimum. Your analysis will work from marketing estimates for average weekday passenger loads between HUB and other cities, and may assume that rates of travel during the hours 7 to 10 a.m. and 5 to 8 p.m. are double those at other times of day.

(a) First describe the arrangement you will use to define and lay out possible schedules on the spreadsheet. That is, describe by example what parameters would be entered into what cells. Describe the manual steps the user might use to find improved routings.

(b) Describe how this problem could be set up or modified so that required number of arrival/departure "gates" at the hub, or other airports could be easily determined.

12. If you are familiar with a computer database management program such as *dBASE* (a registered trademark of Ashton-Tate Corporation), or a similar program:

(a) Describe how you might use it as an aid in partitioning and keeping track of subsystems and parts of a complex system of 10,000 parts, distributed over 50 subsystems.

(b) Recommend a systemwide *part identifying (naming) convention* which will not require the user to remember part numbers and similar data. Assume that subsystems contain quantities of one or more each, of parts from a list of parts which are used only in that subsystem, but that certain parts may be used in several subsystems.

BIBLIOGRAPHY

W. R. Beam, *Command, Control and Communications Systems Engineering*, McGraw-Hill, New York, 1989. This book illustrates many system design principles as they are applied to a specific class of systems.

Gordon Bell and Allen Newell, *Computer Structures: Readings and Examples*, McGraw-Hill, New York, 1971. A comprehensive definition of computer structures. Uses a simple *system descriptive language* which demonstrates both benefits and difficulties of such media.

B. S. Blanchard and W. J. Fabrycky, *Systems Engineering and Analysis*, Prentice-Hall, Englewood Cliffs, NJ, 1981. An excellent introduction to systems engineering.

U. Rembold and R. Dillman, *Computer Aided Design and Manufacturing: Methods and Tools*, Springer-Verlag, New York, 1986 (2d edn.). A wide-ranging treatment which covers current and projected future (robotics, machine vision, etc.) applications.

G. P. Richardson and A. L. Pugh, *Introduction to System Dynamics Modeling with Dynamo*, MIT Press, Cambridge, MA, 1981. Chapters 1 and 2 describe the concepts and methods of system dynamics modeling. The approach is computer-based. Although Dynamo is a programming language designed especially for this type of analysis, methodologies are not language-specific.

E. Teicholz (ed.), *CAD/CAM Handbook*, McGraw-Hill, New York, 1985.

Engineering simulation takes a variety of forms of which *computer simulation* is the most common. This is often described in publications dealing with a technical area (heat transfer, aerodynamics, electrical circuits, etc.). *Hardware-in-the-loop simulation* involves adding some of the system hardware to a computer simulation. *Man-in-the-loop simulation* involves humans who respond to synthetic stimuli to operate the simulated system. A *Simulation Symposium* has been sponsored since 1968 by a varying group of engineering societies including IEEE (New York); its proceedings are available. Training simulation (especially for Defense Department applications) is addressed by an annual *Training Equipment Conference* (Orlando, FL), which also publishes proceedings.

Product-driven Design Objectives

If users' system requirements always fully defined the systems they order, creativity would seldom be called for on the part of system designers. Even in designing systems for such users, however, designers usually have great freedom of design. Designers are even more in charge in commercial system development where systems must be developed speculatively, without an assured market. Here, system objectives must be defined *in toto* by the enterprise and its designers, based on their perceptions of what users need or will buy.

The process of arriving at a system design often consists of satisfying (if possible) *all* of the user's known requirements while at the same time satisfying certain important objectives of the designer and builder. One of these must of course be *profitability*. We shall examaine this and other common objectives of the designer and builder in Chapter 6.

A system designed to *emphasize* all the objectives its owners and designers can think of will most likely turn out "plain vanilla." Compromises must be made to reach *any* difficult objective, and if "all" objectives are equally important, the system will likely be so compromised as to not stand out from its competition. While this may be acceptable for a firm which is guaranteed its market share of business, competitive firms tend to emphasize a limited set of features, in order to capture particular "segments" of the market for that class of systems.

We will distinguish two classes of system design objectives, each of which is present to some degree in almost every system development (Figure 5.1). The first class we will term *product-driven objectives*. These are characterized by focus on the characteristics of the system itself. The second group, of *process-driven objectives* (discussed in Chapter 6), relate to the design and construction processes and to other system features frequently more important to system designers or builders than to buyers and owners.

Most system designs *are* required to balance a variety of different objectives, each with its own priority . . . and a good many of these

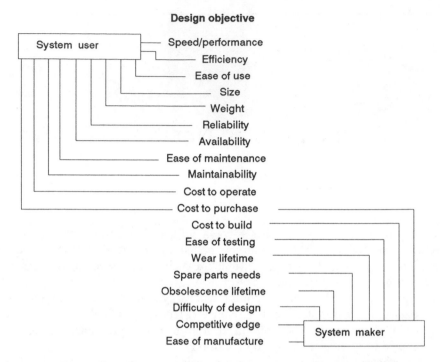

Figure 5.1. The system developer/builder and end user have different interests. Whereas the user is interested in the system itself rather than the resources required to design and build it, the developer is influenced by the facilities available to design and produce. Profitability is also important, as is maintaining a "competitive edge." The design considerations at the top of this list we refer to as *product-based* while those nearer the bottom we call *process-based*.

priorities may be high. To most simply explain design methods for addressing *individual* objectives, however, it will be simpler to envision each objective as if it were very strongly driving the design, that is, if it is of highest priority. We shall discuss the common task of juggling several high-priority design factors in Chapter 7.

5.1 Design for high reliability or error-free operation

5.1.1 Representative requirements

The system designer may be faced with a broad range of requirements addressing system errors and failure. Systems wherein high reliability is demanded may typically be used in operations where safety of life or large economic losses would threaten if system errors or failures occurred. For example, aircraft *flight control system* failure may, if there exists no manual backup, result in loss of the aircraft and possibly the lives of its occupants. Security systems also must be reliable to reduce the exposure for financial or other losses. Even banking systems, if lacking high reliability, can expose the user to a chance of large fiscal losses.

In some systems, occasional errors, errors limited in scope or consequences, even brief unplanned shutdowns, may be acceptable, provided that damage or loss can be strictly limited. For example, navigation systems for aircraft or seagoing vessels invariably display some inaccuracy, but as long as they are within the expected limits, these systems are considered to be operating satisfactorily. Computer systems which experience an unavoidable loss of electric power may not be expected by their owners to continue operating indefinitely on auxiliary power, but they will be required not to lose data *in memory at the time of power loss*.

Users may require systems not to experience unplanned shutdowns exceeding a certain period of time (e.g., not occurring more often than once per month), nor lasting longer than some small time interval, nor overall representing more than a small fraction of the expected operating period. These are common user expectations, even where more critical reliability concerns do not apply. Owners of motor vehicles, for example, should expect occasional vehicle failure but (by law, in many states) cannot expect remuneration from the manufacturer unless the failure frequency and/or repair time exceed certain generous limits.

5.1.2 Sources of system errors and unreliability

Although *randomly occurring errors* and *breakdown* usually affect system users in quite different ways, they share common analytical bases in system design. That is: (1) both are *probabilistic*, and (2) both must be treated *combinatorially*, for systems having more than a single source contributing to error or failure.

Some system errors, of course, are *systematic*, characteristic of the design: a four-digit pocket calculator may be expected to exhibit error of at least 1 part in 10,000 when multiplying or dividing. Though the competent system designer must also appreciate the possibility of systematic errors and know how to eliminate or minimize them in design, random errors and unreliability are unavoidable properties of most systems.

Probabilistic system variables exist for various reasons (Figure 5.2):

- Parameters of system parts are subject to *manufacturing tolerances*. If system error or failure are contributed to by a part, it must be treated in error or reliability analysis. Often, potential for system error or failure because of *initially* incorrect parameters is detected by testing. However, it might not be, if critical *combinations* of operating-parameter values are never exercised during tests. A test of a system operating with reduced supply voltages, or another test at normal

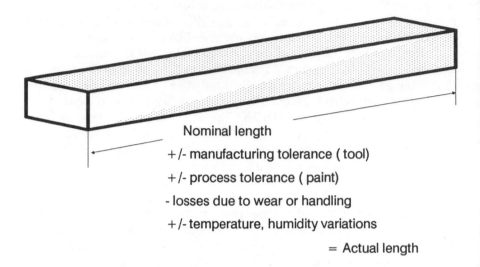

Nominal length

+/- manufacturing tolerance (tool)

+/- process tolerance (paint)

- losses due to wear or handling

+/- temperature, humidity variations

= Actual length

Figure 5.2. The values of system parameters may vary as a result of tolerances in fabricating, assembling, and finishing parts (there are analogies in electronic and other technology). In system operation, parameters may change due to the system's environment and degrade due to wear and aging. When expected changes are beyond what the design can tolerate while delivering acceptable performance, scheduled maintenance must address them.

voltages but increased ambient temperatures may not reveal a failure which would turn up if low voltage and high temperature were applied together.

- System construction and integration processes are subject to variations (soldered joints are notorious) and can harbor latent failures ("time bombs") which can make themselves felt even after successful testing and system delivery.
- Many system components are subject to wear or deterioration with use or age, or in hostile environments. The chance of failure of the component increases with time, but time of its failure cannot be accurately predicted. Change of parameters over time may also be a cause of increased system error.
- Environments in which systems are operated are often subject to variations (temperature, humidity, etc.) which may not be controlled. These effects may include mechanical, electrical, or thermal surges which cause system parts to be overstressed and to fail. Even when there is no failure, system accuracies may be affected, particularly in nondigital systems.

A *correctly designed* system's parts should fail only because of some combination of manufacturing errors, operating stress, wear, or degradation. In complex systems, it becomes impractical to predetermine all possible error or failure mechanisms, since they can depend not only on error or failure in individual parts, but also in parts working in combination. Likewise, it is seldom feasible to monitor *all* stressful environmental conditions, in every nook or cranny of a system, which may cause some part to fail or to contribute to system error. Accordingly, in large systems, system error or failure modes are most often treated *macroscopically*, that is, on a collective basis for the entire system. This should be a valid approach if failure of one or more in a known set of parts can result in a particular mode of system failure, or if system error is a result of variations in operation of certain parts. While the macroscopic approach to reliability enables us to determine approximately how long (on the average) a system may be expected to operate without a failure, it will not help in locating and repairing the *causes* of failures.

Analog and digital systems Systems, or their subsystems, are often classified as *analog* or *digital*, depending on whether internal variables may take on meaningful values in a continuous range, or only discretely many distinguishable values. Errors in analog devices or systems are often represented in terms of maximum expected deviation above or below the correct value, as a percent of full-scale value. Error in digital devices may be described in terms of a device or system output having an *incorrect state* (e.g., a digit error, or a switch being open when it was supposed to be closed). Digital systems and subsystems thus have error characteristics different in nature from those of analog systems or system elements. A simple example: at a 10 percent reduction in power supply voltage (or, say, hydraulic pressure), an analog system will typically produce output reduced proportionately, or a bit worse. A complex digital system, under the same circumstances, may produce correct output, very erroneous output, or quite possibly none at all.

Programmable digital systems, i.e., systems containing computers, are also subject to error or failure due to *incorrectly or incompletely designed software*. Though hardware errors could also be present, digital hardware such as computers are of broad utility and usually subjected to thorough developmental testing. It is appropriate to refer to *errors* in software due to improper design. However, the occurrence of undetected errors in software in operational use has sometimes been (inappropriately) referred to as "software *unreliability*." Software is *not* subject to wear or stress effects. Any system error or failure due to software design or incorrect modification is latent in it, from the time the software was installed or modified. Discovery of error after system installation is invariably due to failure, to anticipate or test for, all conditions experienced in actual operation.

Analog systems have characteristic errors, when they deal with signals or motions comparable in strength to *electrical noise* or *thermal motion*, which is present in the system at any temperature above zero degrees

Kelvin. Though, in fact, the indicating pointer of an analog instrument has a small natural "wobble," these effects are most apparent in radio receiving systems. There, electrical "noise" enters the receiver because its antenna seeks a *thermal equilibrium* with those regions toward which the antenna is directed (Figure 5.3). Additional noise may enter the antenna from operation of arcing or radiating electrical equipment nearby. More noise is added because the receiver itself uses electron motion as a means of amplification; that motion too has thermal fluctuations. Similar error-producing noise also affects high-sensitivity mechanical instruments such as seismometers, which accordingly must be operated distant from highways and other seismic noise sources.

Error metrics System *error* may be defined in terms of expected system performance, usually the values of system outputs. There are usually two kinds of errors, *systematic errors* and *random errors* (Figure 5.4). The former represent errors which occur predictably, though they may depend in complex ways on the system operating conditions. Random errors can be predicted only in a statistical sense, if at all. Systematic errors may be corrected, provided that all parameters on which they depend are measurable. The cost and complexity of correction may be large if a high degree of correction is required. Ultimate accuracy is usually limited by the state of technology or by unpredicted sources of error. For example, the aiming accuracy of ballistic missiles has been improved since the 1950s, by improving the technology of inertial guidance systems which direct the missile. Ultimate accuracy limits may

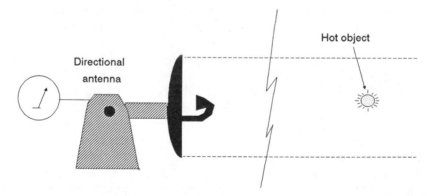

Figure 5.3. Heated objects radiate energy, which is absorbed by cooler bodies nearby and raises their own temperature. If all radiated energy is absorbed nearby, the object eventually reaches thermal equilibrium with its surroundings. The process of *radiation*, along with heat *conduction* in a solid or liquid or *convection*, when a gas moves heat, are responsible for the thermal behavior of most systems. Though an antenna pointed toward a distant star does not absorb enough heat to physically warm it perceptibly, it does introduce noise into a connected radio receiver in proportion to the temperature of the star and how much of the antenna's field of view is filled by the star.

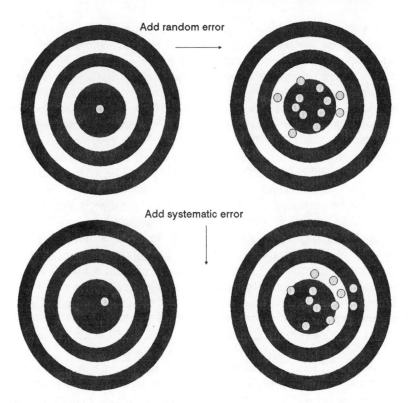

Add random error

Add systematic error

Figure 5.4. A process may have both random and systematic errors, as illustrated by these targets. The gun was aimed to put all shots through center, but systematic error (in this example) biases its aim up and to the right. Random error disperses the pattern symmetrically, on average, in the example shown.

be determined by microvariations in local gravity along the missile trajectory.

Error-limit specification for a one-dimensional system variable may be *absolute*, e.g., "+/– 1 psi," or as a maximum percentage of a reference value "+/– 5 percent of full-scale pressure." The specification "+/–0.1 v +/–3 percent of full-scale voltage" would suggest *two* sources of error in the instrument, one absolute and the other dependent on the full-scale value. Where a measure is a vector having several dimensions, different notation may be used. The term "average miss distance," in a three-dimensional guidance situation, indicates an expected value for distance of closest approach to some objective or aim-point. The term "CEP" (*circular error of probability*) is used to describe military systems which are aimed toward a point in a plane, e.g., a point target on the ground. If a circle of radius R equal to the CEP value is drawn around the aim-point, on average *half* of the attempts to hit the aim-point will fall within the circle, and the other half outside it.

Reliability metrics *Reliability* of a system is usually specified in terms of a *mean time between failures* (MTBF, Figure 5.5), which can be specified in any convenient units, say, hours or years. While it most commonly refers to *operating* time periods rather than elapsed time, the conditions should in each case be clearly stated. MTBF is also the reciprocal of the *probability of system failure per unit time* (assuming that repair took place instantly). Though we are most accustomed to probabilities in the range 0 to 1.0, a *probability per unit time* can be greater than unity, i.e., if the probability of failure of a system per hour were .001, the probability of failure *per year* would be 8.76 and the MTBF would be 1/0.001 = 1000 h.

Additional information is often required to pin down a definition of MTBF. In the system in question, what constitutes "failure?" Surely if failure included *any* tiny deviation from complete and correct system operation, MTBF would be smaller than if failure included only catastrophic or irreparable failure of a major system component. Should we, for example, call an automobile "failed," merely if its radio will not play? Probably not. Hence the importance of defining what constitutes an operational condition of a particular system. Since there is no universal definition of system failure, the acquirer/user and designer must reach accord on what is to be considered system failure.

An additional factor which must be considered is whether a system retains a degree of utility, even when parts of it are inoperative. The system might, for example, be able to perform only the critical important part of its functions, or to perform all its functions at reduced capacity or

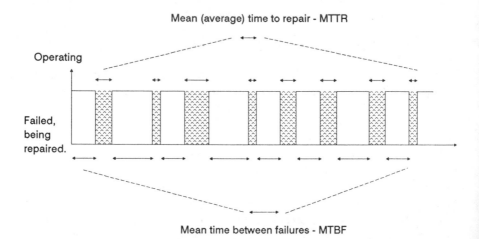

Figure 5.5. The system illustrated operates until it fails. Then it is repaired immediately. The sequence of alternating operating and repair intervals is averaged to yield mean time between failures (MTBF) and mean time to repair (MTTR), as shown.

speed. This leads to a design-dependent specification of *failure*, and to definition of *usable states*, for a particular system. A separate MTBF may be derived, if required, for *each state transition*. This requires the definition of the condition (state) of the system at the time at which measurement of the time-between-failures begins. Where the initial state is a fully operating state, MTBF is usually measured beginning at completion of the maintenance or repair event which last put the system in full operating order.

Where there is system utility in partly failed system states, probabilistic analysis (discussed in the section following) will allow the determination of a separate MTBF for each state transition to some less-useful state. This measures the average time interval, from when the system reaches that state to a point when another failure causes it to go into a less-useful or a non-functional state. If each possible failure occurs via a mechanism independent of the others, an overall MTBF for failures from that starting state can be estimated by taking the reciprocal of the *sum of the reciprocals* of MTBF's connecting that state to more-failed states separated from it by single failures. If there are some failures which *themselves do not further reduce* system utility, as in the simplified automotive failure-state chart of Figure 5.6, these need be considered only if they also alter the likelihood of subsequent failures which *do* reduce it. Thus, the operability of the automobile is not affected by failure of its air conditioner, nor may the likelihood of engine failure be altered by most failures of the air conditioner. Dealing with such issues involves the principles of *conditional probabilities*.

The second important measure used in system reliability has to do with how long the system should *remain* in the failed state. This is usually treated using the measure *mean time to repair* (MTTR), (c.f. Figure 5.5). This is usually taken to be the average time required by a maintenance crew to effect a repair. However, a correct MTTR must include: any delay after failure but before the failure is observed and the repair crew notified, time required for the repair crew to reach the system site and begin repair, and delay before repair parts arrive. The true MTTR starts at the time of failure and continues until the system is again operable. An MTTR estimate may not consider *quality* of service; but, with poor maintenance, a repaired system may be more failure-prone then prior to the "repair."

If actual system repairs are controlled and carried out by others, none of these important constituents of the MTTR can be controlled (or even well estimated) by the system designer. MTTR, accordingly, is usually difficult to estimate. The developer must place strong caveats in any MTTR estimates, if there are uncertainties in the quality and timeliness of maintenance. Sources of MTTR information should include operation of current systems by the user, or operation of similar systems by other

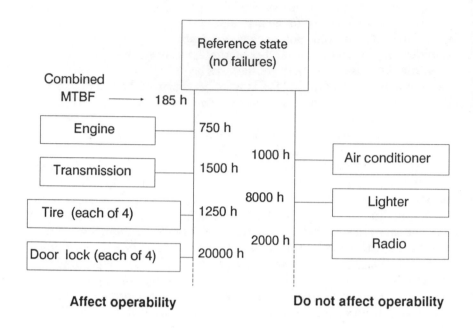

Figure 5.6. This diagram illustrates contributions to MTBF by various major and minor subsystems or parts of an automobile. The overall contributions of the items on the left is equivalent to 185 h net MTBF, and does not include all possible reasons for failure. All parameters are estimated and would in practice depend on the make and age of the vehicle. Most modern automobiles, when new, are probably more reliable than that represented by these numbers.

users having similar activities. Often this is the only kind of information available. After adjustment for known or expected differences between observed operation and the predicted system operation, this analogy-based estimate will almost surely be superior to a purely hypothetical estimate.

MTTR is also difficult to estimate because one must often consider a wide variety of repairs. An "engine repair" on any automobile, for instance, may involve anything from merely tightening an oil-sump bolt to a complete disassembly and reassembly of the engine. Maintenance is greatly simplified, and MTTR greatly reduced, if a system is designed so that its subsystems, or smaller modular elements found to be failed can be: located rapidly, accessed easily, removed without damage, and quickly replaced by spare units which are kept immediately at hand in sufficient quantities to meet maintenance needs. The subsequent step of returning failed subsystems or parts for repair or replacement does not contribute to the measured MTTR. However, the cost of ample spare subsystems or modules may represent a large fraction of the cost of a second system.

This can be offset by repeated use of small numbers of module types, wherever practical.

Practically speaking, the estimation of MTTR is limited to consideration of predictable failures of parts of the system for operation-related causes. Catastrophic incidents such as war damage, vehicle collisions, fires which destroy or damage systems, or sabotage are generally ignored in both MTBF and MTTR determinations.

There are additional subtleties in defining MTBF for a system to be operated intermittently, say, not used evenings or weekends. MTBF may be estimated on the basis of actual operating time, with the argument that the system cannot fail nonoperatively. However this can be misleading since systems often fail because of extra stress during start-up. If a system is kept in partial ("standby") operation, to reduce startup problems, it certainly may fail during periods of nonoperation. It is accordingly important to identify any special *operating conditions* to be assumed in MTBF estimates if the system is not to be operated continuously.

Intermittent operation can reduce MTTR if expected repair times are comparable with operating periods. A repair begun when the system fails may continue during periods of normal nonoperation. In this case, *if the system is not required to finish uncompleted work after repairs have been completed*, part of the time to repair may not inconvenience the user. This, of course, assumes that the repair crew will continue its work during periods when the system is normally not in use.

It should be clear from this discussion that both MTBF and MTTR are subject to application-dependent factors which can strongly influence a system's true utility. Prior to performing or commissioning elaborate MTBF or MTTR computations, system designers should understand all user-controlled factors which will affect these estimates.

Error and reliability analyses Estimates of MTBF are a principal problem of reliability analysis. The usual definition (Figure 5.5) is the expected time for which the system will operate without failure, beginning from a state which could be described as "brand new" or "just repaired."

Most MTBF analyses for systems containing many failable components use a *uniform random failure rate* approximation. That is, the probability that a given system component will fail during a time period dt is assumed proportional to dt. If the MTBF of the component is measured to be T hours, the probability of failure of that component during an hour of operation is assumed to be $1/T$. This may not be accurate, of course, if parts are allowed to wear or degrade to the point of failure before their replacement, neither can it accurately account for "infant mortality," in which parts fail early in system life because of manufacturing defects (discussed later). It is, however, a useful compromise, particularly in the case of large systems in which detailed failure mechanisms of many different parts cannot be accurately estimated.

To be more precise, one should consider for each failure-prone component n in the system, and at a particular time t_n in that component's service life, the probability $P_n(t_n)$ dt, that this nth component will fail during the interval following t_n-t_n+dt. The probability that the *system* will fail during the interval dt is the *sum* of the probabilities of all component-failure combinations which would cause system failure.

If we assume, by way of example, that the failure of *one or more* system components will lead to system failure, we might consider all possible failure combinations, e.g., component 1 fails but no others fail, component 2 fails and no others fail, etc., plus the set of probabilities that components 1 and 2 fail but others do not, and likewise for all combinations of 2, 3, 4, . . .,(n-1) failed components. As often occurs in probability calculations, it is easier to evaluate the probability that *no* component fails in the interval dt, which is simply

$(1 - P_1 dt)(1 - P_2 dt)(. . .)(1 - P_n dt)$.

(Here the P_n are functions of the actual operating life of each part; using a uniform random failure approximation, they are taken to be constants.)

Since we may assume that all $P_i dt$ terms are infinitesimal compared to unity, the probability of non-failure of the system is $1 - \text{Sum}_n(P_n dt)$, and the probability of system failure is to a first approximation $\text{Sum}_n(P_n dt)$, or since dt is the same for all terms, $\text{Sum}_n(P_n)$ dt. If the time t is measured in hours, the probability of failure in a 1-h period is therefore $\text{Sum}_n(P_n)$. The system MTBF at time t is its inverse, or $(1/\text{Sum}_n(P_n))$.

Supposing that a particular system contains 1000 components, each with P_n = 0.0001, and that if any one failed, it would cause system failure. The system MTBF estimate is accordingly $1/(1000 \times 0.0001)$ = $1/0.1$ = 10 h. This underscores need for high-reliability components in complex systems.

Although a single integrated circuit microprocessor may contain 500,000 individual semicondutor components, its MTBF may be 100,000 h. The older discrete-circuit technology which it replaced was far less reliable. von Neumann's 1952 vacuum-tube computer produced MTBF of less than 8 h, based on limited first-hand observations by the author.

Redundancy to improve system reliability If a system is designed so that one or more functional parts can fail *without* causing system failure, the system is said to possess *fault tolerance*. This is typically accomplished by incorporating duplicative components or subsystems, arranged so that failure of one or more may occur without causing system failure. Although any system component might be duplicated, there must be some means to remove the failed component(s) from having undesirable influences on system operation.

System redundancy often involves providing duplicate subsystems, such as computers. For example, in a computer-based system such as aircraft flight control, any "hard" failure, or even a discontinuity in operation, could be catastrophic and life-threatening. In this application, three or sometimes four computers each execute simultaneously the same flight control program. The computer output is a logical combination of outputs of the duplicate computers which is typically the *output of the majority* of computers known to be operable. If a computer fails, it is not sufficient merely that one or more others continue operating. Some form of logic is required to ensure that any output which differs from the majority is effectively disconnected from system output. Early redundancy schemes used logic circuits separate from the computers for this "voting" function. Later, designers realized that this logic itself had become the critical element. Modern multicomputer redundancy schemes often require each computer to continually compare its output with those of all the others, and to *disconnect itself* from further operation after perceiving itself to be in error. For systems having *three* duplicate processors, after the failure of one processor the test is not meaningful, and the system typically fails when a second processor fails. If there are four, the system can operate until three processors have failed.

Most redundancy schemes assume that an element which fails will quickly be repaired or replaced and returned to service. If this can be done *before the final redundant element fails*, the system as a whole should seldom experience a failure, and the resulting MTBF can be very long. In systems such as earth satellites, however, where the failed elements cannot be repaired, different redundancy schemes must be utilized. Suppose, for instance, that we used the same type of four-way redundancy in a satellite computer, each redundant processor having MTBF of 60,000 h. The expected MTBF for failure of a first processor will be 60,000/4, or 15,000 h. With three computers still operating, MTBF for a second failure should be an additional 60,000/3 = 20,000 h. When the third processor fails after an additional expected 60,000/2 = 30,000 h, there may be no way to determine correct operation. This redundancy scheme produces an effective *system* MTBF of 65,000 hours, whereas with only a single computer it should have been 60,000 hours.

In non-repairable systems, such as earth satellites, it is more practical to operate only a single element at a time, providing for detection of its failure by another system element which tests it periodically. When the first redundant unit is found to have failed, it is disabled and a second unit is enabled. This is repeated until all redundant units have been operated and have failed. If there is no deterioration during periods of non-operation, the result is that n redundant units can provide a system MTBF which is n times that of a single unit. However, *continuous* correct operation is not attained, as it is in redundancy schemes based on rapid repair or replacement.

For those systems where it is possible to *repair* failed units, it is relatively easy to show how much the MTBF can be increased through redundancy. Let us assume that we have a twofold-redundant computer installation, which requires only one computer to be operative for correct system operation. We will assume each computer to have an M-hour MTBF, and an R-hour MTTR including any delay in initiating repairs. During a time interval, the probability that the *system* will fail requires the *event* that the second computer fails *under the condition that the first is under repair*. Using the parameters M and R, *each* computer is under repair a fraction of time $R/(M+R)$. Hence the probability at any instant that one computer has failed is $2R/(M+R)$; the factor 2 accounts for the possibility of either computer failing. The probability, per hour of its operation, that a second computer (of which there is only one) will fail is $1/M$. The probability in a one-hour period that the system will fail (i.e., a second computer fails while the first is being repaired) is the product $2R/M(M+R)$. The MTBF for the combination is the inverse of this probability rate, or

$$M(M+R)/2R.$$

If for example M were 1000 h and R, 10 h, the effective MTBF would become 50,500 h, but if R could be reduced to only 1 h, MTBF would increase to over *one-half million* hours. If triple redundancy is used (with only one computer being required, for system operation), MTBF rises even more dramatically. In practice, immediate repair is often infeasible, as, for example, for systems in an aircraft. In this case, the MTTR estimated with the aircraft on the ground should be increased by one-half the mission length.

Diversity as a redundancy scheme   In subsystems where unreliability occurs for natural or otherwise uncontrollable reasons, it may not be possible to increase system reliability by simple redundancy in which identical elements are duplicated. In long-range high-frequency radio communication, for example, a set of identical transmitters and receivers operated between the same locations and at the same frequency, will not overcome excessively weak or noisy propagation in that radio band and signal path. Neither can redundant telephone cables in the same trench or on the same poles safeguard communications against an earthquake, or an auto bringing down a pole and its cables.

Diverse redundant communications (Figure 5.7 is a humorous example) may use different frequencies, modulation techniques, or signal paths: direct radio, radio via satellite, underground cable, microwave relay, or channels following completely different routings. As in the duplicative

Figure 5.7. Redundant systems using *diversity* are designed to rely on different technologies or phenomena, and thus to have different sensitivities to environmental problems. (Whoever uses the redundant communications "system" above has experienced failure of *all* alternatives.)

redundancy discussed in the previous section, means must be provided to select a best alternative among all signals received.

Diversified redundancy is practiced in many other areas. Both wheeled and tracked (e.g., tracked "snow-cat") vehicles complement ordinary automobiles for emergency crews who must seasonally operate in deep snow. A commuter needs alternative ways of getting to work if his or her car breaks down. A variety of output printers may be available on a computer network; though they may be intended for different purposes such as draft printing and final copies, when one's preferred printer breaks down, another can be used to complete an urgent job.

Analysis of the reliability improvement and, where pertinent, performance limitations imposed by diversified redundancy schemes is relatively straightforward, provided the analyst understands the reliability and other characteristics of each available alternative.

5.2 Design for system availability

5.2.1 Representative requirements

Availability, defined in Figure 5.8, measures the degree to which a system can be expected to operate, at those times *when it is needed*. System availability requirements may be based on a variety of different descriptions of system operating periods:

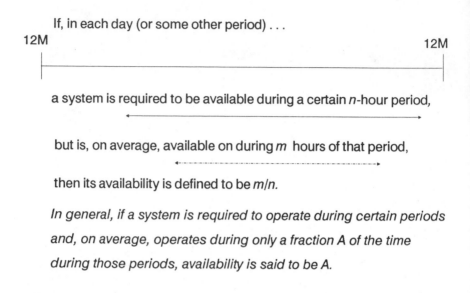

Figure 5.8. Availability is defined in this time sequence.

- Specific daily or weekly time intervals during which the system is required to operate
- The fraction of the total time the system is expected to operate
- Average frequency and duration of operating periods of a system which is to be used intermittently

Availability may be of great importance to a system user in two special cases: (1) where a system must operate continuously, or nearly so, and (2) for a system of an emergency nature, which may be used only infrequently but which *must* operate when required.

5.2.2 Availability and reliability relationships

Availability is closely related to reliability, but with some differences. Whereas reliability is closely related to system-failure events, availability might be better described as related to "system-success" events, i.e., the likelihood that the system will operate when needed. An availability requirement may be coupled to a reliability requirement; for example "during the 8 h operating shift, system failure shall not occur on average more often than once each month."

In systems which are operated continuously, availability can be related directly to MTBF and MTTR. That is, without redundancy, overall availability *cannot exceed MTBF/(MTBF+MTTR)*. (For each MTBF hours of operation, on average MTTR hours of repair—non-operation—is needed.) Hence if experience with an existing system shows that the ratio *MTBF/(MTBF+MTTR)* is less than the availability required, system reliability must be improved in some way, such as through use of fault-tolerant concepts.

Scheduled maintenance Systems having *no* parts subject to wear or other forms of degradation can be operated *without maintenance* until they fail. It may be decided to operate other systems without regular maintenance if (a) the system is unrepairable, (b) system *replacement* costs less than a representative repair, or (c) scheduled maintenance is impractical because of system inaccessibility or for other reasons.

The purpose of *scheduled maintenance* is to eliminate or delay actual system failure by repair or service operations made in advance of actual failure. The same intelligent car owner who would not ignore periodic auto maintenance to replace lubricants, fuel, and coolant may continue operating a cordless electrical appliance until the battery is exhausted or service an expensive television or stereo receiver only "when it breaks." In fact, most of us own many household or personal systems for which scheduled maintenance service is simply not economical.

Scheduled maintenance can improve availability as well as reliability. This may require that maintenance be done without system shutdown, or that time for scheduled maintenance be available within the system's operating schedule. Systems which are required to be available and operating on a 24-h basis should be designed to permit scheduled maintenance without system shutdown, wherever feasible. Certain scheduled maintenance may not be possible without total or partial shutdown, for example, the replacement of bearings in mechanisms. If 100 percent availability is essential, subsystems which need to be shut down for scheduled maintenance may be duplicated.

On-line test and diagnostics A useful complement to scheduled maintenance is a provision in the system for continual (or even *continuous*) testing during system operation. We should be familiar with this as in automobiles which use a voltmeter and/or "idiot light" as a continuous test of condition of the battery and alternator. If a system application demands it, on-line test outputs can be connected to automatic operator alarming, or used to enable subsystems whose function may be to carry out still more rigorous tests on-line, to take a system unit off-line temporarily for more rigorous tests, or to shut down the system and alert users and maintainers.

On-line tests should not noticeably impede or alter operation of the system. Tests which generally *cannot* be carried out on-line include those in which system parameters are altered until the system fails indicatively though harmlessly. For example, an off-line test of electronic equipment may temporarily raise or lower power supply voltage, and check for incorrect operation.

On-line diagnostics equipment now customarily uses computers in the system to evaluate system performance indications and predict the need for repair. After a system failure has occurred, on-line diagnostics may direct maintenance personnel to a failed component. This scheme is employed, for example, in modern aviation-electronics (avionics) systems. A failed component in a radar may be located by the radar's computer, while the aircraft is returning from its sortie. The pilot is notified of a problem, and radios the maintenance operation at the operating base. Maintenance personnel can meet the aircraft when it lands and thereby shorten turnaround time.

Availability analysis Availability analysis may, for example, involve demonstrating that a particular combination of scheduled and unscheduled maintenance of a system will meet the users' availability requirements. Or, it may involve design of system redundancies, diagnostics, and built-in system tests sufficient to satisfy user needs.

Care must be taken to ensure that the definition of availability used is consistent with the intent of the user. Availability is generally defined to be *the portion of the time during which the system is able to be operated, of the entire time during which is it required to be operable.* If a system is required 10 h daily, but because of unscheduled maintenance can actually operate only 9 h per day on average, its availability should be given as 0.9. The fact that it may be operable during most of the remaining 14 h each day may be of no consequence to the user. However, if the user *must* achieve a certain minimum amount of system operation each day (such as production of essential daily computer runs), it will be useful to estimate average *elapsed time* required to obtain this amount of daily operation. This will provide data needed to estimate overtime pay requirements for system operators.

Intermittent and "one-shot" availability Some systems are required to operate intermittently, and perhaps only for brief periods, after long intervals of inactivity. Here availability can be measured by the probability that the system will operate correctly when it is activated. This may be accompanied by the condition that the system be recharged or inspected periodically or after each use.

There are many emergency devices, and special-purpose systems such as military missiles, for which assured availability in an emergency is the principal requirement. Ammunition is probably the most extreme example. Such diverse systems as fire apparatus, *emergency locater transmitters* (ELTs) in aircraft, *intercontinental ballistic missiles* (ICBMs), and building sprinkler systems are useful only if available for operation whenever needed.

Some of these systems can be distinguished by provisions for built-in test or diagnostics. In most cases ample time is available, between operations, for testing and on-line diagnostics. However, for many such systems, there are no operating personnel who are assigned responsibility on a continuous basis.

Although equipment designed for such purposes is always accompanied by recommended test and maintenance procedures, these invariably require users to exercise strict disciplines. Lack of simple but essential maintenance, such as replenishment of evaporated battery electrolyte, is often a cause of failure. The author recalls a dark and stormy evening in a Phoenix, Arizona airline terminal when electric power failure revealed that 7 out of 12 emergency lighting units were inoperable.

Because of the difficulty of enforcing maintenance schedules, systems which minimize frequency of viability testing or which can alert users to their incipient failure have great advantages. Systems which include visual "tell-tales" such as voltage or pressure gauges must, of course, be visited. Local or remote indicating alarms in emergency systems must be tested on a regular basis, which almost invariably requires humans to initiate the operation. However, the alarm may serve to signal the system's impending failure, as in residential smoke detectors, which emit distinctive short blasts of sound as battery life approaches its end.

Detection and false alarm rates as reliability factors Systems which operate automatically upon detection of certain signals or other circumstances exhibit special forms of unreliability. Fire or intrusion alarms are familiar systems of this type, as are computer input devices which read magnetic-ink characters coded onto bank checks.

There are usually two important reliability factors, one related to correctness and a second to errors. Users of automatic character-recognition equipment, for example, are concerned not only with what fraction of characters are read correctly, but also for what fraction a wrong value is *substituted* (substitutional errors). The third possibility is that the system can indicate "don't know." This is much preferred over incorrect substitution since it signals a human to examine further any checks which cannot be read automatically.

Other detectors have the problem of determining not only the nature of an emergency but whether an emergency exists at all. Detectors which sense the presence of smoke or fire may *fail to sense* an emergency which actually exists, or may sense one when none exists. The first characteristic is usually measured as *probability of detection* , i.e. in what fraction of actual emergencies the system responds correctly. The second is characterized by a *false alarm rate*, usually the expected number of false indications per unit time. Clearly, the functioning of such systems can depend greatly on the environments in which they are operated. A smoke detector operated in a short-order kitchen will, in general, produce many more false alarms than the same unit located away from natural smoke sources.

Systems which must provide a high level of correct detection combined with a low false alarm rate can do so only if the situations they are supposed to distinguish have inherent, measurable "features" which distinguish them unambiguously. Automatic handwriting recognition, for example, works poorly because handwritten (cursive) characters are not sufficiently distinctive in shape.

In addition to problems caused by incomplete, ambiguous, or noisy information, systems of this sort often fail for two other reasons: (1) The system must have the ability to detect and to measure *features* at a level of accuracy sufficient to distinguish objects or situations unambiguously. In the real world this is not always possible, in part because new objects or events, not anticipated when the system was designed, may possess features indistinguishable from those previously defined. (2) The system may have insufficient logic and computing power to analyze combinations of features. Aside from these broad design considerations, systems of this sort are likely to be strongly application dependent. For example, where the sole object of interest possesses a unique and detectable feature, along with other features in which it is non-unique, optimum detection may involve concentration solely on that one feature. The best way to locate humpback whales, for example, may be to home on their audible "songs."

Intermittent failures   Reliability analysis does not deal separately with a problem very common in system operation, *intermittent failures* in systems. These are defined as failures which typically occur repeatedly, but not at predictable times, nor are they sustained long enough or under the conditions necessary to speedy location and correction. This type of problem is very common, particularly where parts or subsystems believed to be defective are sent to a remote maintenance facility for repair. Upon receipt and subsequent test of the units, it is often reported "no trouble found."

The most common causes of intermittent failures are different for various classes of systems. In electronic systems they are often caused by poor electric contacts. Their occurrence may be greatly reduced by use of contacts with higher unit pressures, gold plating, or wiping action. In mechanical systems, there are many reasons for intermittent failures, such as friction which prevents moving parts from free operation, partial blockage of a fluid channel by metal chips or sealant materials, or parts which have unplanned freedoms of motion. Local heating of poorly ventilated systems, during extended periods of operation, is a frequent initiator of intermittent failure. Maintenance conditions may not accurately represent those of actual operation, particularly if the equipment is "opened up" for examination.

On the whole, intermittent failures are often conseqences of careless design or manufacture. They can in most cases be controlled by the same good design practices which enhance overall system or subsystem reliability.

Special designs and technology for intermittent or one-shot use   Systems which operate for brief periods, after long periods of inactivity, generally utilize different designs and technology from that used in similar systems operated frequently or continuously. A good example is provided by electric power tools. Major manufacturers may provide three different classes of these tools: (1) for homeowner use, (2) for "light-duty" industrial use, and (3) for "heavy-duty," or continuous use. The first group of tools typically uses only prelubricated sleeve (journal) bearings, the second group substitutes "permanently lubricated" ball bearings for the fastest-rotating parts, while the third may have roller bearings which require occasional regreasing in all bearing locations.

Systems designed for intermittent use are often designed and constructed such that disassembly and repair are impractical. This is characteristic of low-cost consumer goods for which the cost of a typical repair would exceed replacement cost. Although high-quality wrist watches are still designed to be repairable, decreased production costs of electronic watches timed by precision quartz oscillator crystals, and the lack of understanding of their repair by traditional watch repair personnel, have put most in the "throw-away" class.

Systems requiring very high availability on a "one-shot" basis must avoid use of parts having short *shelf life*. Electric batteries are typical. The acid or alkali electrolyte will ultimately deteriorate (though some systems are better than others) and can damage not only the battery but external components as well, in some cases virtually exploding. Battery life may in certain units be extended by an occasional full discharge or by continuous charging at a low current. Batteries used in costly electronic systems such as military missiles are designed to contain the liquid electrolyte initially separated from other components by a membrane. At activation, the membrane is ruptured by acceleration or another means. Operating lifetime is limited to a few minutes.

Electronic circuits which must operate after long periods of inactivity should be designed using components which retain correct operating characteristics under these conditions. The electronic components most subject to deterioration are *electrolytic capacitors* which, like batteries, contain a liquid electrolyte (usually alkaline). Special capacitors using alternative technologies are available. Although semiconductor circuits are far less susceptible to deterioration, low-cost units are usually encapsulated in plastic which will ultimately deteriorate and deposit films of contaminants on the device inside. Circuits encapsulated only in ceramic, glass, and metal enclosures, though substantially more expensive, have much longer lifetimes.

A principal cause of inoperability of *mechanical systems* activated after long periods of nonoperation is deterioration of organic fuel, lubricating fluid or grease. Systems required to operate for only short periods or at light mechanical loads may make better use of stable plastics such as teflon for bearings, or may employ nondeteriorating *solid lubricants* such as graphite or molybdenum disulfide. In some high-speed systems, lubrication is accomplished by gas pressure in the space between moving parts. Small high-speed rotating parts may also be suspended, in a near-vacuum enclosure, using electrostatic fields. In short-operating-cycle systems using liquid fuels, such as certain rockets, the fuel may serve as a poor but adequate lubricant for its pump, thus reducing the number of organic materials involved.

Solid fuels, as used in military rocket engines for missiles of all but the largest sizes, improve system availability. They require little maintenance and no fueling operations before use. However, solid fuel units incorporating volatile solvents will characteristically shrink somewhat with age. These annular fuel elements are intended to burn only on inner surfaces; however, shrinkage gaps between the fuel and its outer housing may permit hot gas to completely surround the element and allow burning over its entire surface. The higher burning rate can cause the unit to explode. One solution is to line the housing with a long-life compressible elastic liner which will expand to seal shrinkage gaps.

5.3 Design for maintainability

5.3.1 Representative requirements

Maintainability is more difficult to define than are reliability and availability. It has no common metrics of its own but often uses MTTR. Its objectives are usually simple: minimization of repair and preventive maintenance time and effort. A related requirement, of increased importance as systems grow more complex, is that average skill level of maintenance personnel *cannot* be expected to increase. Nor should greatly extended training be required.

Maintainability requirements may specify expected and maximum scheduled maintenance and repair time, in which case the resources assumed available should be defined (numbers of technicians and their training, maintenance equipment, spares, supplies, etc.). Where large or heavy systems require maintenance, limits may be placed on maximum weight or size of *line replaceable units* (LRUs) which must be handled manually by maintenance personnel. Maintainability criteria may specify a maximum allowable amount of system disassembly required to gain access to a part or subsystem for its service or replacement. They may also address the ease with which a technician of given weight and body size must be able to access parts of a system requiring physical entry. And, of course, they often required development of tools, maintenance manuals, and training curricula concurrent with system development.

Special maintenance fixtures and tools used for routine maintenance or repairs can greatly facilitate maintainability. If complex systems are to be maintained routinely by user personnel, as required for major military systems, the development and production of maintenance equipment and maintenance training equipment can represent a significant fraction of system acquisition costs.

Where maintenance by user personnel is deemed infeasible, the system contractor may be required to demonstrate its own ability to do so. It may be required to develop a system which ensures timely repair or replacement of defective parts or subsystems, including a parts "trade-in" procedure and pricing basis. For a complex system, this in itself constitutes a major development effort. If the contractor must invest in maintenance equipment and a spare parts inventory, the financial commitment required to support a maintenance agreement may be very large.

Systems which will be installed or used in multiple locations may require both local and central maintenance functions and parts inventories, the central facility typically used for complex or low-frequency operations for which personnel or facilities would be unaffordable if located at every operating location.

5.3.2 Maintenance processes and techniques

The processes required to maintain a system in operating condition depend on the detailed character and scope of the system, the technologies it incorporates, personnel and facilities available to perform maintenance, and the timeliness with which repairs must be completed.

The *maintenance environment* is very important. Systems with open-air or remotely sited system installations may pose special problems for maintenance in all bad weather conditions or pose difficulty reaching maintenance sites. If the site of a large outdoor installation is a major system complex, industrial equipment such as "cherry picker" cranes may be applicable, provided maintainable system parts are within reach of the equipment. Large or remotely sited installations must usually include permanent access provisions, such as gangways or ladders, with provisions for temporarily mounting portable hoists, to move large or heavy parts. Where equipment must be maintained in all weather, design consideration must be given to protection not only of maintenance crews but also equipment, using portable shelters possibly with space heaters and other utilities incorporated. Installations in polar or mountaintop sites usually call for the system to be completely maintained within its enclosure.

Access to maintainable parts is a necessity sometimes overlooked by system designers. Most of us have experienced frustrations while attempting to replace light bulbs in a ceiling fixture or spark plugs in an auto engine where access using ordinary tools was impossible. Although special tools may in certain instances be the sole practical solution, these also add to system costs and the complexity of maintenance. Superior system designs permit access to maintainable parts with *minimal* removal of unaffected parts. Additional removal and replacement steps not only add to MTTR but risk damage or the incorrect reinstallation of parts which did not themselves require maintenance. While it is understood that removal and replacement of failed parts may require effort to disconnect and reconnect the interfacial cables, bolts, plugs, hoses, and the like, routine maintenance should *not* require such efforts.

Access for maintenance can usually be greatly improved by the provision of access doors fastened by reliable, quick-operating, fasteners which demand little physical effort and time for correct functioning. Reliability of a fastener relates to the sureness with which it holds in system operation. However, it applies as well to the ease of unfastening and refastening, and its resistance to incorrect procedures. *Captive fasteners*, which need not be removed to disengage them, are highly desirable since they save maintenance time and avoid use of "baling-wire" and other makeshift fastenings.

Where system parts must be moved for test, repair, or access to other parts, features such as robust sliding tracks and hinged mountings can reduce maintenance problems, especially if units that have been slid out do not fall off the tracks when the limit of motion is reached. Such support schemes should be substantial enough to withstand normal wear and tear but require as few operating steps as consistent with the rigidity required in the operating system. In this respect mobile systems differ enormously from stationary ones. An aircraft or overland vehicle can develop high transient accelerational forces in any direction, requiring that its parts resist all movements.

Maintenance provisions must allow for the presence and use of test instruments and fixtures needed for diagnosis or repair. Electronic circuits which require voltage measurements but lack connection points for test probes may require one maintainer to hold probes in place while a second manipulates the test instrument. This recalls a collection of bad jokes about the numbers of people required to screw in a light bulb, but it is far from humorous.

In a system which contains numbers of subsystems of a similar type, for example, electronic subsystems, maintenance provisions can be incorporated through *design rules* applied to all such subsystems. Where there are special one-of-a-kind subsystems such as an antenna tower, or where many dissimilar subsystems are integrated into a single system, however, maintenance poses special problems.

There are few sources of guidance for design for maintainability. Those old standbys of the designer, previous-generation systems, are often poorly designed from the maintenance standpoint. Good maintainability will not reveal itself at a glance. Special attention to this design area is usually important, using experts where available. A preferred way to achieve good maintainability is by exercising representative maintenance operations on a prototype system. This requires discipline to ensure that a wide enough variety of representative maintenance exercises are practiced, preferably using personnel representative of those who will actually perform the system maintenance.

5.3.3 Special test facilities

Because electronic circuit boards have steadily increased in complexity along with the integrated circuits they contain, testing a single board often involves checking operation of many thousands of circuits. *Automatic test equipment* (ATE) for electronics testing has evolved into a major system area. The designer of systems containing circuit boards may be required to supply, along with the system design, a comprehensive test program to verify correct operation of each board. These may be required to be written in the *Atlas* ATE language for use on the customer's automatic test equipment. This type of equipment, because of its cost and productivity, is normally installed only in major test facilities. In the U.S. Navy, this sort of equipment is located aboard major sea vessels, because of the delay required in returning parts to a land-based repair depot.

Built-in test In some electronic subsystems, built-in test operations are programmed to take place automatically in various subsystems whenever the subsystem computer is turned on and also when it is not otherwise fully occupied. In any digital systems, it is often possible to add a test program as part of a sequence of programs which are executed automatically and sequentially. Thus, implementation of built-in test in this case requires principally the addition of software. For other types of systems, it may require addition of sensors or detectors, to provide indications of system parameters, and sometimes a computer to perform the analysis.

When there is automatic built-in test, it is necessary that the system incorporate some means to pass test results to system operators or maintainers. In systems which are built up from existing subsystems, this may prove very difficult, since many subsystems cannot communicate directly with system operators. Increasingly, modern systems are being designed with built-in test as an intrinsic part of their *architecture*. This implies several design features: (1) a means (usually local) to command testing of any part of the system; (2) a network through which test results can be directed to a maintenance console; and (3) software separate from the operational software, through which these functions can be coordinated.

A long-popular view of built-in test of electronic systems was that a future electronics board should contain a single light-emitting diode lamp indicator, which would indicate if the board had failed or was operating correctly. This is proving to be too simplistic an approach, since more information is available and often more can be used.

Built-in test of mechanical systems may involve features such as continuous testing of lubricating oil to determine if minute metal particles have entered it signifying imminent failure of certain highly stressed parts. Vibration sensors may also be used to determine condition of high-speed machines.

As Figure 5.9 illustrates, built-in test equipment need not be limited to complex computer programs which operate automatically. Many system functions lend themselves to on-line visual observation by operators or maintainers. Since the failure of complex built-in test facilities is a growing possibility, some of the simpler schemes offer significant advantages both in initial price and life cycle cost.

5.4 Design for compatibility with existing systems or interfaces

A frequent design assignment, which provides a major challenge, is a *system enhancement*, in which an *existing* system is extended or modified to meet additional requirements. Rather than working top-down, as is more desirable for new systems, the designer must accommodate an existing system as part of the framework, or as a portion, of the new one. Old technology must be accommodated as well as new. Achieving compatibility between old and new elements often requires very creative design. Advanced system concepts may need to be applied to both original and new parts of the system.

5.4.1 Representative requirements

The simplest requirements of this type are for *additions* to a system in being. Additions may, for example, increase the system throughput or allow the system to serve additional users. Additions may be duplicates of subsystems already in use, with no intent to alter basic system operation. Such assignments are often directed to the original system

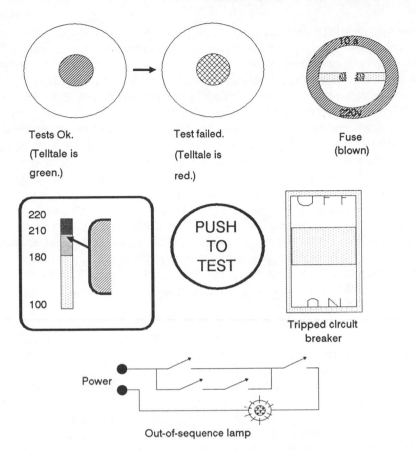

Tests Ok.
(Telltale is
green.)

Test failed.
(Telltale is
red.)

Fuse
(blown)

PUSH
TO
TEST

Tripped circuit
breaker

Power

Out-of-sequence lamp

Figure 5.9. All built-in test equipment need not be complex or automatic. The familiar concepts shown here should not be overlooked in a search for reliability in modern systems.

contractor(s) if experience with the original system proved favorable. If required additions represent a significant growth in system capacity, or features not envisioned in the original architecture, some architecture changes may be required. For example, if a subsystem performance requirement exceeds capabilities of the largest single subsystem unit available, *multiple* subsystem units will be needed, along with consequential changes to system layout.

Another frequent requirement, usually more complex than merely adding duplicate subsystems, is *replacement* of obsolescent system components: e.g., a central computer, data storage unit, electric power source or power distribution facility, or telephone branch-exchange switching equipment. A major objective may be cost reduction, especially if the equipment to be replaced was leased, rather than owned, by the user. In this case, the user's motivation is to replace hardware or software, not to alter system functions. The original system contractor will again ordinarily be favored unless products of competing contractors were stimuli for the replacement. The system architecture may need revision to accommodate, or take better advantage of, replacement subsystem(s).

At the highest level of complexity is the *enhancement* of an existing system, i.e., implementation of *new system functions*. It is least complex in cases where the new functions can be implemented by an *appended* subsystem, connected to the original system via existing interfaces. Providing such interfaces during the original system design, in anticipation of future needs, is often referred to, in either hardware or software, as "providing hooks." In more complex cases, enhancements may alter existing system functions, perhaps requiring new subsystems or software replacing those of the original system. The most complex enhancements are frequently those requiring many small changes throughout existing system functions, particularly if the new objectives are inconsistent with the architectural decisions made for the original design. The problem is made more difficult if the client lacks appreciation of how difficult are such "minor" enhancements.

Let us illustrate problems which will occasionally be encountered in undertaking to modify an existing system. The client may expect the system designer to incorporate as much as possible of current system assets, and reflect the organization's newest way of doing business, in designing the expanded or enhanced system. To this end the client may set meaningful *overall objectives*, for example: new components to be based on up-to-date technology; lower system operating costs, with increased performance and reliability; an ability to expand the system more easily in the future; and minimum disturbance to ongoing, revenue-producing or otherwise important system operations.

We might consider, as an example, a small communications company which has historically served *local* telephone needs of small communities. Its expanded objectives, to be implemented via system changes, might be to grow into a *regional* telephone system while continuing to derive maximum economic benefit from previously installed equipment. It will likely wish to continue operating in ways which are comfortable to both its employees and subscribers.

System requirements generated by conservative, near-term oriented clients may superficially seem to be thorough. On closer examination, however, they may include much unimportant detail while remaining uninformative about top-level issues or future system needs. A client firm may understand its own operations in such inflexible terms that the requirement becomes essentially a "bottom-up" system description. It may ignore, or even be internally inconsistent about, the real top-level issues. Even if a designer carries out top-down design, it may be expected, nonetheless, to follow closely this "bottom-up" system detail.

A further impediment may be that the existing system has, over the years, been highly optimized, in a local sense. New features which could address the uncertainties in evolution and growth will in this case probably appear less optimal to the client. The new, more general top-level system may seem to the client only a minor improvement, though it is likely to cost more than the asset value of existing systems assembled over many years.

Though this is not an actual case, it is typical of problems in dealing with system modification problems, which are often overconstrained and underfunded and may have *no* solutions which are both available to the designer and acceptable to the client.

5.4.2 Design of system enhancements

We will discuss enhancements requiring changes of system architecture, of design, or both. Lesser modifications may be considered subsets of this general case. Approaching any modification from a high-level point of view is usually prudent. At the outset, e.g., on initial inquiry of a prospective client, a system design firm usually cannot immediately determine what parts of the scope of changes required (or desired) will ultimately prove within the client's budget.

Understanding requirements and existing systems In any systems engineering project, *understanding the client's requirements* is an initial, and important, task. The client's requirement may sometimes be stated informally, perhaps merely verbally. Even if documented, when the client's requirement falls short of what is needed to understand a system requirement, a way must be found to bring the designers' application awareness to a level at which system design can proceed intelligently. (*Requirements definition* is itself a complex art, though not a tutorial objective of this book.) Under certain circumstances, a system design firm may be willing, as it must be *able*, to assist a client in completing a requirements definition, possibly as an initial unpaid activity. From the client's own point of view, however, paid assistance by competent consultants or an organization which will not be involved in system design and construction will keep open more client options.

Requirements for a system enhancement must be examined within the context of the *current system*, which will often lack complete or correct description. One may be able to locate only dog-eared instruction manuals for commercially purchased units such as terminals, possibly a crude operator's manual, primitive sketches of interconnecting cables, hoses, etc., and perhaps an assembly drawing or two. If the system includes special software, user documentation will usually prove inadequate for interfacing to that software. If the same software programs were delivered to multiple customers, source descriptions will probably be considered proprietary and not available. Existing software which must be retained, or its functions reproduced in detail in the enhanced system, may represent the single most difficult problem area in system enhancement.

Requirements must be understood through discussion and negotiations with the client. Above all, there should be no misunderstanding as to the scope of the work. If there are uncertainties, they should be addressed at the outset, rather than assuming they will work out favorably. Necessary developments and acquisitions must be defined, required work and associated costs estimated, and a proposal prepared. Proposal development may in some cases be carried out under an initial contract; project continuation may then be contingent on a design and implementation proposal with cost and schedule acceptable to the client. The proposal efforts, that is, preparing a preliminary design, estimating costs, and laying out the proposal itself, may be assumed by the system contractor as an overhead cost. This is usually so, if the contractor carries out many very similar projects for different clients. Otherwise, proposal preparation costs may be shared with the client, or partially compensated by a fee agreed upon in advance.

Architectural extensions The original system architecture (and its detailed design, if possible) should be examined carefully for presence of features ("hooks") intended to aid future enhancements (Figure 5.10). If planned enhancements were not previously implemented, the hooks may have been forgotten by the original designers and may never have been understood by the system owner. If they exist and can be uncovered, they may suggest the best routes to enhancing the system, possibly without architectural changes.

Given need to extend the architecture of the enhanced system, a good system designer will attempt to disturb it as little as possible. New functions will, where feasible, be implemented through *existing* system interfaces, for example, via system buses in the case of digital electronics. If added elements are designed to emulate (i.e., behave like) existing subsystem structures, the original system may not require any architectural changes.

Some revisions of this sort involve adding structure *above* the "top end" of the original, thus making it a subsystem of the new one (Figure 5.11). This works well if the original system has a central point at which it can be fully controlled, and at which information is collected and distributed. The new top-end element can then communicate with and control the original system, and serve as well, as a framework to which new subsystems may be appended. New subsystems may often also be readily appended at the "bottom end" of an existing system, for example, attached to control or communications connection points. However, from this vantage point they may have limited ability to control or to access other

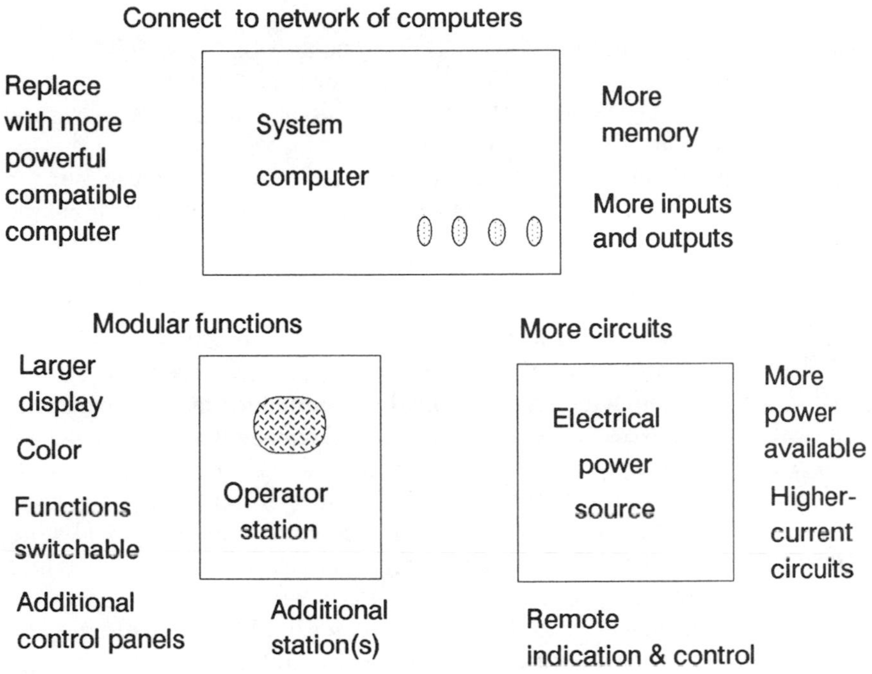

Connect to network of computers

Replace
with more
powerful
compatible
computer

System

computer

More
memory

More inputs
and outputs

Modular functions

Larger
display

Color

Functions
switchable

Operator
station

Additional
control panels

Additional
station(s)

More circuits

Electrical

power

source

More
power
available

Higher-
current
circuits

Remote
indication & control

Figure 5.10. Extensions to a given system are greatly facilitated when hooks (provisions) are created early in the design. The figure represents some of the many directions system extensions or enhancements may take.

Top end

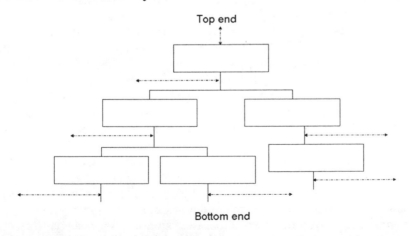

Bottom end

Figure 5.11. Extensions to a system may be connected at any of a variety of existing interfaces, as depicted here. Capability to control or to interact with the original system are usually limited to those already available at the interface. Attachment at the top end of the structure may provide greater control, but may not have access to some of the information available at lower levels in the existing system.

subsystems. (The system described in the appendix was extended from top and bottom.)

Still other access points (Figure 5.11) may usually be found at interfaces at which subsystems were attached in the original system. Additional subsystems can often be attached at these same points, using the original interface definition. Depending on the capabilities which may be accessed and controlled from such points, it may be practical to use such a connection as an interface to subsystems which add new function or couple to other systems.

For enhancements requiring interfacing with *several* existing subsystems, design and implementation problems are minimized if most new interfaces derive only *inputs* from the original system. An example might be a chemical processing plant enhancement to provide *automatic system control*, for a system which was originally designed for manual control at a central location. New inputs could be derived from new sensors attached to various subsystems, with control outputs sent through the manual control interfaces at the control point. This sort of multi-interface class of enhancement may require the enhancement designer to deal with undocumented characteristics of the original system, such as its failure modes, transient behavior, or subsystem delays. These can usually be dealt with by mechanisms and/or programs within an added subsystem. In this example, if the existing system is controllable and its performance is heuristically understood by its operators, techniques such as expert systems, with control theory properly applied to describe system dynamics, should work well.

Interface adaptation As suggested by Figure 5.11, existing interfaces on the present system may be used as "takeoff points" for enhancement. Other interfaces may need to be added where none existed. An interface standard may have been chosen or defined during the original design. Dimensions, voltages, pressures, mechanical tolerances, or other parameters, however, may not conform to any current standard. Where new subsystems are obtained with standardized interfaces, it may be necessary to design interface adaptors. In simplest form, these may consist merely of special connectors, cables, hoses, etc., with an original-system interface at one end and a standard interface at the other. Even such simple adaptors must be described in detail in system documentation and labeled clearly with part number or other identifying information. After a later repair, or replacement using an incorrect interface part, unsafe operation or system damage might occur.

In enhancing a system, more elaborate, "smart" interface adaptors may be required to: convert formats of signals, transform pressure or voltage levels, or convert data to different forms. One-of-a-kind adaptors must be designed, constructed, and tested as carefully as are higher-volume units. Detailed documentation and drawings are required. Adaptors should be fully and carefully tested, prior to attachment to the existing system. The cost of development of special interface adaptors is often underestimated.

If an old system is described as using standard interfaces, it is crucial to verify if the standards have been updated in a compatible way. Where interfaces must be redesigned, serious shortcomings of the original interfaces should, where feasible, be corrected. If, for example, the original interface failed to include proper electrical grounding, this could be added. If the electrical grounding of the original system is questionable, it would be better practice to provide electrical *isolation* between the original and added parts. Carefully isolating the older parts of an enhanced system from the newer parts, to the degree possible while allowing necessary communication, reduces the risk that problems in the older system will cause the added parts to fail.

If *system user interfaces* (i.e., controls and displays) are altered during system enhancement, major rearrangement often appears desirable. However, since user personnel are familiar with the original system, allowable interface alterations should be clearly defined in the contract. It may be better to *add* controls or displays, instead of changing the appearance or function of existing ones. Good human-factors design rules should of course guide modifications.

Occasionally, a new interface must be created *inside* an existing system. This is risky for it may affect the operation or safety of the original system. Merely connecting an additional output to an electronic device will change the magnitude and dynamics of its output and may even damage or destroy it. Cutting mounting holes into a structure to add components, in an aircraft, *or even a building* can weaken the structure disastrously. Modifications which could affect original design parameters of an existing system may invalidate warranties or service contracts between your client and the original manufacturer. In advance of any such modification, the designer should check both technical and legal implications.

Other design considerations The designer of a system enhancement must appreciate the users' view of the system. It is important not to defeat the utility, maintainability, and other features of the preexisting subsystems by careless enhancement design. If, for example, new subsystems will be placed adjacent to original ones, maintenance access must be considered.

Wherever practical, enhancements should complement the original system: its physical appearance, user operating procedures, maintenance operations, failure-indication schemes, etc. Reliability of new system elements should preferably be higher than that of original ones for two important reasons: (1) The enhanced system will have lower overall reliability if any new parts fail, and (2) an enhancement contractor will usually be blamed for any new system problems, whatever their cause.

This emphasizes need for quality in special interface adaptors, connecting cables, and the like.

Choice of system elements should, where possible, limit operational complexities such as numbers of separate maintenance contractors required or programming languages to be dealt with. Unless the client directs otherwise, standard types of subsystems such as power supplies, pumps, and processors should, where practical, be obtained from sources which supplied the originals. New subsystems using expendable supplies such as recording paper, magnetic tape, or plotting pens should accept items currently stocked by the user.

The designer of a system enhancement must establish a close rapport with the existing system and its users, despite annoying shortcomings due to original designers apparently being idiots. Failure to give due respect to a system which has served the client long and well will adversely influence your relationship.

Integration and testing of enhanced systems A new system may be assembled and tested in any preferred sequence. It is common to set up in the client's site and test it operationally. If it is to *replace* an older system, the two may be operated concurrently for an extended time, so the client has "fall-back" until the replacement is fully acceptable.

When enhancements are made to a system used regularly or even continuously by a client, they *must* be tested as fully as possible prior to actual installation. Opportunities to test the overall (enhanced) system may be presented only at times when the original system is not in use. If it is in constant use, a tightly limited, periodic schedule for integration and testing—perhaps time normally devoted to scheduled maintenance—may be assigned. This time *must* be utilized effectively even if in early morning or on weekends. A careful estimate and schedule, allowing time for unexpected problems, should be prepared to assure the client that inconveniences should be temporary. Care must be taken not to alter or damage the original system during installation or testing. A detailed schedule and step-by-step instructions must be prepared in advance, though of course any test schedule may require modification. If activities are to occur in the presence of client personnel, installation and testing personnel must act courteously and discreetly.

When access time is very limited, enhancements should be designed for quick connection and activation at beginning of a test period and disconnection at its close. Though this may be of little continuing value after system installation, the modest cost will in most cases be offset by the more efficient test operation. Where enhancements derive inputs from the existing system, design should make it possible for these to remain connected during system operation. The outputs of the added system elements can be examined to verify their correctness, without disturbing operations.

User personnel experienced in operation and maintenance of the existing system should be involved in the integration and testing operations to the degree their work schedules will permit. This should be characterized as part of user training. They will become familiar with the enhanced system, operate it, and assist in trouble-shooting, with their knowledge of the existing system.

5.5 Design for ease of use

Until recent years, relatively little design attention was paid to ease of use. A sad example is the typewriter, whose 100-year-old keyboard was laid out specifically to *slow* typing. Typists could jam the mechanisms of early typewriters, had they incorporated user-friendly key layout. A retired U.S. Navy officer designed and demonstrated the *Dvorak keyboard,* almost 50 years ago. Dvorak's design has been found near-optimal in later studies. But neither his nor any other improved keyboard design has yet made headway against the inefficient "QWERTY" keyboard.

The lesson is that most people enjoy the familiar, and avoid the unfamiliar. Much more recently, the U.S. Air Force acquired the F-16, in which the familiar "joystick" was replaced by a 6-in handle ideally located for right-hand operation (Figure 5.12). To add to its novelty, the stick doesn't move—force applied is converted to an electric signal fed through wires into an all-electronic flight control system (hence the term "fly-by-wire"). Although one four-star general with thousands of flying hours was said to have immediately disliked the control, young pilots with less experience and bias have adapted easily.

In some instances, system operators actually relish the job security implied by a difficult-to-use operator interface. The early brute-force controls of trucks and construction machinery can, with hydraulic assistance, be operated by a petite woman or even a child. However, many system operators of either sex take pride in work which a newcomer may find difficult.

Although physical effort required to operate modern systems is usually small, system complexity and performance have continued to advance so rapidly that *sensory and cognitive limitations* of human operators must be considered in modern system design. The term *user-friendly* describes systems which attempt not to harass or confuse users. Some even provide operating instructions on request.

5.5.1 Representative requirements

There are three distinct areas in which systems may, physically or psychologically, challenge users. The first is requirement for high *physical effort*. Since lower required physical effort reduces operator fatigue,

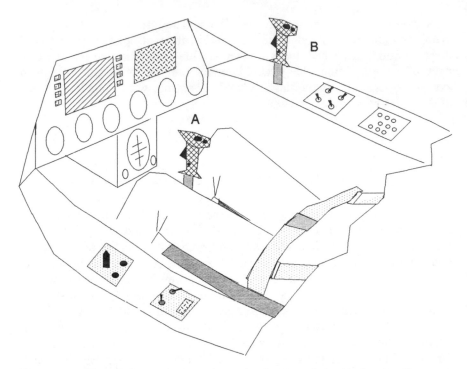

Figure 5.12. Though they never appear together in one cockpit, this drawing illustrates both (A) the traditional "joystick" and (B) the right-hand short stick used in U.S. F-16 fighters and a few gliders. A benefit of the center-mounted stick is that it can be operated using either hand. The side stick, however, is convenient in cockpits where there is little room for movement.

it *is* cost-effective to use hydraulic or electromagnetic assists when heavy work must be done. A system requirement or associated standards may cite maximum effort and movement required to actuate a (mechanical) system, but details are typically left to the designer.

The second area relates to *sensorimotor activities* by system users. Such complex activities typically require the operator to sense need for some change in a system and to make the change in a timely way with minimal effort or likelihood of error. The field of *human factors engineering* arose in the 1950s to address needed improvements in user interfaces. Human factors originally addressed matters such as location and size of controls and the force or delicacy and amount of motion required for their adjustment. It also related to location and size of displays which must be sensed or read, quantities of textual information displayed, and the minimum practical size of alphabetic or numeric characters which may normally be read at a given distance. Since the early years, the practice of human factors engineering as a specialty appears to have reduced. However, its standards for knobs, switches, and displays have been incorporated into modern system components. Design engineers today have access to handbooks which describe human factors design practices.

Requirements may cite standards of human factors design, for example, military standards such as MIL-STD-2167A for software development, among many others. Today's system designer still has *carte blanche* for control and display layout, and design engineers having awareness of good human factors design are still relatively rare.

The third challenging interface area deals with design of *cognitive tasks* involving mental activity, involved in system operation. Cognitive abilities of humans vary widely and involve a variety of different skills in perception, analysis, formulation, and execution of intellectual activity. Yet there are still no established norms of human cognitive performance against which systems can be measured. The predictable result is that most system designers ignore the cognitive workloads their systems demand of users. Where cognitive activity is considered, is most often described in terms of *decisions* which operators must make, with little attention given to processes which can be employed or data which is needed. Cognitive tasks may, at best, be allocated between operators in ways which require little interoperator coordination. Though operator training to carry out cognitive tasks is clearly important, most training regimens use rote methods.

Many activities previously carried out as human cognitive tasks have been assumed by computers. More are being taken on by automation, with each new generation of systems. Automatic aircraft landing systems, as in Boeing 747 and Airbus 300 aircraft, are representative. Implementation of this kind of automation tends to suppress concern about designing systems having reduced operator cognitive-task difficulty or stress. The other side of this issue, however, is *how far to go*. What tasks should be reserved to humans, and for what *automated* tasks should humans serve as backups for system computers? The complexity of some of the simplest real-world cognitive tasks is still beyond the abilities of computers or the programs which operate them. Though we may be making some headway through research in artificial intelligence, we have hardly begun to understand human cognitive abilities. Meanwhile, rapid evolution of personal-computer programs demonstrates the salient differences between well-designed user interfaces and poor ones. Such features as pull-down "windows," "help" functions, and graphics pointing devices represent some present high-water marks in cognitive interfacing.

Requirements for cognitive interfaces between systems and their users are hardly ever addressed directly. Such requirements may be couched indirectly such as requiring interactive access to tables or data used by system operators. Human factors data suggests some of the limitations to human perception. But as yet there are no standards which can be applied broadly to *design* of cognitive tasks.

Design of operator and maintenance tasks System operation increasingly requires operators to carry out complex mental exercises, in response to observations of system operating parameters or of the external environment, or both. These tasks range in complexity and time-urgency from rapid responses to system emergencies to system planning activities. Modern system technologies, principally in programmable systems, allow these tasks to be carefully designed, and supported by the system with information, suggestions, or automated operation.

Where a task is an emergency response to a situation such as loss of engine power in an aircraft, the response proposed by the system developer is, most often, timely actions by the operator. Traditional and accepted training is by memorizing simple symptoms to which apply nonvarying sequences of actions. These may be referred to as *preplanned responses*. Little attention may be paid to the correctness of problem identification since any delay in response may prove fatal.

In a complex system, more routine operator tasks, such as adjusting the system to accommodate changes in input or environment, may be required. In these cases the activity required by the user or operator can be described as a cognitive task. Often such tasks consist of six main steps:

- Examination of the system state
- Decision whether or not action is needed
- Development of (at least one) action, with alternatives if time permits
- Assessment of alternatives
- Selection of an alternative
- Execution of the task

Often these processes overlap. An experienced operator may bypass some of the steps, based on experience with similar situations. Many of these cognitive steps may be *computer-aided* with a computer subsystem recommending actions or executing a selected course and ultimately communicating with the rest of the system.

As systems become still more complex, and the consequences of operator actions grow in importance, additional effort must be placed on painstaking *design* of cognitive tasks required of system users or operators. This field is at present very immature, and it is impractical to offer universal rules of cognitive task design. However, cognitive task design requires answers to a number of important questions. These may guide the design process.

- **What, specifically, must be done?** In short, what *is* the task? In the past, cognitive tasks have not been articulated. They may involve spatial pattern recognition, examination of written text, a search for data within large files, etc., etc. Often they involve a sequence of different activities for which different designers may have varied opinions as to the most timely or accurate approaches. As in other systems engineering activities, here it should be possible to decompose a task into constituent subtasks, whose characteristics and needs can then be further defined.

- **Within what environment is the task to be performed?** A task may be performed under high-G acceleration, in cramped or overheated quarters, after many hours on duty, or accompanied by other tasks which compete for attention. In these cases less can be expected from system operators.
- **What are the success criteria or, how detailed and accurate a response or action is expected, and what are the penalties for error?** A coarse response will usually be available more rapidly than a detailed one. If only a detailed response would be meaningful, the system can help to enforce an orderly decision process.
- **How long a time is available to complete the task and what are the penalties for delay?** The means provided to support performing the task must be compatible with its time-urgency. It makes little sense to provide elaborate computer display options for a task which must be completed in no more than a few seconds.
- **What information is needed by the operator(s)?** These may include information in all forms, including that retained in the memory of the performer, indicator lights, displays, the contents of a computer database, etc.
- **What human errors are likely in carrying out the task?** Consideration of possible errors, particularly those which carry high penalties, should help formulate the way the task should be set up.
- **What division of responsibility can be made between several human operators, or between humans and computers?** Certain subtasks such as recalling large amounts of dynamically changing data are far easier for computers than for humans. Other tasks such as recognition of patterns within an image are at present easier for humans than computers.
- **What are the most appropriate measures of performance in the task?** This may relate to trade-off between speed and accuracy. If there are especially demanding areas of performance, special operator screening may be needed. (Screening tests would presumably be based on tasks similar to those required in system operation.)
- **What human resources will be available to operate the system?** Cognitive tasks must not demand more from system users than their native abilities can provide, after training.

Because we currently lack other bases for evaluation, comparison of cognitive task designs is best carried out experimentally, using subjects from a representative group, provided with representative training for the task. If the task is time-urgent, requires precision, or is to be performed under particular conditions, these factors should be introduced or simulated realistically.

A process for assessment of complex user interfaces Any design must be tested and assessed. Whereas entirely automated systems can be tested using technical means such as meters, systems in which human operators are "in the loop" require more appropriate testing. The best-accepted current technique for assessment of complex user interface designs involves *real-time simulation of system operation*. This simulation is sequenced through representative operational scenarios while human subjects, acting as operators, react to simulated external events and initiate system activities.

The simulation must be implemented with realism appropriate to the tests to be carried out. User-interface hardware and software should be in forms which allow rearrangement of controls and displays, of control functions, the information displayed, and display modes. Design and testing personnel, including behavioral scientists, execute human roles external to the system. Events and operator actions are captured using instrument, television, and voice recordings, while the observers make time-stamped observations. Following the session, subjects may be interviewed using prepared questionnaires.

The simulated system and interfaces are initially set up in a baseline configuration derived from experience with previous systems of similar type. The experimental subjects, who must be representative of user personnel, are given a standardized training curriculum in system operation. One or more system scenarios, which may run many minutes or hours, are run repeatedly to capture responses, reactions, and "learning curves" of several groups of test subjects. The baseline configuration is then changed to what is expected to be an improved configuration and the simulations are rerun. The process may be continued to evaluate two, three or more system configurations, which may differ significantly or merely in detail. Similar data is captured for each system configuration, where possible, so that dissimilar results do not need to be compared. A measure of the relative utility for each system configuration is derived from analysis of operators' responses, response delays, accuracy and error, along with operators' subjective views learned through carefully structured interviews of test subjects.

This highly experimental approach is unquestionably so costly and time-consuming that it is unlikely to be used except for high-dollar-value systems. There is no assurance that an optimum will be discovered, or that interface problems can be discovered and repaired. In the absence of any proven theoretical design methods, however, it is far superior than the mere application of rules of thumb. It has been applied to the detailed design of many military system interfaces and somewhat less formally in automotibile and test equipment design.

5.6 Design for system or personnel safety

At least three types of safety concerns arise in connection with systems: protection of the system itself, protection of systems with which it interfaces, and protection of operators or bystanders within range of harm by the system through either normal or incorrect operation.

The system itself may be threatened if while in operation it is exposed to forces, temperatures, voltages, flammable vapors, or other influences capable of damaging it, either catastrophically or over an extended time. When the operating environment will include these threats, the system must be designed to counter them or be isolated from them by distance or by some sort of protective covering or shelter.

Systems may also pose threats to themselves. For example, most propulsion engines are able to be operated at high speeds which, especially when output loads are small, may very quickly cause physical damage. Many electric devices operate using voltages which, if applied improperly to other parts of the device, could produce arcing and permanent damage. Protection of a system against its internal threats is more subtle. Since the threat cannot be removed, it must be controlled. Engines may be equipped with governors which preclude them from operating above certain speeds. Thermal equipment such as furnaces and water heaters must include limit switches, which open electric circuits if operating temperatures exceed preset limits.

A principal difficulty with such protection is that it usually imposes limitations on system performance. In some cases it may be more important to retain the potential for exceptional performance, even though it is accompanied by risk of damage. For example, an aircraft may suffer main wing spar failure if the pilot pulls sharply back on the control stick or yoke *at normal cruising speed*. Pilots are supposed to be trained to be aware of this risk. At the same time, the aircraft retains the potential to make a maximum-g maneuver if that were the only alternative to a disastrous midair collision.

Safety threats are very commonly posed by systems to other systems. A small private aircraft landing, say, a quarter-mile behind a wide-body jet transport encounters turbulent air created by the larger aircraft. This can, almost instantly, roll the smaller plane sufficiently to cause it to become uncontrollable and crash. The only solution to this threat is an operational one: If you're a little guy, stay well behind the big boys. There is little the wide-body's designers *could* do, and probably nothing they *should* do, since this threat is well known and there is no reason to maintain close aircraft spacing during landing.

Many safety threats posed by systems to other systems have nothing to do with their design but are inherent in their application and environment. Consider, for example, operation on a busy highway of a tank truck containing liquified explosive gas. Unless special care is taken in design, many systems may pose threats to those nearby. Aircraft or automobile parts designed so that they are likely to detach and fall from the vehicle pose very serious threats. Highway trucks carrying heaped-up loads of gravel without the covers that are now mandated in a number of states are capable of seriously damaging other vehicles and endangering life.

5.6.1 Threats to humans

Many systems pose a variety of possible threats to their operators, service personnel, or innocent bystanders. Aside from the obvious threat of having the system or some part of it fall on you, most electronic or electrical systems employ lethal voltages. Safety precautions here usually take the form of interlocks which automatically remove voltage inside cabinets containing electrical apparatus when their doors or covers are opened. In some cases this may not be adequate, for there may be large storage capacitors (usually associated with dc power supplies) which retain a lethal charge for seconds or longer.

A system designer is usually expected to take "reasonable steps" to protect both users of potentially dangerous equipment, and bystanders. Service personnel often choose to operate equipment when high voltages are about or must do so to carry out tests. Although their training emphasizes the dangers, accidental electrocution does occur with regularity, principally because of carelessness though sometimes because of failure of safety devices.

There are many other system operations which may threaten human safety. Equipment which operates at high temperature or pressure is potentially hazardous, the more so if poorly maintained or in a deteriorated condition.

Some relatively modern threats are considered especially pernicious because their effects are not immediately apparent. One class of these are purely chemical in nature, certain organic products being toxic or acting as carcinogens (cancer-inducing agents) even in minute quantities. Effects of some of these may be delayed for years. Other subtle threats are posed by electromagnetic radiation, which is usually characterized as ionizing radiation and nonionizing radiation, depending on the wavelength (or photon energy). Microwave radiation is nonionizing but is well known as a heating source and has been suspected of causing damage in certain organs (eye, testicles) which have little self-cooling capacity via blood flow. Though it is widely believed that microwave radiation at high peak but low *average* power (e.g., less than a few milliwatts per square centimeter) has no deleterious effects on humans, there are occasional claims otherwise. Since radio, television, and radar signals produce large peak (but lower average) radiated power throughout much of the world, there continues to be interest in any possible long-term effects. Certain ionizing radiation, which includes X-rays and radiation from radioactive materials, is known to be carcinogenic, even though it can also be used to destroy cancer cells.

In light of strong present concerns, and the potential for astronomic debts due to successful negligence lawsuits, it is appropriate, and in most cases essential, for *any* system involving mechanical, electrical, electronic, or chemical components to be subjected to a safety audit. This is usually best done by safety experts rather than the system designers, though the designers' cooperation is also needed to explain system operation. Fortunately, most safety corrections to system design can be in the form of add-ons, though attention to safe designs from the first will save time and added costs later. Safety must also be exercised in system development and testing where risks are increased because many safety features are disabled or will be added later.

5.6.2 Life-cycle safety considerations

Safety considerations are important throughout the life cycle of systems. If a system is safe at the time of delivery, it may not be so after many years of operation and possibly after several modification cycles. Improper or postponed maintenance is a major cause of unsafe systems. As systems become older, attention to operator training in safety matters often becomes ignored.

The system designer may have little potential for influencing safe system operation many years after a system was put into service. However, steps can be taken during design which will assist in maintaining system safety through its useful life. Below is a partial listing. Detailed safety design will depend considerably on the type of system, and its applications.

- Clearly marked *warning labels* or plates should be installed on the system in regular or constant view of system users.
- System parts subject to wear should where feasible signal their own wear, as, for example, vehicular brake linings which produce audible "squeals" on brake applications long before lining end-of-life. Visual monitoring schemes may include features such as the sunken ribs in automotive tires, whose appearance at the tread surface is an indication of tire wear.
- User manuals and plates attached to the system should contain instructions for regular inspection of system parts whose failure could impose safety threats. Operation and maintenance manuals should continue to be easily available throughout system lifetime.
- *Inspection ports* and other conveniences should make it simple and quick for busy users or maintainers to inspect safety-related parts subject to

deterioration or damage. These facilities must, however, be simple to use and their use should be enforced if feasible.

- Intelligent application of *factors of safety* (i.e., factors by which system-parameter operating limits are multiplied to determine failure limits) should be applied to stressed system elements. For example, certain parts may be *intended to fail* to avoid failure of other parts which would pose difficult safety problems. This is the principle of electric fuses or circuit breakers and of shear pins on drive shafts of outboard motors. Such *sacrificial parts* pose certain problems if replacements are not readily at hand.

- *Safety equipment* must itself be designed to especially high standards. Flimsy escape ladders which corrode quickly to uselessness, escape door handles which break off when operated urgently, or fire extinguishers which refuse to operate in an emergency are worse than none at all.

Designers and manufacturers of systems having the potential for liability in private or class-action lawsuits also can protect themselves to some degree through liability insurance. In light of the large amounts of litigation promoted by an army of lawyers, such protection has become an essential to doing business in many jurisdictions.

5.7 Design for long or warranted useful life

Designers of low-cost and expendable consumer goods are often accused of creating products which are designed to wear out or fail quickly. It is indeed frustrating when a product fails (usually, just after the warranty period is over) in some minor way though unrepairably. Repair cost of high-technology products frequently exceeds their replacement cost.

No physical system is likely to be operated forever. However, competent designers must create designs which, *with necessary periodic maintenance*, can be expected to yield service lifetimes *appropriate to their cost and function*. At the end of a product or system's useful lifetime, its purchaser should believe that it has been cost-effective. Otherwise, its replacement will, if practical, be obtained from another source.

5.7.1 Representative requirements

The most common lifetime requirement applied to a system may specify a "service lifetime" in terms of accumulated hours of operation, or *operation count* if that makes more sense. Certain regular maintenance operations may be stipulated as conditions of a warranty. Where a product or system is of a class expected to be used intermittently or rarely, both *nonoperating* and operating lifetimes (or alternatively, *minimum num-*

ber of acceptable operations) may be specified. A minimum *maintenance interval* may also be specified ("2000 hours between major overhauls," for example).

5.7.2 Design precautions

The components and subsystems which make up a system will generally fall into several classes: (1) ones not subject to significant wear or deterioration, (2) ones which deteriorate due to factors *other than* operating stresses, and (3) those whose useful lifetime is a function of operating stresses. In mechanical subsystems, the first group will usually include heavy nonmoving parts such as case frames or supports. The second may include less durable nonmoving parts subject to corrosion. The third group includes most moving parts. In systems which are comprised principally of electronic components, incidental mechanical moving parts (disk storage drives in computers, for example, or electric switches) frequently have the shortest expected lifetimes, followed by those electronic components operating at high temperature and power dissipation. In stationary equipment of most types, especially if used in noncorrosive indoor environments, housings and supports will have effectively infinite expected lifetimes.

When a system is to be designed to operate or be available for a specified lifetime, the designer should select components proven to yield at least that lifetime within the operating environment of the system. The ratio of component cost to lifetime may often be a guide in part selection. That is, if expected lifetime of one component barely meets system requirements, but a component with substantially longer expected lifetime is obtainable for a few percent more, the superior component may increase system reliability sufficiently to warrant its selection.

If the *only available* component for a certain application is predicted to yield shorter than the required lifetime, it must be arranged for easy replacement. If its expected lifetime is only a fraction of that required, the component may be listed for replacement during scheduled maintenance, at a frequency sufficient to avoid in-service failures. Finally, if the failure of the component may induce other problems, a replacement may be installed with easy change-over. (We are reminded here of toilet stalls having an extra roll of paper in reserve.)

Manufacturing defects, infant mortality Many products intended for commercial, industrial, or military use are delivered with a warranty which provides no-cost replacement if the product fails within some initial use interval. In the case of electronic equipment, this period is usually 1 year. If a light-duty product is subjected in normal use to an unusually rigorous operating environment such as that of continuous industrial use, its warranty may be invalidated.

A principal intent of such warranties is to replace components which have inherent defects ("infant mortality") caused in manufacture but not revealed by testing. Many small products are tested only representatively: A small fraction of units is tested, based on the manufacturer's experience with product reliability and the expectations of purchasers. Since individual unit testing can significantly increase the cost of products, some manufacturers test little, provide a warranty, and readily replace items found faulty by purchasers.

A system designer faced with stringent system reliability or lifetime requirements thus may need to require that certain components be 100 percent tested by the manufacturer, even subjected to stressed testing (heat, cold, vibration, and shock—while operating or idle). Often testing is least costly if performed by the component manufacturer. If the manufacturer cannot provide adequate tests, the system developer must perform them. Yet another option is testing by an independent testing laboratory.

Manufacturing defects usually account for most system failures early in system life. Where it is important to avoid this problem, the best approach is to operate failure-prone components under conditions representing those in system operation. This must be done for a sufficient time to reveal manufacturing defects. Unfortunately it is not always clear how long such a "burn-in" cycle should be. Since most components deteriorate at a rate which increases exponentially with absolute temperature, high-temperature burn-in while subjecting the component to elevated operating voltage, pressure, etc., may be used to shorten burn-in time. If these conditions are applied, however, for some parts this will shorten remaining lifetime. Many systems include materials such as elastomers, grease, plastics, or electrolytes, all of which decompose rapidly at elevated temperature. High-temperature burn-in should be used sparingly to prevent materially reducing the useful lifetimes of system components.

Infant mortality may also be attacked by the less drastic but more time-consuming procedure of operating system or subsystems, prior to integration, for periods of hours, days, or even months. Although operation during system integration and testing should also reveal infant-mortality problems, these will further complicate integration and test procedures.

Rapidly degradable materials   Many materials from which systems are constructed exhibit natural modes of degradation. In particular, most organic materials, including elastomers and structural plastics, among many others have more limited shelf and operational lifetimes than do most metals. Although most organic materials contain carbon, hydrogen, and oxygen, some, because of composition and processing, are incompletely formed chemically. They may include mobile gaseous or liquid components, which evolve gradually (usually producing distinctive odors). Since high temperatures produce more rapid decomposition, low-temperature storage is a means to extend storage life. This also works for deteriorable inorganic materials. Since most materials become stiff or brittle at low temperature, cold-stored parts should be brought to room temperature before their installation or use. Repeated temperature cycling to very low temperatures should be avoided because thermal stresses may cause material fracture.

Practically any unprotected material will deteriorate with time, whether on the warehouse shelf or in operation. In many cases, covering the surface with a recommended coating (a metallic plating, or perhaps paint) will greatly increase utility. Certain high-strength aluminum alloys containing significant amounts of iron may fail catastrophically when cyclically stressed. These *stress corrosion* failures begin at surface cracks. Parts fabricated with these alloys should be inspected for surface cracks, filed free of surface nicks and scratches which could become failure sites, and protected after manufacture by flexible surface coating.

The coatings we refer to collectively as "paints" constitute a complex variety of solutes, mixtures, and suspensions, each with a particular chemical basis (alkyd, epoxy, etc.) and unique properties. Some coatings are formulated to prevent oxidation of the underlying surface by themselves capturing oxygen. Other coatings may include fungicides, to prevent growth of fungi on systems used in tropical climates. Some coatings can provide limited lubrication ability for parts which must move freely on occasion but with light mechanical loads.

Most materials relied upon for mechanical strength lose their strength as operating temperature rises. In metals and many alloys this may be caused by nearing the melting point and is reversible if no damage occurs. In other materials, irreversible chemical or crystalline changes will occur. Electronic materials, for the most part, must operate within a limited range of temperatures: At high temperatures semiconductors short-circuit. At very low temperatures they do not conduct, switch, or amplify. Magnetic materials relying on storing a magnetized state will operate only *below* a characteristic *Curie temperature*. Above this, they will no longer retain magnetism. Momentary excursion above Curie temperature will cause the magnetism to be erased. Curie temperature is often not far above room temperature. Magnetic devices can pose system problems when in thermally stressed environments. There may be no affordable or acceptable alternatives to maintaining reasonable temperatures.

A system designer should exert caution in selection of unproven "miracle materials." While new materials such as high-flux magnetic materials, advanced fiber-composite structural materials, and new lubricants may indeed perform as advertised, there may be secondary characteristics which suit them poorly in certain applications. A high *vapor pressure* means that the material (the volatile components of it) evaporates. This may not be acceptable in spacecraft, for example. Temperature limits of materials must match expectations for the system operating environment.

Materials are often sensitive to *stress rate*. Applying mechanical stress to materials *gradually* may produce a different result than if the same peak stress is applied abruptly. (It will usually be stronger and more brittle at higher stress rates.) The operating temperature of materials in a system is usually above that of the system environment because of heat generated internally or absorbed from the sun or nearby heat sources. As with other detailed design features, the systems designer may not specify the materials used. The system designer must, however, understand principles of material degradation and any special complications in a system's working environment. Specifications prepared for detail designers should contain realistic criteria for their materials selection.

Design of nonmaintainable systems Systems for certain applications or environments may be designed to operate without maintenance. A prime example is earth satellites, which must operate unattended for periods of many years to justify the enormous costs of raising them into orbit. Maintenance would be impractically difficult *and* expensive. Certain highly complex and expensive devices such as gyroscopes are hermetically sealed to render them free from contaminants; maintenance, if required, must be done in the same laboratory conditions in which they are initially assembled. Residential air-conditioning compressors are likewise hermetically sealed (welded into a steel housing) and are often treated as nonmaintainable. Other devices or systems are designed for unmaintained operation, to allow simplified design or construction and a reduced selling price.

Designs intended to deliver some specific minimum operating lifetime without maintenance may employ different technologies from those of comparable systems which can be maintained. For example, nonmaintained *lubricants* must retain lubricating properties for long periods and must have low volatility if they can escape into the atmosphere or outer space. Components may be *derated*, i.e., limited to operating stresses less than those for which they were designed. Many failure or degradation mechanisms are nonlinear functions of operating stress (e.g., of voltage, pressure, or temperature). Even modest derating can greatly lengthen useful component life. The incandescent light bulb is an example, its life being limited by evaporation of its tungsten filament. Though household lamps are frequently designed for 1000-h useful life, even modest reduction of operating voltage can increase predicted lifetime by orders of magnitude (Figure 5.13). There are usually penalties in derating. For example, electronic switching devices will not operate at as high switching speeds if supply voltage is reduced to derate the unit. Degradation due to high operating temperatures can be reduced for many system

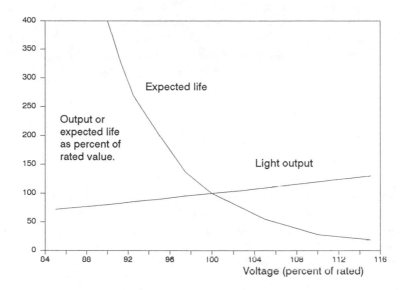

Figure 5.13. Variation of light output and operating lifetime of a gas-filled incandescent lamp wih tungsten filament, as a function of operating voltage. All quantities are measured in percent of nominal values. (Adapted from *Standard Handbook for Electrical Engineers*, McGraw-Hill.)

components merely by better removal of *system-generated* heat. Although the system designer should seldom venture far into details of component lifetime limiting processes, good awareness of the quantitative importance of heat and other stresses must guide the design of systems intended for unmaintained operation.

5.8 Design to allow easy system growth or enhancement

Ability to readily expand, or add new functions to, a system is very important in applications where systems of a particular class address wide ranges of users' needs. This is characteristic of many computer, manufacturing, communication, or transportation systems. Although *system architecture* is the initial vehicle for definition of expansion and enhancement features, the system design implementation must follow through with the details required for the actual expansion or functional growth. The designer usually strives to provide maximum future expansion or enhancement potential at minimal initial investment, for few system buyers wish to invest in unused system capabilities.

5.8.1 Representative requirements

The most common requirements are those anticipating additional *system capacity* or *throughput*. Users will seldom predict expansion beyond a doubling or trebling of their initial estimate. Any growth requirement often includes a caveat against high cost of preparing for future growth. Even where not explicit, such warnings may be implicit in a competitive systems development environment.

Thus, a designer is often faced with the enigma of designing a system which is optimal for current needs but which can also expand easily to meet future needs, again optimally. When the chips are down, the design must meet current requirements against current competition. Those features included to support future growth can represent only a small portion of the system's initial cost.

Practically speaking, this often equates to supporting growth in those system capabilities where it is economical to do so, while ignoring future needs which would significantly increase current costs. The designer may identify certain system features, for example, where only complete redesign would permit expansion of system capacity beyond some modest limit.

Requirements seldom address *future* system enhancements in a meaningful way. The user will, in most cases, have identified *all known requirements* and may assume the position that there *are* no others. An exception sometimes occurs if the user finds the full set of current requirements to result in an unaffordable system. Here the designer must reduce the system scope to meet available budget, while leaving the other options open. In the Department of Defense, early planning for future system requirements has been called *Pre-Planned Product Improvement* ("P-cubed I"). This worthy concept, however, has failed to achieve the popularity it deserves. Frequently the Congress, or the Administration perceive it more economical to delay a system's acquisition until *all* projected requirements can be afforded. Since projected requirements will continue to evolve, the result may be an indefinite delay in initiating the system development.

5.8.2 Growth and enhancement features

A typical provision for future growth is expansion of *throughput capacity* of the system beyond the current needs. In a few instances this may require no current additional hardware but merely an architectural provision for future addition. In some cases, additions are made only to basic and relatively low-cost/low-technology system elements, such as power (or other utility) supply, extra conductors in cables, or unused control panel areas.

If these provisions are to be realistic, both architecture and the current design must be laid out, at least at broad levels of detail, with the eventual system cabability in mind. Locations and functions of additional subsystems and their interfaces to the original system need to be described. Even though future growth may not take the precise forms at first envisioned, the simple provisions made should allow the actual growth to be accommodated more readily than had the advance work not been done. Where growth requires addition of duplicated subsystems, a scheme whereby these added components can be connected should, if possible, be used in the original design.

Planning for enhancements is most difficult if the designer is unaware of the directions which will be taken by the client's activities. Some guidance may be obtained from knowledge of similar organizations having larger or more advanced operations than those of the client.

Where the system is being designed speculatively as a commercial offering, and the system requirements are arrived at in part through market survey, many features beyond the acceptable cost range or the reach of current technology may be proposed. These can often be used as a basis for ranking the most desirable future enhancements. This can then guide the system architect toward a system compatible with all or most of the proposed features.

5.9 Additional product-driven objectives

Special requirements may arise in applications or environments highly sensitive to certain system parameters. These requirements may pose special challenges to the designer, especially since they are often added to the more common requirements. Examples are systems whose designs under normal circumstances would be straightforward but which are required in addition to operate in a scorching desert or thousands of feet deep in the ocean. Users are often shocked at the large increase in system costs imposed by additional requirements of this sort. More than one system firm, ignorant of the true difficulties to be faced, has lost its virtue in such an endeavor.

5.9.1 Special power or energy requirements

Many special systems have stringent power or energy requirements. These are usually based on system portability or operation at sites remote from sources of power or fuel.

Earth satellites have traditionally employed semiconductor diodes (*solar cells*) which convert photon energy in sunlight to direct-current electricity. For high-orbiting satellites, this energy source is virtually continuous, but in low-orbit satellites it must be buffered by energy storage cells. These energy converters transform 10 to 20 percent of the sun's radiation impinging on their surface into electrical energy, if they are mechanically oriented to remain continuously normal to the direction of the sun's rays. Hundreds or even thousands of individual silicon cells must be connected together electrically, to multiply voltage and current to operating level for powering a system or charging rechargeable batteries. This complex wiring network requires protection against device and connection failures.

Portable systems of many kinds, including satellites, use nonrechargeable (primary) or rechargeable (secondary) electrolytic storage batteries. Either type is heavier, *per watt-hour stored*, than most other sources of electrical power. Cost of providing power with nonrechargeable batteries, over the lifetime of a system such as a portable two-way radio, will be thousands of times greater than if it were deriving power from a commercial power plant. Recharging of a secondary cell requires access to some source of power and is typically no more than 50 percent efficient. In each repeated discharge-charge cycle, the battery will retain slightly less useable energy. For most rechargeable cells, a few hundred substantial discharges is representative useful lifetime.

Propulsive mechanical energy, or electrical energy for onboard systems may, in some systems, be generated from hydrocarbon or other fuels. In small units generating a few kilowatts or less, this is very costly, and system life often short. One of the most popular fuel combinations, in terms of energy available per pound, is combination of hydrogen and oxygen. This combination appears in such applications as NASA's space shuttle and *fuel cells*. Fuel cells are primary batteries which combine gaseous hydrogen and oxygen to produce electric power for some spacecraft.

5.9.2 System weight

A premium is placed on low system weight in aircraft and spacecraft systems and subsystems. For a given type of aircraft, one may estimate increased aircraft cost per additional pound of weight to be carried. The estimate comes to upwards of $1000 per pound for high-performance aircraft. For a given class of avionics, the cost per pound of the avionics systems themselves is comparable. Traditionally aircraft and spacecraft design have required subsystems of small physical size. This has become easier to achieve due to large-scale electronic circuit integration and is seldom the pacing problem it was in years past.

Given an airframe design, subsystem weight can be traded for greater fuel capacity, therefore longer range. In spacecraft, lighter weight means smaller launch costs, or higher attainable orbits. Because of the limited varieties of expendable booster rockets available, geosynchronous

(22,000-mi altitude) orbits, for example, can only be attained using a given booster only if the all-up spacecraft weight is under a limit imposed by the particular launching rocket.

When necessary, weight reduction of systems is accomplished by techniques such as:

- Removal of all unneeded system parts
- Use of strong but light materials such as titanium
- Replacement of low-frequency transformers containing heavy iron cores by smaller ceramic transformers operating at higher frequency
- Replacement of long copper-conductor signal cabling with fiber-optic cables carrying many signals in a single fiber
- Operating propulsion plants at higher temperatures at which thermodynamic efficiency is higher
- Making use of subsystem enclosures to add strength or rigidity to the system's main structure

5.9.3 System robustness

Systems destined for operation in harsh environments should have special requirements for robustness. These usually treat environmental hazards such as vibration, shock, and temperature extremes but may also include operation at high altitudes or resistance to physical abuse. Traditionally, mobile systems' robustness was tested using vibration tables shaking the system sinusoidally for each of a set of frequencies and amplitudes, along one axis at a time. Recently, robustness tests have been modified to better simulate operational environments, using combinations of heat or cold, vibrations, and shocks applied for time periods representative of system application.

The design of any system which must operate successfully in a moving vehicle *must* take this operating environment into consideration. Even systems which do survive rough shipping and handling experiences may fail after only brief periods of operation on a moving platform. The problem is intensified by transient "shocks" which can excite natural vibrations of the equipment over broad frequency ranges. Although it is often possible to "ruggedize" equipment by adding stiffening or elastic mounting and beefing up equipment housings and attachments, this may double or treble costs of normal commercial equipment without providing any guarantee of improved robustness.

Systems shipped from manufacturer to user require protection against shock and vibration in shipping. In handling by shipper personnel, a label marked "fragile" is sometimes an invitation to release frustrations. Provisions for vibration or shock resistance generally take the form of elastic mountings (a standard hardware item), plus rigid mounting and locking features on plugs, hoses, circuit cards, and the like. Individual parts inside a system must be designed to be sufficiently rigid and strong that they will not flex destructively. One hazard, often overlooked in this design area, is the possibility that system parts will collide with one another in flexing and vibrating excursions atop their elastic mounts. Where strong vibrations are likely in operation, flexible parts should be supported along their length, so that natural resonances will be increased and those resonances damped in intensity.

The most highly ruggedized electronic systems are often immobilized, by surrounding them ("encapsulation") with a poured-in thermosetting plastic material. This is normally a good thermal insulator. Its dielectric constant is also high, and circuits sensitive to excess capacitance will behave differently after encapsulation. The cost of ruggedization thus involves certain electrical and thermal trade-offs.

Damage during system delivery can be minimized by proper design of shipping containers and internal supports. Filling all empty space in a container with commercially available foam-plastic fragments is excellent with lightweight systems. Heavy equipment requires a suspension structure to space the equipment from the walls of the container and to provide a well-damped (nonresonant) flexible support.

5.9.4 Controlling system vulnerabilities

System vulnerability, when addressed by user requirements, usually refers to system behavior in the presence of unintentional hazards or of intentionally hostile ones. Much electronic equipment, for example, is sensitive to presence nearby of high-power microwave heating equipment. The powerful signals can enter sensitive receiver circuits and cause them to malfunction or to be damaged. Communications equipment may be adversely affected by strong signals *outside* the frequency bands in which the system is intended to operate. Military communication systems, radars, and radionavigation systems may also be subject to enemy countermeasures in the form of jamming. If sensitive equipment has no built-in *counter-countermeasures* features, jamming may render it virtually useless.

Mechanical systems have major vulnerabilities when in environments containing dust or chemical vapors, or exposed to mechanical shocks. Elements such as *gyroscopes* operate with such precision that one dust particle can cause significant error. High-density magnetic storage disk drives have a similar problem and are also acceleration-sensitive. If subsystems such as these are sufficiently compact, an entire unit may be hermetically sealed in a metal, ceramic, or glass enclosure, which may form a part of its support structure as well.

A common vulnerability of electronic equipment, too often overlooked in design, is sensitivity to incorrect supply voltage levels, or to short pulses in electric supply voltage caused by switches being opened. Electron devices which operate at very low power levels are especially sensitive to voltage, or may be destroyed by minute static-electricity charges when being handled by service technicians in a low-humidity environment.

In general, vulnerabilities are both *application* and *technology* dependent. A system may be fully acceptable in one environment or application but useless in a second, depending on both the nature and degree of hazards. Technological dependencies, in particular, must be addressed with special expertise which should be tapped when special requirements arise or sensitive technology is to be utilized. Often hazards can be removed, or systems or their parts protected at little cost, if the problems are recognized in early design.

EXERCISES

1. Integrated circuits increase system mean time between failures (MTBF) because they reduce the numbers of interconnections. Given that the MTBF of a single soldered joint is 10,000,000 h, and that of a semiconductor package is 100,000 h, compare the net MTBF of a complex integrated circuit contained in one semiconductor package and connected to a system with 180 soldered connections, with an equivalent circuit comprising 900 simpler integrated circuits whose packages have 200,000-h MTBF's, connected by 12,000 soldered connections.

2. A system which has no provision for maintenance, for example, an earth satellite, may use redundancy by including redundant copies of components with short useful lifetimes, using each one in sequence until all have been used up. If the probability of failure of each such unit is independent of the state of the other copies and of the time when it is turned on, then probability theory indicates that the expected MTBF of the system with N duplicate copies (one of which is needed for system operation) is just N times that if only a single copy were included. Considering any kinds of component you wish, suggest rational reasons *based on conditional probability* as to why the MTBF with N duplicated components could be more or less than N times individual MTBF.

3. Describe five or more partly-failed states of an automobile which can reduce but will not destroy its ability to provide transportation. Identify also one or more such states in which *multiple* failures of similar components would further reduce utility.

4. (a) Using your own automotive experiences as a basis, roughly estimate the MTBF in *working hours* for failures due to the following: tire damage or leak; loss of coolant for any reason; and engine failure for any reason other than loss of coolant, *in the absence of scheduled maintenance*. (Where you lack statistical data, you should make rough estimates of failure interval and convert to working hours.)

(b) Make another estimate of each failure MTBF, based on *annual* scheduled maintenance which inspects parts but does not repair them unless they show signs of significant wear.

(c) Repeat the estimate, on the same basis as (b), but with *monthly* scheduled maintenance.

(Explain your rationale based on what maintenance may uncover in each case, the wear-out periods for certain components, and the likelihood of unpredictable failure modes versus wear-out modes.)

5. Consider automobile tires. Assume: (1) End of tire life occurs predictably after 600 h of driving; (2) in addition, there is a probability for each tire of 0.0002 per hour of operation that unpredictable failure will occur; and (3) there is an additional total probability of 0.01 that a tire will fail at some time within its first hundred hours of operation, due to manufacturing defects.

(a) Taking all three factors into account, plot MTBF of the vehicle against a single tire failure, as a function of operating lifetime of the vehicle from time of purchase, assuming that no tire fails. In particular, note that limited wear-out lifetime acts to limit the probabilistically determined MTBF to *no more than the remaining wear-out life.*

(b) If you replace immediately any tire which fails for any reason (including wearout) by a new tire, determine by properly averaging over the four tires (which will fail and be replaced at different times) the average MTBF between tire failures. (You may prefer to set up a simple Monte Carlo computer analysis to solve this problem.)

6. Extend the analysis provided in the text for dual-redundancy-with-repair MTBF value in terms of M and R the MTBF and MTTR of a single subsystem, respectively. Consider triple and quadruple redundancy. You may then be able to write down the general case of N-fold redundancy by inspection.

7. *Diversity* is used as a redundancy technique in many everyday ways. For example, for commuting to the office, an individual might be able to use a car pool, bus, hitchhiking, or a personal auto. One of these will usually be designated a *primary mode* while the others rank lower in desirability. Identify at least five regular personal or business functions for which you use alternatives of different kinds only if the primary choice is not available.

For each *function*, characterize its characteristics or parameters which are important to you (time, cost, convenience, satisfaction, etc.). Rank or measure *each* alternative used for that function, by *each* such factor. Combine the factors in a logical manner which is applicable to all, and rank the alternatives as to desirability.

8. Estimate values of MTTR for a single automobile tire failure, under the conditions *with* and *without* a spare tire carried in the car. Assuming 500 h of driving a car per year and an MTBF, per tire, of 30,000 miles, by what percentage does having the spare improve the car's availability neglecting other sources of nonavailability?

9. Many first-order reliability analyses do not consider conditional probability of a failure if another failure has occurred in the system and the related part(s) replaced. For example, installation of a new pump with tighter seals may raise

system pressure and cause leaks or breaks unlikely at the lower pressure experienced prior to pump replacement. Examine systems of kinds that are familiar to you for the potentiality of this type of *related failure*. If possible, discover potential failure mechanisms of "chain-reaction" or "domino" types in which one failure may create another repeatedly. (One broad example of this sort occurs in mechanical structures where breakage of one supporting member increases the loads on remaining members.)

10. Give examples of several *on-line tests* you use to check correct operation of simple systems which, in the absence of action on your part, cannot be determined to be operating or to be operating *correctly*. (A simple example: revving the engine of an auto which has no tachometer and which is inaudible at idle speed.)

11. If you are sufficiently familiar with computers, give this one a try: What on-line test sequence could a program use to check a personal computer's memory for ability to store 0 or 1 in any bit, using only the time spent actually waiting for keyboard input to cycle through memory positions? How much otherwise unused information storage will be required to perform the test? How will you test memory locations containing the test data or program, or other programs or data which should not be altered?

12. Consider either your television set, personal computer, or automobile—whichever you know best—as a system you would require to respond immediately from a cold start. Describe the operator sequence required for starting and the internal startup activities (steps) involved. What happens which does not occur during continuous operation? What can go wrong, or what delay may be imposed, during each step? Then *rank* the steps in the sequence from the weakest to the most reliable (or error-prone).

13. Discuss three purposely lifetime-limited devices or systems you use at work or home (select devices more complex than, say, no-deposit, no-return bottles, please!). For each, describe if and how the design seems to have been optimized for its short-term use. For each, how long do you believe the product could remain in normal storage if it were required to operate satisfactorily when put to use?

14. Examine your home or automobile, identifying and describing features which have been added to enhance *maintainability*. Organize these features into *classes*, for example: features allowing commonality in handling maintenance problems; features for convenience; features reducing force or effort required for maintenance.

15. Examine your home or automobile, critiquing its maintainability and suggesting changes or additions to improve its maintainability. *How* will each change improve maintainability? For each, estimate cost-effectiveness on some basis you believe reasonable.

16. Make a list of different products (or systems) whose designs have been created to be compatible with systems *designed and built by others*. (Do not repeat minor variations of the same sort.) For each, cite the objective(s) of compatibility as you understand it, or as explained by its designer. For each, also discuss limits of compatibility which prevent operation or reduce effec-

tiveness under certain described conditions. (Example: Telephone handsets, once solely the property of *American Telephone and Telegraph Corporation* within its network, are now sold to end users by many firms. Some cannot operate on pulse-dial-only networks, and voice distortion in some is worse than required by internal AT&T standards.)

17. Discuss *all major steps* which will need to be taken to expand a one-story residence to two stories, increasing the number of its rooms and floor area by factors of 2. Assume that *in its original design* there was no contemplation of such a modification. Identify those components of material or labor cost which should double, more than double or less than double, compared to their current cost for a one-story structure, and explain why.

18. Plan a useful or desirable enhancement to a building or system with which you are well familiar: First describe it *prior to* the enhancement; then lay out the requirements which enhancements are to satisfy, as specifically and succinctly as you can. Finally, locate and describe those elements of the *architecture* which are not well suited to enhancement without significant changes.

19. Assume you are a U.S. systems engineer and must adapt a set of interfaces to couple to a system whose interfaces do not conform to current U.S. standards. Describe how you would handle each situation below, including if need be: the description of a specialist to whom you could address important questions, and the specific question(s) you would ask, or information you would provide, prior to turning the problem over to the specialist.
(a) The system has a strange-looking power plug, and you discover the notation 220-V 50⁻ on an information placard.
(b) There is a missing cover screw, and none of the ones your technician has in his collection will fit.
(c) Cooling liquid supply and drain tubes, not of standard sizes, are evident on the front face of the subsystem, whereas in the system you want to interface it with, only cooling air is required.
(d) The maintenance manual is printed in German; important diagrams, whose purpose is clear to you, refer to "DIN-" items which are unfamiliar to you.
(e) The unit appears to require four electric input connections to produce output. Your system has provisions to connect to only three of them.

20. Characterize what you consider to be the 10 most important *ease-of-use* features commonly found in automobiles, and describe in what ways each contributes to comfort, safety, health, or convenience.

21. (Multiple system operator coordination) Consider a real-time system which requires two operators at separate terminals which are required to be located across a room, so that they are out of sight and cannot hear one another. For some of their activities the operators must act in concert, executing related parts of rapid sequences of steps, based on data presented only to one operator or the other, in accordance with the system's design. Consider a variety of *conditions* (at least five) for which operators might need to coordinate views or activities. For each, state performance measures which should be important to the operation. Again for each, suggest some type of system change implementation which could help in coordination.

22. (Here you describe a *cognitive task*.) A telephone operator, working at a shopping mall, is required to provide telephone numbers to callers requesting information on products sold by stores in the mall. The operator must use a special directory organized by category, like the familiar yellow pages directory, listing in this case merchandise categories, all stores selling each category in alphabetic order, and their phone numbers. A second directory lists stores by name and indicates their mall location (how to reach them) and their hours of operation if nonstandard.

Provide *explicit* instructions to the operator for responding to those calls asking for help in locating a particular item of merchandise. If you find it useful, sketch out formats for each of the two directories the operator will use.

23. Make a list of at least 10 specific occupational safety hazards in the use of office systems which are known to you. Specify both the potential damage and the reason it might occur. For the most serious five of these hazards, examine and suggest system solutions which should materially reduce the hazard. If your solution also involves some lesser hazard, describe it and explain why it is more desirable.

24. If a system is to be delivered with a (real or implied) warranty of certain minimum operating life (design lifetime), any parts which could wear out or will be expended must be made sufficiently robust or redundant to last for at least the warranteed lifetime.
(a) Identify classes of vehicle parts or materials (found in some particular vehicle type, such as automobiles or aircraft) which *suffer wear or deterioration or are expendable*. Distinguish the classes of parts or materials by *common replenishment or reserve strategies*. (An example. Fuels: Provide adequate reserve storage, in a sealed tank resistant to attack by the fuel or its decomposition components.)
(b) Describe a replenishment or reserve strategy for each class defined in (a).

25. Based on any reference(s) you select, or your independent views, discuss the *definition* of "infant mortality" in system reliability. How *should* it be specified for a given part or subsystem, based on reliability data for that part or subsystem? Under what conditions might system users or maintainers decide that a given failure had occurred because of inherent or manufacturing defects?

26. Computer architectures often set limits to growth of a computer "family." Select a popular personal computer or mainframe architecture. Within the range of computers which provide software compatibility, determine the ratio of available top-end to low-end processing power and of maximum to minimum available memory capacities. Determine what aspects of the architecture limit each end of this power and memory range.

27. Make a list of the *enhancement features* available for a popular personal computer; for each enhancement, describe how the system is enhanced, and also any limitations which are imposed by the enhancement (for example, incompatibilities of an enhanced system with certain applications).

28. System *vulnerabilities* must be defined with respect to certain *threats* posed to the system. Certain wrist watches are vulnerable to water if it merely splashes on the case; others can sustain immersion at hundreds of feet depth.

Select and study some particular type of *residential intrusion alarm system*, for example, one operated by window-open and door-open sensors or with other (specific) sensors. Describe *vulnerabilities* of the system to a stealthy intruder. That is, what specific steps could a knowledgeable and determined but *unarmed* intruder take to avoid detection by the system? For the system examined, how might its vulnerabilities be reduced?

BIBLIOGRAPHY

M. B. Beyer (ed.) *Encyclopedia of Materials Science and Engineering*, (nine volumes), MIT Press/Pergamon, Cambridge, MA. Material properties by class and application. Automotive to electromagnetic and more. (Systems engineers need to understand the physical limitations of their systems.)

B. S. Blanchard, *Logistics Engineering and Management*, (3d edn.), Prentice-Hall, Englewood Cliffs, NJ, 1986. An outstanding book on the bases and practices of system ownership. Important to the designer as well.

B. S. Dhillon, "Reliability Engineering in System Design," Van Nostrand 1983. The essential amount of math without reaching for "elegance." Has extensive bibliographies. Includes safety considerations.

E. M. Dougherty, *Human Reliability Analysis*, Wiley, New York, 1988. Deals with *how well* humans carry out complex tasks.

Elodyne Corp., *Weather Data Handbook for HVAC (Heating, Ventilating & Air Conditioning) and Cooling Equipment Design*, McGraw-Hill, New York, 1980. Gives year-round weather data for the United States.

D. G. Fink and H. W. Beaty (Eds.), *Standard Handbook for Electrical Engineers*, (12th edn.), McGraw-Hill, New York, 1987. This is a quality professional engineering handbook, edited by a longtime friend of the author. (Systems engineers often use handbooks such as this, for information in design disciplines other than their own.)

W. Hammer, *Operational Safety Management and Engineering*, Prentice-Hall, Englewood Cliffs, NJ, 1985. A manual and guide oriented to the system operator's safety.

C. M. Harris (ed.), *Shock and Vibration Handbook*, (3rd edn), McGraw-Hill, New York, 1988. Most reliability problems appear when systems are vibrated, shocked and/or heated. Might as well learn how to do it.

F. E. McElroy (ed-in-chief), *Accident Prevention Manual for Industrial Operations*, National Safety Council, Washington, D. C., 1980. Aimed strongly at the designer.

J. Moraal and K.-F. Kraiss (eds.), *Manned System Design*, Plenum Press/NATO Scientific Affairs Division, New York, 1981. Deals with human resources utilization, simulation and interface design.

V. P. Nelson and B. D. Carroll (eds.), *Tutorial: Fault-tolerant Computing*, IEEE Computer Society Press, Washington, D. C., 1987 (Order No. 677). Fault-tolerance approached from hardware, software, and system viewpoints.

W. L. Pritchard and J. A. Sciulli, *Satellite Communication Systems Engineering*, Prentice-Hall, Englewood Cliffs, NJ, 1986. A good example of a book written by and for systems engineers. Is direct and to the point. Has the basic math needed, describes system limitations and practical problems.

J. Rasmussen, *Information Processing and Human-machine Interaction - an Approach to Cognitive Engineering*, North Holland, New York, 1986. Introduces a number of concepts in how system users carry out complex tasks.

M. S. Sanders and E. J. McCormick, *Human Factors in Engineering Design*, (6th Edn), McGraw-Hill, New York, 1987. Includes much of what the system designer needs to know to satisfy human users' needs and limitations. (Mostly descriptive rather than mathematical.)

Process-driven Design Objectives

In actual system design and construction, the users' requirements are never the only constraints placed on architecture and design. The designer-manufacturer has certain needs which cannot be ignored: need to make an adequate profit, need to keep designers fully occupied, need to make use of costly test facilities or production machinery. The designer-manufacturer is also resource-limited vis-a-vis people, equipment, processes, and capital available directly, and other resources available through subcontractors. Most system acquiring organizations expect system completion within some reasonable time—months or years, depending on system novelty and complexity. This sets constraints on the use of newly available or untested system elements or methodologies. All these constraints influence system design, in different ways.

6.1 Cost as a design driver

System costs are often separated *by the developer* into two components: *recurring* and *nonrecurring* costs. Nonrecurring costs are those which do not depend on the *numbers of systems* to be built. Recurring costs are those incurred *for each copy* of the system. Design costs are, of course, part of nonrecurring costs, which also include *tooling* costs.

System owners have a somewhat similar perspective. They typically separate costs into *acquisition costs* and *operating costs*. Acquisition cost is paid initially for system design, construction, and installation. Operating costs are commonly presented on an annual basis, although a unit of function (e.g., a ton of ore processed) or a single transaction (e.g., an insurance benefit processed) may alternatively be used as a basis. Cost in a bank system might be expressed in cents per check cleared. Cost of volume manufacturing systems, likewise, might be expressed in terms of cost per part produced. In either case, of course, a transaction (or production) rate should also be specified, because at low operating volumes most large systems are not cost-effective.

The various cost components are related. In particular, *additional design costs* will usually be incurred if there is design effort spent toward reducing either manufacturing costs or users' operating costs. Or, in system development, nonrecurring effort and cost spent on preparing a design for manufacture typically saves recurring costs.

It is characteristic of physical systems to operate within limits imposed by their design or operating environment. A commercial aircraft can fly economically only at about 90 percent of the velocity of sound (i.e., Mach 0.9). At that speed, attempts to increase cruising speed only marginally incur disproportionately large increases in fuel consumption, engine thrust, and other design and operating factors. Likewise, a given class of manufacturing equipment can operate satisfactorily up to a certain production rate, above which reliability, maintenance, useful lifetime, or quality of product output begin to degrade rapidly.

Design efforts required to extend system capability beyond limits characteristic of the technology and state of the art will disproportionately increase design costs. Figure 6.1, representing a hypothetical performance parameter with an upper limit, portrays this fact-of-life problem. As one approaches any theoretical or practical limit to cost or performance, even marginal improvement can raise costs far out of proportion to benefits.

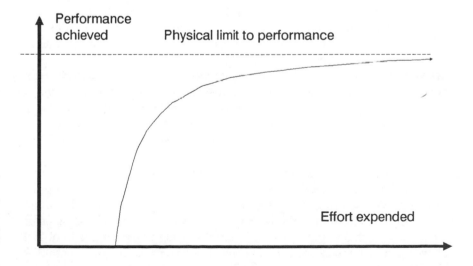

Figure 6.1. A law of "vanishing returns" governs most attempts to obtain the last increment of performance theoretically possible from a system.

While, for tutorial purposes, we may fearlessly illustrate this relationship, in practice these difficulties are seldom obvious in advance. The greater experience the developer can bring to bear, the better will they be anticipated. The shrewdest design managers are those able to separate uninformed conservatism from an appreciation of technical difficulty, i.e., if limits are imagined rather than fundamental, a different approach could transcend the supposed limits.

One is unlikely to use the same design and manufacturing methods for systems expected to be built in quantities of one or of one million. An essential and difficult problem, for both developers of commercial products and for government contractors, is estimating the quantity to be produced, possibly even before a prototype has been built. The experience of the developer or the availability of sound market advice may avert overproduction or investment in tooling for unrealistic production levels. Market-estimate errors are frequent, even in experienced organizations. Estimating is especially difficult if a system or product has many competitors, even more so if the competition are different in their system concepts or pricing structures. Many manufacturers of large systems (aircraft, computers, etc.) plan production and pricing from fairly conservative sales volume estimates.

6.1.1 Economy and diseconomy of scale in design

Economy of scale most often refers to manufacturing cost versus volume relationships, which should offer *lower unit cost at larger quantity*. This is normally to be expected, since most aspects of manufacturing can be reduced in unit cost when volumes are high. Fast-food stores are eminent examples of economy of scale and have proved that quality can be maintained in the process.

However, there may also be economies, or in some instance diseconomies, of scale associated with design processes. For example, a system firm may over a period of time receive contracts to design systems which differ only slightly from one another. The firm can take good advantage of this situation by laying out design activity such that unnecessary redesign and related efforts are not done.

In some organizations, resource economies are achieved through a "matrixed" design organization (Figure 6.2), wherein specialized design knowledge resides in individuals, organized in discipline-based groups, whose services are available to any design project, as required. A negative aspect of this organization is that the design team itself may lack coherence. A way to offset this, in a matrixed organization, is to establish a full-time *design cadre* for each major project, so that from one-quarter to one-half of the designers will stay with the project long-term.

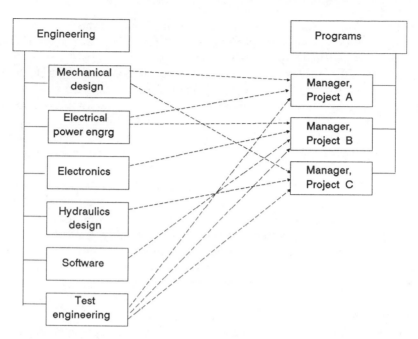

Figure 6.2. In a matrixed engineering organization, most technical specialists are attached permanently to a discipline-structured organization. They support programs through their temporary assignment to project organizations.

Many design activities yield benefits from larger project size. Where small projects may not be able to justify full-time use of certain specialists, larger projects can do so. Where small projects must use off-the-shelf solutions to design problems or standard design tools, larger ones may afford a special solution or tool which is more economical or better suited to their needs. Large military projects, for example, have sponsored new programming languages or new computer designs. Major new aircraft have sometimes brought along an engine more economical (in various ways) than any off-the-shelf engines.

There can also be *dis*economies of scale associated with large design projects. This is particularly noticeable in large software design projects. It is, no doubt, also true in other large efforts requiring extensive coupling and integration. Diseconomy of scale comes about principally because of need for close coordination and for extensive communication between many individuals or groups. Where a lone designer may require little interaction with others from start to finish of a multiyear project, the same task when attempted by 100 designers may require (for comparable results) at least an order of magnitude more design hours.

The diseconomy may to some degree be reduced by relaxation of unreasonable design schedules. The most effective approach is to start by creating an architecture which allows work to proceed in smaller, parallel, and largely independent activities. (Recall, from Chapter 4, that the fundamental objective of good system partitioning is a design which *can* be carried out efficiently in parts.)

In the special case of software, recurrent production cost of the software is negligible. Copies can be produced for little more than the cost of magnetic storage media on which to send them or communications to transmit them. This poses difficult problems to software developers. There are no reproduction-based economies at the overall product level. A thief can create and distribute copies of an expensive product with little investment.

An economy most software firms practice is to reuse software components designed for previous systems. This software may be modified to meet new needs without incurring the expense of complete redesign. It has the further advantage of being demonstrably correct—at least before modification. Even limited modification of complex programs exposes the risk of errors which adversely affect cost or schedule. Reuse of portions of a software program developed for a client who paid the development costs could raise ethical questions unless the developer's practices are made clear in contract agreements.

Because of the potentially long useful life and low reproduction cost of software designs, it becomes especially important to establish software licensing or ownership in any contract where delivery of software is anticipated. For example, certain clients (the U.S. government is often one) may expect to "own" software developed under contract, or even to license use of it by third parties. In such situations ownership and responsibility for correction of errors should be spelled out clearly.

6.1.2 Economy of scale in manufacture

Economy of scale in manufacture is most readily achieved if the rate of production can be established by the manufacturer rather than by a purchaser. In low-rate production of a volume-produced system, scheduling of production may be by *blocks*, each containing a certain number of systems. At higher production rates, production volume will be measured in the daily, weekly, or monthly output of completed systems.

Subsystems and parts are typically acquired from subcontractors in block orders matching the system builder's needs, including requirements for spares along with the original production. It is of course desirable that subcontractors be assigned firm quantities and schedules as far in advance as feasible, in order to minimmize costs.

Economies of scale are greatest when a single manufacturer produces the entire quantity of some subsystem, or part, on a schedule which is favorable in respect to utilization of the manufacturer's invested capital. However, a noncompeting subcontractor may be inclined to view his customer as "captive," and adjust prices upward. Because of this, large system integration firms customarily seek *alternative sources* for most parts or subsystems. At high volume, the average cost per unit purchased may be little more, if requirements are fulfilled by two or even three subcontractors. This won't happen if the product is available from only one source. In this case, the system developer is somewhat at the mercy of the subcontractor, unless it has the role of sole *customer*. In this case, success of the system house and subcontractor are tightly coupled. Certain large system firms maintain limited component or subsystem development and manufacturing capability in-house to provide plausible competition for outside subcontractors.

Manufacturing costs can be controlled early in system design. This is particularly important at low system volumes for which separate manufacturing engineering activity is uneconomical. The primary precaution is to use conventional design methods and proven system components. Construction of systems using exotic materials, for example, not only poses problems in obtaining materials but also in fabricating material into parts, attaching the parts, developing protective surface treatments, etc. If parts or subsystems are unconventional, subcontractors naturally raise prices to offset uncertainty. Where nonconventional manufacturing is required, a capable system firm often undertakes it in-house to limit the uncertainties with which it must deal.

System operating economy is often a design objective for systems filling well-understood applications. A fuel-economical automobile, low heat loss in a commercial building, or a photocopier with reduced maintenance requirements, all have sales appeal. Though minimum operating cost is not always of first importance, it may be essential not to breach traditional cost levels by which competitors' systems are measured. Cost of expendables will earn low scores for a system out of line with costs for competing systems.

Although *economy of scale* is most often associated with mass production of consumer products, similar economies can be practiced in lower-volume manufacturing by using components or subsystems repeatedly within one system. This should have additional benefits in reduced maintenance cost and increased availability and is worthy of careful study during the system design.

Diseconomy of scale *can* occur in manufacturing. It often appears as a result of commitment to manufacturing processes or equipment designed for higher rates of production than can be supported. If production must be increased above the rate at which the process, assembly line, etc., being used is optimal, the manufacturer must (1) commit a portion of production to a less-automated or manual process, (2) find subcontractor(s) willing and able to take on the excess production at an acceptable cost, or (3) install a second production line (Figure 6.3). Installing addi-

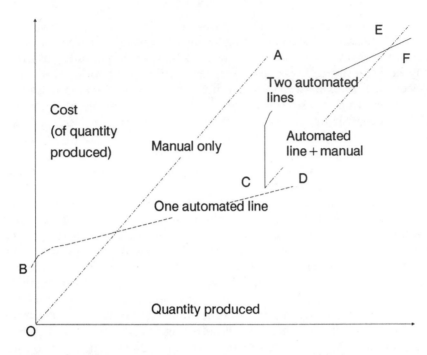

Figure 6.3. The relationship between cost and quantity produced is one part of the data needed to assess marketability of a product. *O* to *A* represents cost of manual production (for total quantity). *B* to *D* represents cost of automated production using a single automated line or processing unit. When quantity rises above its capacity (at *D*), the curve *C* to *F* could be followed by adding manual production. If a second automated line or unit is purchased, the cost curve *B-C-F* can be followed. It will be uneconomical until output rises high enough to obtain efficient use of the second line or unit.

tional production equipment should cause only short-term boosts in production costs provided that the production rate is steadily increasing. If there can be no assurance that current production rates will continue long enough to amortize tooling investments, there is no incentive for additional investment. Unless the factory facilities are in operation a large fraction of 24 h every day, it cannot be said that production has reached its limit. *Temporary* peaks in production rate requirement can usually be met by increasing daily or weekly hours of operation.

6.1.3 Design to cost

Design to cost refers to systems whose *requirements* include *strict system cost limits*. The requirements must in this case also include flexibility, for example, a priority list, with regard to system function or performance requirements. The design to cost concept apparently took form first in the U.S. Department of Defense, continually plagued by escalating system costs during the 8- to 12-year acquisition cycles characteristic of its major systems. While DoD system *users* (the uniformed military services) generate system requirements, funds to pay for acquisitions are controlled by the Congress, and by the Office of the Secretary of Defense.

The basic design to cost concept calls for trading away *desired* system features if projected system cost is expected to exceed a design-to-cost limit. The incentives of this concept are seldom appreciated by either users or system developers. System utility could drop below some practical threshold to achieve the desired design cost, thus rendering the completed system of no utility. Contractors also feel little incentive to do an exceptional design job, on projects for which the profits could vanish completely.

Design to cost has definite appeal to taxpayers since it implies "buy only what you can afford." There is great difficulty, especially for military systems, in establishing *what is needed*, in the absolute sense implied. There is always, however, a lower limit. If the resulting system can do no more (or do a job no better) than an *existing* system, its marginal utility is zero.

The most important problem arises if design to cost results in a system design only marginally more capable than some existing system. Both Congress and defense industry executives shun projects which would result in system replacements which are little or no improvement—as well they should. With military systems, a second important factor is *military competitiveness*. Though new system **A** may not be a large improvement over our existing system **B**, it may be superior to an adversary's system which is *also* superior to our existing system **B**. Should we produce it? In military systems, performance differences of as little as 10 to 20 percent in range of a missile or maneuverability of a fighter aircraft, are measurable. This difference is easily offset, however, by inexperience in system operation or by maintenance problems.

6.1.4 Cost-effective designs

A *cost-effective design* is said to be achieved by a system if the system cost is effectively repaid by profit or savings through use of the system over the system's lifetime. Marginal cost-effectiveness is nil if system operation yields a less positive financial result, as compared to existing systems. If competitors offer systems for the same application, a user's system acquisitions will frequently be guided by the net financial benefits expected to be realized using each competitive system.

Cost-effective designs should not disregard other system considerations, such as reliability or availability. A system that is less reliable in use or difficult to maintain may be cost-effective by virtue of a low initial cost or operating costs. The designer of a system which is intended to "make money for" owners must be driven largely, if not entirely, by the cost-effectiveness objective. Features "nice to have" but which reduce system cost-effectiveness should, at least by that measure, be omitted.

Actual cost-effectiveness of complex systems is often impossible to estimate accurately, even if the system characteristics are fully known. One reason is that the system itself may play only one part in determining user revenues. A second is that the user may not employ the system in its most cost-effective way. . . for example, employing more operators than the minimum needed. Another is that a new system may require changes in user operations for which neither the system developer nor the user has a valid experience base. Estimates of cost-effectiveness *in advance of system installation and actual operation* are necessarily speculative, and often reflect unwarranted optimism or pessimism on the parts of user or developer.

6.2 Design for high-volume or specialized production

The term *mass production* came into use long after the start of the industrial age. It is applied most often to assembly operations wherein each production worker is responsible for only one or a few tasks out of the long sequence. An objective of mass production was to reduce required levels of worker expertise. The broadly knowledgeable craftsman of the previous century gave way to an assembler who might be required only to insert a part or tighten a screw.

A feature often identified with mass production is the *assembly line*, on which the products (Henry Ford's Model T automobiles were among the first examples) move on a mechanical conveyor. As they pass by, a series of assemblers attach parts in correct sequence, completing the product by the time it reaches the end of the line. Assembly lines have been used also for food products (butchering of food animals, canning, bottling), for consumer appliances and other products made in large enough quantities to justify the investment. Assembly-line-like factories have even built prefabricated homes. These are mounted temporarily on wheels and moved to their sites in major sections small enough to travel by highway.

Not all high-volume production features assembly lines or conveyors. High-volume residential construction may be carried out on-site, using coordinated specialist teams with a prepackaged collection of components to complete each part of the construction. Each crew moves along a street of homes which differ little, if at all, in interior arrangement. William Levitt, in his Levittown communities in Pennsylvania and New York, was among the first to produce, in rapid order, thousands of homes. To the surprise of many doubters, properly-managed volume production of most products yields higher average quality than can be produced using lower volume processes, and at lower cost.

6.2.1 Characteristics of volume production

Materials and components employed in mass-produced products or systems are typically selected to be of greater uniformity and often of higher quality than those used in unit production. Whereas a carpenter installing wood framing in a single house must search for and select the straightest lumber for door and window framing, in mass production this is too time-consuming. The standards used for all mass-production components or subsystems expect full interchangeability.

Mass manufacture often requires special production machinery. Aside from conveyor belts, transport trays, or wooden pallets on which products are mounted for movement within a factory, fabrication and assembly tools take different form than those used for manual production. Most screwdrivers and wrenches are motorized. *Numerically controlled machine tools* used in volume production may do the work of lathes, milling cutters, shapers, punches, or drills. Their operations to produce a part are controlled by a digital program in memory or read from a paper tape. These devices require operator assistance only for setup and finished-part removal.

In manufacture of sheet metal parts such as automobile bodies, robot spot welders can produce rows of welds far more uniformly than typical of manual operation. Other robots may assemble parts. To apply this level of automation, parts to be assembled, and their assembly jigs, must generally be far more accurate than was the practice in hand assembly or manual welding. The resulting product is also more precise and uniform.

Manufacturing processes for high-volume production often aim at minimizing time required to perform operations. This reduces both factory area and inventory investment (of parts and incomplete products) which must be maintained. If a mass-manufactured system requires parts with unusual or time-consuming production processes, these items may best be produced elsewhere or contracted out.

6.2.2 Labor utilization

Labor utilization in high-volume production is required to be efficient, since reducing manual steps (touch labor) is a major motivation. Increased production rate and higher product quality are others. In practice, efficient labor utilization is achieved only during continuous production. A temporary shortfall of assembly parts or the need to rearrange a production line for a model change can briefly be very costly.

Automobile manufacturers may temporarily dismiss assembly workers when converting to new models. Automobile workers, most unionized, have been paid adequately in the United States to maintain a comfortable lifestyle. A recent development has been expectations of younger workers for more challenging, less routine work. This has in a few firms caused reversal of long-term trends toward greater division of labor. Some have returned to an assembly mode whereby a small group of workers is responsible for each vehicle.

In the 1980s industrial robots took over most automobile welding and painting operations once done by humans. As routine jobs can increasingly be handled by automation, jobs for humans will require higher skills. While demand for human welders shrinks, more workers who can monitor and maintain welding robots are needed. The bottom line—profitability— will of course dictate which jobs will be performed by humans and which by robots.

6.2.3 Quality assurance in volume production

Volume production introduced new challenges in quality control, now more often called *quality assurance*. With low-volume production, conditions under which each product unit was produced were often very different. It was thus not so meaningful to apply statistical methods, or predict learning curves.

A science-based process of testing became feasible with high-volume production. Initial samples of a product are evaluated carefully to verify product correctness and establish an initial quality baseline. If early production units are faulty, of course, that problem must be fixed before manufacturing continues.

In production under uniform conditions, it is likely that units produced in immediate sequence will possess common characteristics. This of course works both ways. If an employee new on the job goofs when assembling one unit, he or she may make the same error repeatedly until corrected. Batches of parts made from the same barrel of plastic powder should have similar chemical and mechanical characteristics. Though mass-production processes do have the potential for uniformly bad as well as uniformly good results, it is far less difficult to find and repair a common error than to handle the same number of faulty units with more random errors.

Before mass production, for high-quality products it was common to test each unit (100 percent testing). The greater uniformity of mass-produced products led to development of testing techniques referred to as *statistical sampling*. Depending on the product and manufacturing conditions, much less than 100 percent product testing may be required to verify a probability of *very near 100 percent* acceptable products. This is particularly valuable if the act of testing can damage or even destroy the product.

In sample testing, the assumption is made that products in a given "batch" have statistically similar characteristics. A small fraction ("sample") of the product batch is selected at random and tested (Figure

MIL-STD-105D

Example: normal general inspection

Assume batch is acceptable in this case

if no more than 2.5% of parts are defective.

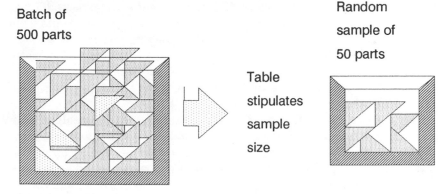

Batch of
500 parts

Random

sample of

50 parts

Table

stipulates

sample

size

Table stipulates:

Test sample parts....

- Accept batch if 2 or fewer samples are defective.

- Reject batch if 3 or more samples are defective.

Figure 6.4. A statistical sampling process which can be used for testing parts received from a supplier. *MIL-STD-105D* provides tables which contain results of statistical computations and make it unnecessary for quality assurance personnel to perform the computations.

6.4). The results of these tests determine the actions which follow. If the test of each of the sampled items is satisfactory, the entire batch may be accepted without additional testing. If some or all of the sample tests fail, additional sample quantities may be tested. If it yields excessive numbers of unacceptable units, the entire batch may be rejected, and later destroyed or perhaps reworked if feasible. Probability theory allows one to say, with a certain degree of *confidence*, that a certain fraction of the remainder of the batch will meet the test criteria. Applicability depends on the assumption that the entire batch was produced under the same conditions and that the selection is random.

There is no magic to statistical sampling. Sample testing can never guarantee the conformance of *untested* products to the test specification. Only its own testing can guarantee that each unit conforms to the specification. However, assume a product parameter has specified limits of 50 ± 5, and that sample testing of significantly many sample units shows a normal distribution about the value 51.2 with standard deviation of 0.8. If you knew that the entire batch had been produced under the same conditions, you should have good confidence in the batch.

Statistical sampling is most useful when applied to production processes which can be controlled very accurately. In volume production of large-scale integrated-circuit electronics, the number of process parameters is so great that no more than about 25 percent of units are expected to pass an initial test. Accordingly, 100 percent testing is required. Units which fail the most stringent tests but pass a less exacting test are often sold as derated devices. Microprocessor chips specified as capable of operation at very high clock frequencies, and those for somewhat lower clock frequencies, may merely be different parts of the statistical distribution for a single chip design.

Where products are produced by the integration or assembly of many parts, a proper *quality-assurance program* deals with tests at every level of integration. The overall objective is to meet product quality objectives, at the *lowest total cost*. If a costly assembly cannot be reworked but must be scrapped if deficient, some of its parts may be more carefully tested than those of assemblies which can easily be tested and, if necessary, repaired. Some parts may be rejected during assembly itself. For example, human assemblers will quickly reject screws having incorrect slots; this ad hoc testing is, however, not recommended if using current robot assemblers.

6.2.4 Design for volume production

Designing for high-volume production is not necessarily more difficult than for modest volumes, but it is assuredly *different*. One can imagine some of these differences by comparing automobile versus private aircraft design and manufacture. While automotive sheet metal is formed by huge hydraulic presses, aircraft sheet metal may be formed by hand, over wooden molds. Larger parts may be riveted together from several pieces of sheet metal, each having two-dimensional shape formed on a simple rolling tool. In volume production, spot-welding is used to assemble automotive sheet metal parts. Riveting is an alternative more suitable for low-rate production (though the rivets are mass-produced). In addition, inspection of riveted assemblies is more straightforward.

For large-scale production, many odd-shaped solid parts are produced by *casting*, or by newer processes such as *powder metallurgy*, in which compacted shapes are heated just below the melting point. At lower production volumes, complex parts are often designed to be machined from solid blocks, or bolted or welded together from flat plates, tubes, rods, or machined shapes. Though processes differ significantly at different production volumes, thoughtful design is universally important.

Some manufacturing processes lend themselves well to volume production and less so to small-scale production. Below are some specific examples:

- *Rolling* or *extrusion* of metal tubes
- *Winding* of metal coil springs
- *Casting* of solid bases or frames
- *Injection molding* of plastic housings
- *Drawing* of thin wire (from larger-diameter rod)
- *Photolithographic reproduction* of electronic circuit boards or chips (Though a volume-production technology, this has been applied economically to a total production run of only a few dozen units.)

In most instances, the volume-produced part is superior to its low-volume alternatives. In some there is no low-volume counterpart; in this case, the system designer chooses an off-the-shelf volume-manufactured product, adapting the system design to available parts as necessary.

Different experience and a different design philosophy are often required to design high-volume parts, using special volume-oriented processes. Design of composite glass-, organic-, carbon-, or metal-fiber aircraft (or automobile) structures, for example, is vastly different from design of sheet metal structures. In either case, there are design rules to be used. For the more common processes which must be relied upon in low-volume production, these rules are generic. Though most can be found in design handbooks, working with an experienced designer is a valuable prerequisite to independent design practice.

Design rules and practices for application using volume-production machinery and special processes must often be learned from the manufacturer of the machinery or proprietary process. Special system components to be built using processes such as glass lamination, large-volume plastic extrusions, or forging are commonly produced by a specialty subcontractor. Design details will ordinarily be handled by that contractor. The system designer must know the capabilities and limitations of the processes.

Even in volume production, nonunique system parts should where available be selected from the product lines of volume producers. Multiple sources are of course very desirable to guarantee competition. This second-sourcing is good practice even for low-volume or one-of-a-kind systems. It is less meaningful if only a single purchase order will be issued, regardless of the quantity ordered.

6.2.5 Flexible manufacturing for midscale production

Small scale production may involve much handwork. Volume production can afford automation and special tooling. What about the middle ground, where there may be total production of 2 to 100 units? This area has traditionally been difficult: You couldn't afford an investment in volume-production tooling, nor could you tolerate the product variations in separately hand-produced systems.

Recently, a concept termed *flexible manufacturing* has been promoted. An automated plant is supposed to be able to produce, economically, a small quantity of parts for each of many different system designs. The key to flexible manufacturing is use of computerization to schedule, set up, and control production sequences using automated and semi-automated production machinery. With the proper type of equipment, it should be possible for a single computer-controlled lathe to produce small quantities of a wide variety of turned parts to higher standards of uniformity and quality than would be possible with low-volume handwork. Other types of tools would likewise perform a few repeats of each operation.

If fully implemented, flexible manufacturing could employ automation in the distribution of raw material, automatic transfer of production and setup instructions to machines and their human supervisors, automated transfer of parts in process between machines, and overall optimization of production schedule.

Whereas traditional manufacturing is either organized into banks of operator controlled machines, or production lines designed to produce a specific set of operations in sequence, flexible manufacturing requires a different organization, usually thought of as a *manufacturing cell*. Rather than containing many production tools of one type, as is common in traditional production, this cell may contain only one each of a variety of computer-controlled tools. Such a cell would be able to take on a wide variety of jobs, up to limiting part dimensions which depend on the size of tools included. *Precise on-line measurement* is an essential to achieve speed with precision. Simplified setup procedure, requiring little human effort, is likewise essential. Both require more than merely good software control programs.

Final assembly and testing is a high-cost activity for low-volume systems. These activities contain high intellectual labor content. When humans must learn a process, there is no substitute for repetition. This is a problem in low-volume system production. In low-volume production testing, it proves useful to record (video tape, digital data, etc.) the integration and testing experiences of a first system. With a well-designed data management scheme, this information can be made available to guide subsequent integration of similar systems, extending or replacing the information as new lessons are learned or new conclusions reached. The objective, as it must be in any flexible manufacturing scheme, is to build a rapid learning curve using data derived from slightly different experiences *in the same environment*. *Process*, rather than product, repetition is the key to progress in this environment.

6.3 Design for testing

Testing takes place throughout the life cycle of a system. During manufacturing and system integration phases, it is exclusively the developer's problem. During system operation, test execution is the user's or maintainer's problem, but test definition and system facilities needed to execute tests should have been dealt with by the developer. Too often, dealing with testing issues has been postponed until late in design. It is then virtually impossible to add architectural features which aid in testing. Most appropriately, testing should be addressed from the earliest architectural and design phases.

There are many different testing environments and objectives. *Development testing* is intended to address the correctness of design and implementation by verifying correct values of operating parameters and correct function sequences or responses to specific test inputs. *Manufacturing testing* deals with departures of parameters from design values in each duplicated system.

In *maintenance testing* emphasis is on detecting changes in system performance parameters which signal impending failures or locating sources of failure after it has occured. *Operational testing* is similar to development testing but deals with the system's performance within its operating environment.

If a system design is required to be adapted for easy testing, it may therefore be necessary to address it from any or all of these different testing points of view. Each requires a different philosophy, different measures, and different tools.

6.3.1 Testing strategies

Design for testing is most often intended to aid in maintenance testing. There is always a need to minimize the numbers, complexity, and duration of tests necessary to predict or to locate any known or potential cause of system failure.

A presumably effective but highly impractical testing strategy might call for measuring *each* parameter of *every* part in the system, while in the system state(s) in which failure was detected. Given an arbitrary order of part tests, on average one should expect to test *one-half of all the parts* in every case.

Instead of testing only at this microscopic level of detail, it is far more efficient to employ some sort of hierarchical, top-down strategy, beginning with tests which are *indicators* of failure location. Often failure can be instantly localized to one subsystem if other subsystems are behaving properly. If the failure appears more general, one typically begins with utility subsystems such as power sources or converters. Instead of testing every component in a power supply which converts ac power to dc, a first test should be to measure average value and time-variation of its output voltage. If results are within expected limits, one might turn attention to other subsystems such as control or switching centers.

In general, the most successful overall problem-*locating* strategy is to develop, then trace through, a series of top-down *inferences*, based on failure *symptoms*. This may in some cases lead quickly to discovery of failure causes. In complex systems, failure symptoms are often ambiguous, in that they could result from many different causes. More subtle approaches considering *combinations of symptoms* should then be used, even though it implies tests to locate other failure symptoms. A good inference-based testing strategy requires a process which is consistent and can reach any failure source in a reasonable number of steps. Rasmussen (J. P. Rasmussen, "Information Processing and Human-Machine Interaction: An Approach to Cognitive Engineering," Elsevier, New York) deals at length with maintenance test strategies.

In designing a system test process, one generally should develop a set of test strategies, beginning from a sound top-level approach. Detailed test requirements, and later test definitions, can then be developed for the particular system. As its design takes on detailed form, it may be necessary to revise either the design, the testing processes, or both. Design for test thus demands activity in early design plus more detailed test development as design and construction proceed.

6.3.2 Combining production and maintenance test facilities

A good design-for-test principle, *where applicable*, is to treat development and operational maintenance tests as two sides of the same coin. With attention on testing beginning from initial design, a fair portion of needed maintenance test facilities can be designed into subsystems. In the most advanced examples, each subsystem will include built-in test capabilities adequate not only to test itself, but certain other subsystems as well. In less elaborate examples, all test connection points required for maintenance testing are provided for, beginning with prototype system designs. Test capability and knowledge will generally increase throughout the development cycle rather than appearing an afterthought. Prototypes, of course, may be equipped with some test features which are unnecessary in production systems.

There are significant benefits if "cradle-to-grave" testing is taken to be an essential element of system architecture and design. This philosophy has characterized mainframe computer architecture for two decades but has not yet penetrated so strongly into mechanical or electromechanical systems, where built-in computers (nowadays, always central to testing) may still be considered alien technology.

6.3.3 Testing homogeneous versus eclectic systems

It is easier to evolve a unified testing scheme for a system if all subsystems employ the same technologies and are designed, concurrently, to the same standards. This is characteristic of all-digital electronic systems. For example, a standardized *test port* may be specified in each subsystem, from which location an *external* test-managing subsystem must be able to examine *all* functional characteristics of that subsystem. If the subsystem is a collection of digital hardware, this might require only software.

If it is a jet engine, however, this strategy may call for addition of many electromechanical sensors *not* required to monitor normal engine operation. Even for such an eclectic system, the best compromise will usually be a testing architecture which uses some of the same interfaces for monitoring routine operation as were used for testing during system integration. Even test software can be altered to adjust to availability of a different sensor set.

Complex technologies used in modern systems may require production tests which are at the frontiers of feasibility. This is surely true of *very-large-scale integration* (VLSI), which puts hundreds of thousands of electronic circuits in a tiny device with the order of 100 external connections. It has been found necessary to *build in* test circuits on the chip, in order to fully test these devices. A similar *integral test capability* is applicable to many other systems and subsystems and should increasingly be used in future hardware systems of all types.

Details of testing Though details of testing a particular subsystem may be below the systems designer's level of interest, the comprehensiveness of the testing should not be. Work in integrated circuit design has suggested certain useful criteria. Software provides what is as yet a puzzling case.

Integrated circuits consist mainly of massively repetitive elements (transistors or gates), connected together in a variety of ways. Some, but not all, of the circuit elements called "flops" or "latches" can retain either of two binary states. Other elements do not possess this storage character but can only respond directly to inputs. Complete logical testing of the circuits can be carried out *only* if the state of each of the flops can be *externally* and independently controlled. *Externally controlled testing must have this capability*, to permit complete testing.

Although the design of test software to run exhaustive tests of a complex chip is not trivial, it can be done. *Software,* however, may be so complex that exhaustive tests are impractical. The number of possible binary parameters in a modest software program may be in the thousands (each numerical constant may, for example, be represented by 32 bits!), whereas the number of flops in an integrated circuit microprocessor may be only in the hundreds. One must also consider that one microprocessor chip design may represent $50 million or more in revenues for its designer, while a software program of the *same logical complexity* will typically be far less valuable.

Up to the present there has been little attention to design of software explicitly for testing. Use of high-level programming languages such as Fortran, "C," or Ada, though not eliminating the need for testing, can materially reduce the incidence of error in typical programs. Formal *independent validation* procedures, where software is studied in detail by individuals other than those who created it, will further reduce imbedded errors in delivered software. No known test method, however, can be proven to locate software errors. There are two approaches to reducing errors. One is to test, and the other is to take steps which reduce error likelihood. It appears that, at least for software, the second is, and will remain, more cost-effective.

6.3.4 Statistical versus worst-case design

Traditionally, engineering designs assumed system components to have *exact* parameter values. This was a consequence of design analysis too complex to encourage evaluation of system performance for all possible parameter values. Though computers have made it possible to examine larger design parameter spaces with little time or effort, few designers attempt exhaustive analyses.

An alternative which has achieved limited popularity is *worst-case design*. Here, analysis is performed with each parameter assigned an extreme value which will most badly corrupt system performance. If, for example, a component is specified as 100 ± 10 percent, it will be assigned the values 90 and 110 for the analysis. Analyses will be made for each combination of parameter values, that is, 2^n cases if there are n parameters. That analysis yielding the worst system performance will be selected. Parameter tolerances may then be tightened to bring performance within specification.

Statistical design considers measured or assumed statistical distribution of each system parameter. Its objective is usually to find the most generous parameter tolerances which will yield a given percentage of systems whose performance can be expected to lie within required limits.

Exact parameter values are unrealistic since all physical elements have manufacturing tolerances. Exhaustive, system-wide parametric analysis is usually wasteful, since parameters interact with one another mainly in a local sense. In this respect, as we shall see, there are situations where parameter ratios, sums, or differences control system parameters. In electrical feedback amplifiers, amplifier gain (output divided by input) is determined principally by the ratio of values of two feedback elements.

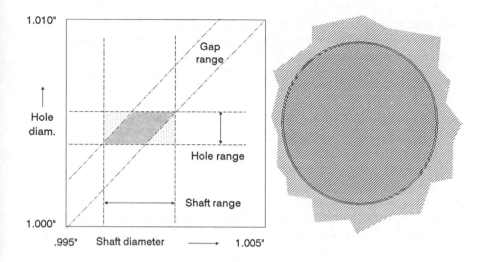

Figure 6.5. This illustrates the simple statistical design example in the text. The rectangular shaded region represents the space of all probable shaft and hole diameters with the suggested tolerance limits. The inner portion represents that fraction of shafts and holes satisfying the requirement of .005- to .007-in diameter difference.

Let us consider an even simpler situation, a 1 in (nominal) diameter round shaft which must fit into a round hole in second part of the system (Figure 6.5). The shaft and the part containing the hole are to be supplied as separate items, more than likely from different sources. From a system point of view, let's presume that our main concern is the *fit* of the shaft in the hole. Let's assume we have established that, for proper system operation, the shaft's diameter must be between .005 in and .007 in *less* than that of the hole. Let us suppose also that the *only available* precision-ground shafting has diameter 1 in, +.000 in, −.003 in, i.e., it may be any value between .997 in and 1.000 in. Our next question: What hole diameter should we specify? If we attempt to apply *worst-case design* and set the criterion that *any shaft* must fit *any hole*, there is *no* hole diameter and tolerance which would suit. In this instance we must select some subset of shafts having a narrower distribution of diameters, for worst-case design to be meaningful.

What about statistical design? If we specified the hole to be *exactly* 1.045 in in diameter (forget the cost!), the shaft tolerances would leave gaps ranging from .0045 in to .0075 in. If shaft diameters were uniformly distributed over 0.997 in to 1.000 in, we might expect that, on average, *two of every three* shafts would fit. In the figure the region of interest can be represented by a horizontal line at 1.045 in hole diameter. Since the gap range (.005 in to .007 in) is .002 in, 2/3 of the range of shaft diameters will satisfy it, as expected.

A more practical hole diameter specification of 1.045 in ± .001 in can be characterized by a pair or horizontal lines at 1.035 in and 1.055 in. The region which satisfies our fit criterion, now in two dimensions, is a shaded parallelogram. The lighter-shaded rectangular region represents the two-dimensional space of all shaft and hole diameters. Assuming uniform distribution of hole and shaft diameters within the region, we can again expect that 2/3 of all shafts will fit the holes. We observe, by this statistical design approach, that we can do as well using ± .001 in hole tolerance as with an exact hole diameter.

Where several individual component parameters combine, in this or similar manner, to yield the value of an important system parameter, a combination with *all* worst-case parameter values is unlikely. We can take advantage of this by statistical design methods.

In either statistical or worst-case design, one must determine consequences of the deviation of parameter values. Correct use of worst-case design *guarantees* that a system built using any combination of the variable elements should work correctly. Statistical design only promises, in a statistical sense, that some statistically determined fraction of systems will work correctly with random parts selection. While this seems a dubious guarantee as compared to that available from worst-case design, purchase cost of more precise components may more than offset the cost of reworking a small fraction of systems which fail to meet specifications.

One might presume that a real system containing thousands of imprecise parameters will produce impossibly complex worst-case or statistical-design criteria. This is almost never the case. In well-designed systems, as part of decomposition and interface definition, attention is paid to isolating critical parameters and variables from one another. This is done in ways characteristic of the technology. Universal joints on the axles or drive shaft of an auto not only provide flexibility for adapting to road bumps, they also compensate for spacing or alignment errors between the engine, transmission, and axle. In electronics, voltage regulation is almost always used to maintain power supply voltages within narrow limits, minimizing transistor circuits' high sensitivity to supply voltage. Feedback processes are used to overcome variations in amplification between transistors used in different examples of the same amplifier design. In hydraulic systems, sections of flexible hose avoid need for alignment accuracy.

With attention to isolating critical system variables, either statistical or worst-case design reduces largely to satisfying *local* criteria. Statistical design results in lower system cost before test and any necessary rework because wider tolerances can be accepted. Test and rework costs must, however, be added. In many parts of systems or subsystems, careful attention to design can reduce the impact of relaxed tolerances and

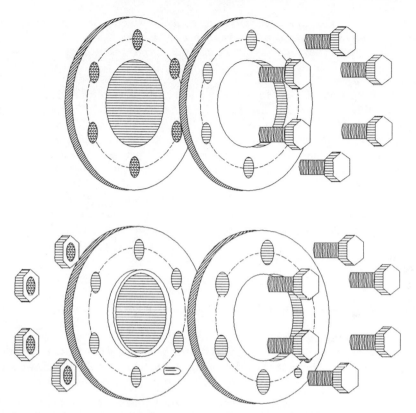

Figure 6.6. At top, a costly way to achieve accurate alignment of the two flanges, by six sets of precision holes, one of which must be threaded. Below, a less expensive design using a raised lip on the left flange, with a pin and a single alignment hole for angular alignment. The letters correspond to the key parts mentioned in the text.

produce a better product at lower cost. In Figure 6.6, the upper coupling of two flanges relies for accurate alignment on accurate location of six bolt holes in one flange and six tapped holes in the opposite flange. That shown below uses a simple flange at its inner diameter for centering. The single *pin* shown, with a matching hole in the opposite flange, is used only for rotational alignment. The six bolts in oversized holes, plus threaded nuts, now serve to clamp the two flanges together. The flange and pin are alignment members not required to oppose operating stress.

EXERCISES

1. Using a 4500-lb (empty weight), six-passenger automobile as your example, examine how its designer could make design trades which transferred cost, i.e., reducing operating costs while increasing acquisition cost, while not altering performance. Aside from, say, a lightweight diesel power plant, two

other aspects of design and construction could be effective in that transformation: reducing wind resistance or vehicle weight.

Estimate wind resistance saving as follows: *Aerodynamic drag* in pounds is given by the relationship

Drag $= 0.5\, c_d\, V^2\, \rho\, A$.

Assume drag coefficient c_d to be 0.4, vehicle cross section A of 30 ft^2, velocity V of 65 mi/h = 95 ft/s, and air density $\rho = 0.0025$ (slugs/ft^3). Determine the power going into aerodynamic drag in lb-ft/s, dividing by 550 to convert to horsepower. Estimate what this means in cost of fuel. Then, estimate the reduction in fuel usage in 20,000 miles of driving per year, if wind resistance could be halved. Convert this to a *Net Present Value*, assuming the vehicle has a 10-year useful life.

For weight reduction, assume that by redesign you can replace 1/3 of the weight of the vehicle with aluminum weighing only 33 percent as much as the steel from which is it now built. Assuming that the fuel usage not attributed to wind drag is proportional, estimate how much less fuel will be used per year if carrying the same passenger/luggage load of 400 lb.

2. Describe three examples which illustrate that design or production cost of achieving improvement in a performance parameter grows faster than linear with performance. If possible, provide quantitative data to prove your points. (Example: Seeking weight reduction in a structural member at constant dimensions and strength might involve a more expensive material, say aluminum alloy rather than steel, along with more costly fabrication. The next step might be composite materials, stronger yet more expensive to design and produce. Beyond this point a structural redesign would be required.)

3. Demonstrate the reality of *economies of scale* in manufacturing by providing three different examples of product or system types in which there are (or have at some time been) both high-rate and low-rate production of similar items. In each case, attempt to determine the sources of economy which have been exploited.

4. For some part, product, or device familiar to you, estimate the ratio of recurring production costs, in quantity of one versus large-scale production. To do this, break down the cost into purchased components plus labor cost items. You might assume for simplicity that you can save 30 percent on purchased items by very large purchases. Estimate hours of labor required for single unit production; estimate large-scale production labor per unit by estimating numbers of workers on an assembly line.

5. Open up the housing of a personal computer and describe the ways in which economy of scale in production seem to have been applied. (Note: Units with unfamiliar brand names may have been produced in smaller quantities than name-brand units, but their producers achieve economies of scale through purchase of major components and subsystems from quantity producers. Studying sources of parts, identified by manufacturer's markings, tells much.)

6. Examine four familiar products and discuss how their designers might trade capabilities for lower cost by a design to cost approach. That is, what features could be removed or scaled back in performance to save production cost, while addressing a reasonable market of customers satisfied with lesser features or

performance? (The objective is to assess how much one would have to give up in order to save a certain amount.)

7. To ensure its market success, a new system must be cost-effective as regards the costs of buying and using it versus its *utility*. Utility may be measured in terms of the cost of accomplishing the same results in a traditional way. This could also be applied in advance of development to estimate the acceptable cost for a system.

(a) A proposed system using *voice and handwriting recognition technology* would permit an executive to dictate correspondence, edit a typed draft by hand or by dictated correction, and receive ready-to-send letterhead and envelope, all without help of a secretary. Reference to correspondent addresses and other previously referenced data would be automatic. Estimate the utility (value) of this scheme, per single-page piece of correspondence prepared, assuming that it would add 2 minutes time to that normally required of the executive. (The basis should be the cost of a secretary doing the same job, using a word processor. Assume annual cost of 2000-h employment of a secretary, including overhead, to be $50,000, and that of the executive $200,000. It will help to lay out a scenario for preparing a piece of correspondence, with estimated timing for each step.)

(b) Analyze the added utility which might be achieved if the new system could respond more quickly than the secretary for priority correspondence. Develop from this a plot along with rationale for added utility if completion time is less than now required by the secretary undisturbed by other activities.

8. Develop and explain a rationale for cost-effectiveness based on utility for a trash-compaction system able to compress refuse to 20 percent of its uncompressed volume, packaging it in a stackable shape.

9. Select some complex volume-produced product whose technology is not alien to you. Identify and describe its design features which indicate that the design was intended for large-scale production. To validate your argument, explain how the same features could have been implemented in low-volume production.

10. Examine a complex product whose technology you understand. Identify what you believe to be opportunities to alter its design to take better advantage of the following volume-production techniques: injection molded plastic forming, die-cast metal part forming, automated spot-welding, cutting or forming many flat parts in a single operation, or automated assembly using robots.

11. Describe a set of production tests which would in your opinion be desirable to apply to 100 percent of production of each of the following products or systems.
(a) Automotive racing tires
(b) Tennis racquets intended for professional or top amateur use
(c) Video games intended for coin operation by the public
(d) Top-range automobiles built by main-line manufacturers

12. Specify in detail an *operational testing* strategy which could be used by an owner to verify satisfactory operation of a conventional tone-dialing telephone handset, not requiring use of measuring instruments.

13. Based on your own experiences, develop a list of features which would make passenger automobiles easier for owners to test. So this will not merely be a "wish list," for each item add a brief proposal describing a way in which the feature might be implemented.

14. A mechanical detail of a precision system involves a short axle of 1-in nominal diameter running inside a roller bearing of 3-in nominal outer diameter. Proper fit, for the application, calls for clearance within the range 0.002 in to 0.004 in between the axle and bearing. The bearing is a purchased item, the axle is to be manufactured in-house. It is intended that the bearing be replaceable by any spare meeting the bearing manufacturer's specification in which inner diameter is toleranced to nominal diameter ±0.001 in. (Presume that out-of-round errors have been taken into account.)
(a) What dimension and tolerances on the axle would ensure that any replacement bearing will fit with the required precision?
(b) Finished (i.e., with ground/polished diameter) axle stock is to be purchased rather than machining the axle to size from larger stock. The stock is available in 1 in + 0.000 in, −0.005 in diameter; each piece of axle stock has negligible taper. For simplicity, assume both axle stock and bearing tolerances are uniformly distributed over their ranges. What fraction of the pieces of axle stock should meet our specifications if required to be used with any bearing?
(c) (example of one form of statistical design) Assume that we don't require bearings to be replaceable. What fraction of axle stock pieces and bearings could be matched in pairs so that *the combination* meets the specified range of clearances?

15. An electric circuit powered by a nominal 9.0-V battery uses a 100-Ω and 200-Ω resistor in series, to produce a nominal 3.0-V output across the 100-Ω resistor. Available resistors are uniformly distributed in value over ±10 percent tolerance. Spare resistors are replaced only in pairs: one 100-Ω and one 200-Ω unit in each package. They are selected from the 10-percent tolerance resistors by measuring their values. If the ratio of resistor values must be 2 ± 2 percent, what fraction of 100-Ω and of 200-Ω (10-percent tolerance) resistors will be usable for assembling pairs of spares? (Hint: Create a drawing similar to Figure 6.5.)

BIBLIOGRAPHY

B. Beizer, *Software System Testing and Quality Assurance*, Van Nostrand Reinhold, New York, 1984.

T. Eschenbach, *Cases in Engineering Economy*, Wiley, New York, 1989. Though not limited to system developments, the case studies in this book deal with real-world situations. (Knowledge of the elements of engineering economy is assumed.)

D. A. Hounshell, *From the American System to Mass Production, 1800-1937: the Development of Manufacturing Technology in the U.S.*, Johns Hopkins University Press, Baltimore, 1984. Provides perspective for the present-day U. S. industrial environment.

A. Jamieson, *Introduction to Quality Control*, Reston Publishing Co. (Prentice-Hall), Reston, VA, 1982. This is the standard statistics-based approach. Unfortunately, texts in this field stop at statistical estimation and do not deal with the forensics required to locate and solve problems.

System Engineering Management Guide, developed by Lockheed Missiles and Space Co., published by Defense Systems Management College, Ft. Belvoir, VA, 22060, 1983. Describes design-to-cost and related acquisition concepts.

K. Rathmill (ed.), *Flexible Manufacturing Systems: Proceedings of 2nd/5th International Conference,* North Holland, New York, 1983 and 1986. Snapshots of a fast-moving field.

G. Salvendy (ed.) *Handbook of Industrial Engineering,* Wiley, New York, 1982. Emphasis is on manufacturing, system operation, and logistics.

J. A. White, *Production Handbook,* (4th edn.), Wiley, New York, 1987 . Provides the system architect or design engineer with a good view of the traditional production environment and its issues.

Synthesis and Analysis in System Design

System design can be described as a combination of four kinds of activity, each applied repeatedly during design activity:

1. **Synthesis:** New combinations of elements are arranged to carry out desired system objectives.
2. **Analysis:** Mathematical or other models (analogues) are created, whereby the most interesting fraction of potential design parameter values or options can be selected for further evaluation.
3. **Experimentation and prototyping:** System aspects which cannot be adequately worked out or demonstrated from a combination of historical evidence and analysis may be demonstrated experimentally. This may involve development of a complete or partial prototype.
4. **Design documentation:** Elements of the design are organized and described verbally and/or diagrammatically to the degree needed to build or otherwise acquire it, assemble, and test it; documentation may take the form of ordered verbal descriptions, sketches, and more formal design drawings, computer programs and data, and sometimes working prototypes or scale models.

This chapter deals at length only with the first two of these activities. Prototyping was discussed in earlier chapters. Other experimental activities carried out in support of design are strongly dependent on traditional engineering and scientific disciplines involved. We do not discuss them here.

7.1 Synthesis in System Design

Synthesis is by far the most difficult of these activities to describe or teach. It involves innovation, even if a system is a painstaking copy of another. Invention, which is approximately a synonym for innovation, is an *unprecedented* combination, of *known* elements, to perform a *known* or a *new* function. It will be observed that in the fine arts the same definition might also be used. When Leonardo da Vinci created the *Mona Lisa*, his paint and canvas were not new ingredients, though the result was new. Synthesis, however, is not entirely innovation or invention, since the system designer often repeats synthesis activities performed earlier, to meet small changes in requirements, or to produce a system similar to others previously designed.

The traditional way to teach synthesis is by teaching traditional practices and techniques, something about the history of evolution of the field, and then assign synthesis tasks under supervision of an expert. Ergo, one teaches artists how to mix and apply paint and recognize color, texture, composition, etc., shows them appropriate art history, and has them create paintings under supervision. Many noted artists obtained this sort of background informally, as apprentices. However, only a tiny fraction of all those exposed to this training ever become world-class artists.

Synthesis aspects of system design are similar to art. One can do little in college, beyond dealing with technique and background of the medium. However, by undertaking major private commissions, noted designers teaching and doing design work in schools of architecture provide the sort of design experience which is seldom possible in systems engineering. Put simply, major systems engineering projects are not carried out in universities. This seems to explain why the most productive design engineers, particularly system designers, have emerged from industrial practice rather than from universities.

7.1.1 A synthesis procedure for personal or small-team use

Most designers will acknowledge that they use a definite procedure for creating a design, though most would also immediately question its applicability to others. Most, however, are probably similar. If you haven't developed such a process for yourself, you can and should do so.

A principal task, in synthesis, is to study a *specific, well-understood requirement*, and evaluate the *widest reasonable variety of element combinations which can satisfy it*. Combinations which perform poorly are rejected without regret, those which perform well are studied in more detail. If known combinations meet the requirements or come near to doing so, these may be used as jumping-off points, or as standards for comparison. *Innovative* solutions to requirements which can be satisfied by well-known combinations of elements require more extreme departures from tradition. Most particularly, the innovative designer must become aware of any newly available elements, their characteristics, and limitations. With this knowledge plus that pertinent to the application, the synthesis process may be begun, as follows:

Capture the requirements   Knowing the results one is seeking is absolutely essential. Don't even try to synthesize solutions if you don't *know* the problem. If system requirements are incomplete, and you *must* come up with an architecture, add your own reasonable requirements, perhaps devising *specific* requirements for system adaptabilities or flexibilities of particular sorts. (These should challenge your innovation, while broadening system utility.) As with other parts of the synthesis exercise, it helps to write down requirements, preferably in abbreviated form and on a single sheet of paper. Examine and refine them, adding to the list when it helps clarify the design problem.

Study earlier solutions or near-solutions if any are known   If you are dealing with requirements which can be satisfied, or nearly so, by earlier systems, list and make note of the elements of these earlier systems and of their major shortcomings. It is important to examine the *widest variety* of such systems, rather than a large number of systems which differ only slightly. If systems differ only slightly, just list the differences. Your objective is to build a compact list of design clues and jumping-off places. It is also helpful to list the *features* found in these earlier solutions, in an abbreviated form. Make a rough breakdown of what they consisted of and their principal characteristics.

If this system requirement could not have been possible or likely until recently, because of social or technological changes, there should be little value in an involved search for precursors. Or, if there is no way to get access to pertinent known or suspected data, it will be better to press on with what can be obtained. It is foolish to proceed in an information void if the only information is proprietary and unavailable to you. But in this situation, it may be useful to proceed, while continuing to test one's own conclusions against the record of others' *actions*. For example, if a certain approach appears excellent, but it is known that a competitor suggested it but did not follow up, one should attempt to discover reasons for their lack of success. (Information on unsuccessful work by others, even though unpublished, may prove relatively easy to obtain.)

Discover and examine potential new elements   Important elements of a design may be: subsystems, components, special arrangements or structures, means of connection, operating sequences or other features which bear on utility, cost, or functional requirements. You should discover and list new elements which *have not been used* in previous solutions to this and similar requirements. The reason they were not used is most often because they were unknown when the earlier solutions evolved. One or more new elements, combined with older ones, is a principal recipe for invention. This is not to say that valuable unused combinations of *well-known* elements do not exist. In a well worked over design field, however, most progress is made by adding something new.

Potential new elements should be grouped according to where they might fit into the design. That is, depending on the requirement, you might classify new elements as operating methods, materials, components, subsystems, or structural schemes, relating them to places where they may be applicable in the required system.

Examine useful combinations Not all combinations discovered will be responsive to your requirements. Discard, as rapidly as possible, elements and combinations which are in no way responsive. Don't be surprised the first few times you go through such an intellectual exercise if everything on your initial opportunity lists gets ruled out.

The most skillful and prolific inventor/designers are not conservative in their choices—either of old combinations or of new elements. Limiting the number of options to those which seem appropriate on initial examination may appear to shorten the design effort. But it may also result in your overlooking a particularly important possibility. My personal rule is to dredge up all the elements which may possibly fit together, until the final ones are becoming silly or far-fetched. Whenever possible, I attempt to classify existing systems, system features, and new system elements into groups depending on their possible function in the system (input schemes, signaling schemes, etc.).

Check for adherence to requirements Even though you should have a specific set of requirements, it may be useful at this point to list the requirements in abbreviated form, as a basis for matching them with elements uncovered in examining existing systems and new elements. Then, collect lists of all the elements which might help to fulfill each requirement item. Abbreviation is useful at this stage. You will want to build temporary lists containing many elements. You will need to be able to scan these rapidly by eye. Some individuals can do much of this analysis without recourse to paper and a pencil. For complex systems, however, a "scratchpad mentality" is best.

At this time you may observe that there are few elements listed which address certain of your requirement items. This is normal. If there are only unique solutions, this could spur you to seek alternatives, particularly for requirement items which are central or unavoidable. Some of the most important innovations may come at this time, when you examine *why* a certain list is so short. Perhaps you have been too closed-minded to acknowledge that there *are* other possibilities.

Select and test combinations The moment (more likely, the hour or week) of truth is at hand. You must next explore *specific combinations* from the lists of solution elements. If you have characterized the requirements well, combination examination may consist of systematically selecting various combinations of elements which satisfy all requirements, and open-mindedly examining the system thus defined. If any of your lists are especially long, you may wish to write down a priority number next to each solution element on a given list, based on experience with and promise shown by that particular element.

If the lists of elements are not too lengthy, you may explore the solution space by considering *all possible combinations* of the elements listed. The number of requirement items used in this exercise might at first be limited so that less essential items are not considered until later. There are good reasons for forcing an exhaustive examination of your options. Most of us have biases or limitations to our knowledge which unconsciously limit our choices. By forcing ourselves to examine first elements, then combinations, which might not pass a bias test, we will be more likely to come up with a worthwhile system concept.

Select, test, improve the winners   Certain combinations may stand out as most desirable in the exercise above. These should be further refined. If you keep *all interesting solutions* alive at this point, it should not be necessary to repeat the entire process. However, learning some new element not previously considered should occasion reexamination of elements or combinations rejected earlier. Should it prove exceptionally useful, and system design commitments have not yet been firmed up, a major change of design may be reasonable. (Recall, we are referring to the early, conceptual phase of work.) However, every system must have the design firmed up at some point. Beyond this, opportunity for redesign should be limited to details which will not adversely affect the design investment, or to unforeseen problems which cannot otherwise be avoided.

The author can attest to the utility of a process of this kind. He has used it to create patented inventions and to develop new and successful systems. A note of caution, however. For any particular problem, the amount and kind of effort placed on each phase of activity must differ. In a personal activity this sort of flexibility is easy to achieve. However, design exercises requiring coordination of hundreds of people must be planned and executed in a more structured way. For this reason, *initial design* of systems, subsystems, or components almost always involves only one person or at most a small, close-knit team.

If several individuals are able to work cooperatively together in an innovative design effort, each must understand and be familiar with the overall problem, rather than being primarily a specialist interested in only one aspect of the problem. The team leader should be a coordinator and also a contributor, not just a manager. One-third to one-half the time may be spent in collective discussions by some or all of the team members. The group process called *brainstorming* which involves coming up with innovative alternatives, is often viewed as of low productivity. However, when participated in by innovative individuals, it *is* an efficient way to generate system solution-element alternatives. It is less efficient as the vehicle for examining system alternatives, but again useful in discussing and refining the proposed system solutions.

7.1.2 A simple example: The better mousetrap

Illustrating a synthesis process by an example incurs risks that the reader will either (1) endeavor to follow each step slavishly, or (2) dismiss the procedure as worthless because the example is unrelated to his or her problems. A short example will, we believe, demonstrate what can be done using an orderly synthesis process. The problem is that old idiom, the "better mousetrap," only this time we are referring to a real mousetrap. We apologize to readers to whom *Mus musculus* is a dear pet, and to others who may never have shared their home with the little critters.

What sort of "better" mousetrap do we want? Since we have no system client or sponsor, we will make up our own requirements. There are some major decisions, for example:

- Method: Is a clean kill preferred, with no small cries in the night, or shall we capture them unharmed for removal to another home?
- Disposal: Shall we throw both mouse and trap away, or reuse the latter? Can the user tolerate dealing with mutilation and death? If there is to be a captive, what to do with it?
- Effectiveness: Here there's little uncertainty. It must be attractive to the victim, and certain in its action once the little fellow has committed to approach or entry.
- Cost: Must be little enough to be disposable, if a killing trap. Otherwise, it must be reusable.
- Ease of use: No question . . . Must be readily operable by a small, old, young, tired, weak, or sleepy person of either sex.

There are two important options among these requirements: (1) kill versus capture, and (2) reuse versus disposable. It will greatly simplify the design analysis if we select a particular combination. However it may undesirably limit the number of options worth pursuing. For present purposes, however, we will proceed with a kill/reuse requirement.

Only one time-honored solution is widely recognized in the United States, a simple mousetrap (Figure 7.1) containing:

(a) A choking/crushing wire U-shaped member powered by

(b) A heavy spring wound from the same wire. This is captured in ready position by

(c) A second straight wire, which is in turn held in position by

(d) A metal trigger strip on which a small piece of bait (cheese?) can be attached, and

(e) a wooden base holding the elements **a** through **d** in relationship.

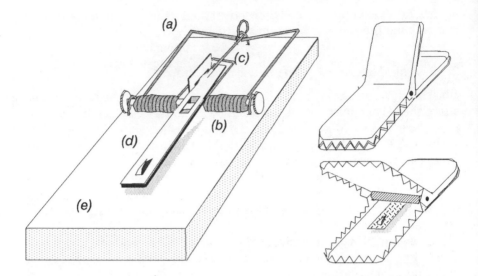

Figure 7.1. The traditional mousetrap is simple, cheap, and fairly effective. The idiom "build a better mousetrap" may, however, be difficult to appreciate, in light of the long time this particular design has persisted. The letters identify the key parts mentioned in the text.

Many variants of this open baited trap have been marketed. One available during the author's childhood, also pictured, used two members formed from metal sheet, connected by a strong spring. A baited metal trigger, when disturbed by the eager rodent, was intended to cause the spring to release and the jaws to shut, though with much less force than in the traditional trap. When tripped, this trap provided its human operator with a convenient grip. When squeezed, it released the dead mouse neatly.

A lack of recent progress in mousetrap design may be attributed largely to commercial pest control organizations but is more likely due to fewer entry locations in a well-sealed modern home. But if one has a mouse in the house, one is driven to remove it, even more so for dreaded (and much larger) *Rattus rattus*.

Since this well-known existing solution does not meet our *reuse* criterion, perhaps we have had poor market advice. However, food-chain industries dealing with grain often have mouse problems in more serious quantities. The $1 (perhaps more these days) cost of "old reliable" (from Figure 7.1) may be considered excessive if substantial numbers of mice have to be eliminated periodically.

This is the point to break down the *elements* of a reusable mousetrap. They seem to be:

- A *bait* or some other attraction (which, preferably, will remain effective even after several mice are trapped)

- A *triggering mechanism* responsive to the victim's presence or actions
- A *capturing/retention* or *killing mechanism*
- A *release mechanism* operable only by a human user
- A *housing* or *mounting structure* integrating these features to produce an economical product

If we intended to trap many mice in a killing trap during a single period of unattended operation, many kill mechanisms must be provided. The number provided will increase the trap's cost and limit its effectiveness. The sole alternative here would require the trap to be able to reset itself. There is thus an advantage in trapping mice in such a way that they move, under their own power, into a retention area.

There again appear to be major alternative approaches here: (1) in which the effort of the mouse itself resets the trap for the next, or (2) some mechanical or electrical device does the resetting. Even though our original requirement did not mention a resetting mechanism, we may wish to add it as a sixth essential element (i.e., for a *reusable, high-capacity* mousetrap).

Lest we depart too early from the objective of examining alternative elements, we should explore each of them. Consider the bait requirement. Some research may be needed to determine what attracts mice, or what would tend to frighten them away; ultrasonic emitters are now being sold as pest deterrents, though how can one be certain that they work?

Cheese, the traditional bait, is presumably effective because of its aroma, but traditional traps often lose this bait without tripping. The bait problem, if examined, breaks down further into a number of subproblems which must be explored: natural versus synthetic bait materials, food versus hormonal attractants, bait held so as to be visible but not edible, bait which simply attracts versus bait affixed to a triggering or trapping mechanism.

Likewise, each of the other elements must be examined. Since some of them may relate closely to others (triggering and trapping mechanisms, for example), it will be desirable to examine specific combinations of the two elements.

Certain elements, in this case particularly the *housing or mounting structure*, are seen to be of little concern in the early consideration of combinations. We must design the housing best suited to support the selected combination of other elements. However, if it is feasible to manufacture certain elements integral with the housing (a retention area, for example), we might wish to examine it sooner.

By examining these requirements both open-mindedly and with knowledge of new possibilities, it is likely that *almost anyone* could come up with a better mousetrap, one which would perhaps be patentable. Financial success, however, would require the ability to build and to sell the product at low enough cost, in sufficient quantity and with enough profit to justify the business operation.

The rest is left to the reader. . .

7.2 Requirements analysis

Of all areas of systems *analysis*, requirements analysis is probably the most important. Without sound requirements analysis, no system can expect to meet user needs.

A friend from the author's Pentagon period, *Leonard Weisberg* (now a vice president of Honeywell Corporation) coined what he referred to jokingly as "Weisberg's uncertainty principle," quoted below.

> *A system development has three major objective areas: system cost, system performance, and delivery schedule. All cannot be simultaneously specified with precision; if any two are intended to be fixed, the third is indeterminate.*

Readers may observe that this is a parody on the *Heisenberg uncertainty principle* of quantum physics. Its message is real, however. Satisfying user requirements is always a matter of compromise. Gains in one area often require concessions in others. Extending schedules may permit cost savings because smaller design teams interact more efficiently. Tightening performance requirements narrows the range of system alternatives which can meet them and can ultimately make a system infeasible.

For a system designer, perhaps the most unforgiveable sin is *not to understand* where design margins are available for trade-off, or where trade-offs could or should be made. As a trivial example, consider the airplane builder who may pay premium prices for miniaturized airborne electronics, while overlooking nonstructural metal aircraft parts which could be lightened by drilling holes through them. This kind of "knowledge gap" may occur in any complex system development wherein managers and management processes are unable to operate effectively *across technical disciplines*.

System requirements analysis and development is not generally taken to be part of the system architecture and design process. However, it is at least an essential precursor. When a system is developed commercially for speculative sale to yet-unidentified customers, a significant amount of requirements analysis is necessary prior to starting system design. In the case of a system to be developed for a client whose analytical resources are limited, the eventual system developer, or perhaps some third-party organization, may be involved in lengthy analyses to define and refine system requirements.

Whereas system design analyses deal with how the system will perform, requirements development deals principally with *what the system must do* to satisfy client needs. Therefore, a major part of the work involves finding out what those needs really are, or what they may evolve into, in the future.

A typical first step is to understand how the client currently performs the tasks for which a new or modified system is desired (if indeed there is precursor business activity). If the problem is one involving system *change* or *enhancement,* it will be necessary to understand in detail how the current system works, how it is used, and user perception of its problems. If there is *no* present system, or where a largely manual system is to be replaced by an automated one, the problem will be more difficult. An exception here is a client entering a new field of activity, who looks to the system developer for requirements knowhow.)

The outcome of requirements analysis, in most instances, will be a *preliminary system specification* defining the desired system characteristics. Accompanying the specification should be a *separate* document describing the work which the system contractor is required to perform. (This may be referred to as a *statement of work,* or SOW.) Items to be delivered (the system, documentation, training courses, accessories, spare parts, supplies, or whatever) and their expected delivery date should be clearly defined in the statement of work. A *specimen contract* should also be prepared, though typically the final contract is subject to negotiation between buyer and seller. In some instances the initial statement of work may cover only system development, with system construction to be performed through a later contract. In this case, the user should in most cases require some form of cost estimate for the complete system, lest it be left with a large cost of development with an unaffordable sole-source system purchase.

From these documents a prospective system developer (or seller) should be able to understand the requirements and other conditions well enough to develop a proposal and firm price for the required system. If the user lacks the resources to produce a good requirement definition, it is almost always more satisfactory to employ different independent organizations for requirements analysis and the system development. During system installation and shakedown, the requirements developer can continue to assist the user.

7.2.1 Qualitative requirements analysis

Much of the analysis during requirements definition is descriptive or qualitative: describing functional operations the system is to perform and desired system features, defining system operating conditions and limits. This demands analysis which is mainly nonmathematical yet every bit as important as establishing proper numerical parameters.

The completed requirement *must be realistic.* In developing a system requirement within a large organization, personnel in different suborganizations often specify what amount to *different systems*. Combined requirements may also add up to an infeasible, internally inconsistent, and unaffordable requirement. Failure to remove these from the final requirement document may result an in abortive development, with unhappy clients. Hence upper management and key staff members in the user organization should be called on as required to help develop needed compromises.

In the qualitative parts of requirements analysis, the essential objective is to determine what kind of system the client will need. It is not always feasible for any particular system to satisfy *every* user requirement and meet all of the user's cost and schedule preferences. If it is likely that not all requirements can be satisifed, they should be *ranked*, individually or in groups of similar importance, as guidance to bidders.

Requirements statements must take on a variety of forms depending on how they will be used. When prepared for a commercial client, one might go so far as to specify particular system components by manufacturer and model number, perhaps suggesting equivalents from several suppliers. If the client is a government body, the requirement may be expected to follow standard practices and be suitable for use in a *request for bid*, including the proposed statement of work.

Requirements studies for commercial clients often recommend specific off-the-shelf solutions when available. Most government agencies, however, must acquire systems *competitively*. Their requirements usually are not permitted to specify a particular make or model, even with the condition " . . . similar to . . .". They must usually issue requirements which *can* be satisfied by several competing suppliers. Occasionally only one system offeror can satisfy special government requirements. In this case, laws may require that a persuasive *sole-source justification* be supplied.

7.2.2 Quantitative requirements analysis

Some requirements can be completed only after extensive quantitative analysis, experiment, or simulation. A new military aircraft or missile, for instance, will often be expected to be able to successfully attack and destroy potential adversary systems, say, a particular model of fighter aircraft. Adversary systems may be available for detailed study only if, somehow, one has fallen into the client's hands. More often, adversary system capabilities are derived or estimated from sparse reports or observations.

The requirement becomes mainly one of *outperforming* a sophisticated competing system. Sometimes the competing system is of similar kind, but often of another kind. An air-to-air missile would outperform an adversary's air-to-air missile if one's own missile had a longer useful range, faster time to target, and greater likelihood of destroying or damaging its target. Each of these factors poses complex issues. Each can be attacked by mathematical analysis but may in addition require simulation, or perhaps even full-scale prototype testing, to convince critics of its effectiveness.

The less knowledge available about the competition, which represents part of any competing system's *operating environment,* the greater is the demand by clients for thorough examination of requirements *prior* to their commitment to system development. This is a psychological phenomenon rather than a scientific judgment!

If too little information exists to permit defining a system which can guarantee competitive success, there are several options. One is *not to develop* a system. If this is unacceptable, another is to characterize requirements generically, then develop the system, using the most advanced technology available, to optimize or maximize values of key system performance parameters. This strategy is applied in development of many military systems. It is also the *principal* competitive strategy of many technology-based commercial system organizations.

A competitive strategy can succeed or fail because of events not predicted and not controllable by the system acquirer. The U.S.S.R. developed the MIG-25 *Foxbat* aircraft, as a long-range interceptor to counter the B-70 being developed in the United States. The B-70 was never put into production, while the Foxbat was. The utility achieved by this expensive Soviet aircraft (which was put to use as a high-altitude reconnaissance aircraft) must certainly have been less than expected.

This is an example of uncertainty in future *system environment.* Although military developments often have this uncertainty, it occurs elsewhere as well. Requirements for extremely large crude-oil tank ships in the 1970s, based on future oil usage predictions at the time, resulted in building large fleets. Many such ships have been idle for long periods. The projection of office space requirements in many major cities has stimulated "requirement" for massive investment in new office buildings. Competing developers often outstrip real growth in need for office space, leading to high vacancy rates.

Establishing *performance* requirements is an important, usually very quantitative, part of requirements analysis. It can be carried out by estimating the required system performance parameters on the basis of all available knowledge of the system environment. If the system environment has not previously been quantified, this may be part of the job.

It is frequently necessary to obtain requirements by surveying different parts of the user organzation and combining their estimated requirements. Individual estimates will contain subjective biases, which the requirements analyst must explore and perhaps compensate in the final system requirement.

Competitive requirements expressed in operational terms ("outfly that MIG," "beat IBM's price") cannot usually be responded to by a system designer. Wherever feasible, requirements must state specific objective performance or cost. For example, "maximum range is to be 22 nautical miles." Often, however, setting all performance objectives as absolute values may not best describe users' objectives. To capture them more accurately, it may be necessary to create combinations requirements. These may relate two or more objective values, in which case they could be referred to as *figures of merit*. Where two parameters are involved such as speed and altitude (of an aircraft) for example, it may suffice to prepare a two-dimensional chart showing unacceptable and acceptable *regimes* of operation as two-dimensional regions.

Many details of quantitative analysis, simulation, and experimentation for requirements development are very much like those used in system design, which are discussed later in this chapter and in Chapter 8.

7.3 Feasibility analysis

Feasibility analysis is essential when dealing with new system designs or unprecedented system capabilities. It is, typically, the attempt to prove analytically that a desired system capability *can be achieved* using a proposed system (which may or may not yet exist).

Here are some typical examples of such analyses:

- Establishing the maximum range, against a particular kind of target, of a proposed radar
- Showing that a racing car could achieve and operate for a sustained period at some particular speed
- Determining that a transport aircraft design could be operated profitably on routes of certain lengths, carrying a given passenger load
- Showing that a new computer design could execute a program which performs some required function, within limits of available memory resources and in less than some given time
- Demonstrating that an automobile tire design could operate without tread wearout for some required distance, at given speed and load

Feasibility analyses deal often with *physical limitations*. The issue is not only whether some particular capability is *possible*, but possible for a certain time, distance, etc. Applied physics and applied mathematics are the two most important tools for feasibility analysis. However, chemistry, biology, and other applied sciences must often be called upon.

It is enigmatic in the development of *software-programmed systems* that the question of feasibility is seldom asked. One can *program* any sequence of logical steps *which one can describe*, with no concern for physical limitations. Thus in software *anything* appears feasible, *given a powerful enough computer with sufficient storage, input, and output capacity.* It appears that only requirements on time and computer resources need to be verified through feasibility study. Practically, however, there may be no computer sufficiently capable which can fit into the system physically. For some demanding requirements, for example, replacing a traditional wind tunnel with equivalent computational power, available computers are *orders of magnitude* less capable than required for real-time simulation of an aircraft in a wind tunnel.

An acceptance of almost universal computational feasibility is a byproduct of our having come to view computing like a utility. If, on the other hand, we would ask into the feasibility of "a computer which can think like a human," the situation is totally different. Here we lack the knowledge needed to broadly define "thinking" in computer terms.

Because of the heavy dependence of feasibility analysis on modeling of the physical world, *applied physicists* have traditionally had the best overall training for this activity. Many engineers also have little difficulty handling it, except where unfamiliar or abstract concepts or relationships must be dealt with. Inexperienced analysts may become more concerned with selecting mathematical methods rather than devising physical models, with logical rather than physical problems, or with optimization rather than synthesis.

7.3.1 Level of analytical detail

Feasibility analyses can often be carried out at any of a wide range of *level of detail*. The models used will depend to a large degree on accuracy needed. Even where highest available accuracy is required, expert analysts first examine simpler models and their solutions. These may be referred to as "ball-park estimates" or "back-of-the-envelope" calculations. They could also be thought of as prototypes for full-scale analysis.

Without such initial efforts, there is great risk that a detailed analysis requiring months of development and weeks of computer time may not be properly designed. Unless the nature of a solution is appreciated, it is easy to select a mathematical model which does not fit the problem.

In this initial exploration, the analyst attempts to simplify, though in general this involves running the risk of introducing fundamental errors. For example, situations involving objects on or near the earth's surface might be modeled using the assumption that, insofar as operation of the system is concerned, the earth's surface might as well be flat. Simple cartesian coordinates then replace the true, but much more complex spherical coordinates. This is not always meaningful, however. One may need to know from how far out at sea one could see a lighthouse rising 100 ft above mean sea level. Proper analytical taste involves using appropriate assumptions at appropriate times.

In any system analysis, the analyst's judgment is critically important. Here are examples of the sort of judgment calls which the analyst must make, usually without assistance:

- Should I carry out this particular computation at single or double precision?
- Should those effects which I previously found to be minor be included in a new analysis with some changed conditions?
- Should I always recompute this from the beginning, even though the initial conditions appear to be unimportant to my analysis?
- Should I include these effects, which I believe are second-order, in the new analysis. . . or will the added complexity take too much time and make the answers difficult to understand?
- Is it time to discontinue analysis based only on hypotheses about the system environment, and attempt to locate some *real* data?

Aside from the purely analytical aspects of the problem, the capable systems analyst must be able to present results in forms readily understood by others who lack his or her analytical abilities. Early systems analyses such as feasibility analysis are subject to error and misinterpretation. The skilled analyst must also painstakingly verify any unusual or unexpected results before exposing them to decision makers. An exciting (though incorrect) result has, more than once, been the basis for expensive though ill-fated projects.

7.3.2 Subjective aspects of analysis: Validating results

There are often questions raised as to the validity of a system analysis. This is especially so if it deals with controversial subjects about which limited real knowledge exists. Questions most often come from individuals or groups in favor of or opposed to the objective implied by the particular analysis. Feasibility analyses seldom avoid this scrutiny and may require exceptional taste both in development and release of their conclusions.

As an example, an analysis to predict the success of a strategic missile defense system in defeating large numbers of missiles and decoys must be sensitive to the numbers of missile warheads used in an attack. Almost any active defense system will be saturated and overcome by sufficiently intense attack. Analysts assuming a lower-level attack than those suggested by critics of the defensive system will surely find their work in question. The most objective analytical approach in this instance would be to examine *various levels* of attacks. It should determine how well the system could deal with each, and develop a realistic view of the saturation limits (if any) of the system. This is unlikely to silence critics, and might even provide them fuel, but it is *objective*. A consequence of public debate of military system analytical results is that it reveals *any* weaknesses to friends and adversaries alike.

Where practical, it is valuable to use two or more independent individuals or teams to carry out analysis against common objectives. If different approaches are used and similar conclusions reached, it will strengthen the conclusions. If teams reach different conclusions, using different approaches, comparison of their work may suggest better approaches to resolve areas of uncertainty. Success of independent analysis requires that competing groups communicate only formally and through some appropriate individual without an ax to grind.

7.3.3 Adding "real data" to theoretical analysis

A point is eventually reached in analytical development at which analysis alone will prove no more. Beyond this point, subsystem experiments, subscale system models, simplified or partially populated versions of the proposed system may be used as a basis for gathering additional data. These experiments must be coordinated closely with analytical issues and questions, and should be directed toward resolving issues which cannot be dealt with analytically. If critics are known, criticisms pertinent to the system operation should be addressed.

When a decision has been made to carry out experimental work to resolve system issues, there is a risk that the demonstration will be used, even enlarged, to serve entirely disjoint management objectives, possibly to a point where it is no longer able to address the system issues. Control of such an experimental program should be retained by a system design organization rather than elsewhere.

The most convincing validation is full-scale demonstration that the proposed system will perform as required, though this may be unaffordable or otherwise impractical. When a convincing demonstration is not possible, analysts may often extend the credibility of analytical efforts, by extrapolating data from earlier trials or field experiences. Even limited data, if well documented and understood, can often be extrapolated with high confidence. Modest amounts of "real data" may be equivalent to infinite amounts of speculation. Its effective application can often convert a theoretical exercise into a solidly based demonstration.

7.4 Trade-off analysis

Evaluating trade-offs is an essential feature of systems engineering. The objective of a trade-off, in its broadest form, is to alter certain system parameters so as to optimize other parameters, meanwhile constraining still other parameters. In designing a transport aircraft, one might "trade off" passenger capacity against maximum useful range, to find the most economical design suitable for routes of 200 to 600 mi.

Although optimization is usually implied, trade-offs are almost never straightforward exercises in optimization. Attempts to fit the problem to some optimization method, e.g., by linearizing the analysis or omitting secondary variables are usually ill-advised. One reason is that usually only a small number of parameter values make sense. Thus the trade-off process involves enumeration of the interesting parameters combinations, with evaluation for each combination. The results may be merely examined, or built into a table or a plot, depending on how they will be used.

Trade-off analysis is often complex. Many quantitative and qualitative system parameters whose relationships may be complex or indirect may need to be defined and their values selected. Trade-offs may involve applying sophisticated disciplines in heuristic ways which must be based largely on experience. It is especially difficult if a proposed system has no comparable precedents. One may find oneself trying to achieve simultaneously multiple optimizations each involving large and overlapping sets of system variables. Criteria for optimality may be fragmentary and altered as a result of the analysis.

Many experienced analysts, faced by such hordes of uncertainties, take shelter in applying familiar analysis methods even though they may be quite wrong. A system designer, therefore, must be prepared to establish criteria, define the model, and perhaps dictate methods to be used for analysis. This requires of course that the designer be widely familiar with applicable methodologies, their capabilities and their limitations. . . and that he or she be unafraid to challenge, and to learn from, analytical specialists.

7.4.1 Trade-off objectives and design parameters

Trade-off analysis can seldom be completed by a single specialist, for it usually spans *many* knowledge or experience specialties. Assigning a particular trade-off problem to a group of individual specialists often produces only the unchangeable views of the most outspoken member(s) of the team. It is more productive to use specialists as resource individuals while vesting responsibility for the trade-off analysis in an experienced generalist.

In its most common form, trade-off analysis begins with a preliminary system description containing *design parameters* which can be varied. There may exist alternative system descriptions, each with its own design parameters. No matter how complex the alternatives and design parameters, there should, however, be a clear and uniform definition of *objectives* for the trade-off.

Objectives, in this case, are usually *desired system capabilities* or *performance parameters*, each typically a function of many system design parameters. Some objectives may take on discretely many values, e.g., the number of passengers in a vehicle, while others may be continuous, such as the vehicle's speed. In the latter case, trade-off analysis is usually simplified by assigning only a limited number of values to the parameter, e.g., cruising speeds of 0.75, 0.8, or 0.85 Mach. It is unlikely that this will cause important options to be ignored since the trade-off can be "fine-tuned," if desired, *after* finding the most favorable solution with the discrete parameter set.

Objectives often specify some threshold, or range of some system capability, e.g."cruising range no less than 2500 nautical miles," "speed in level flight at least Mach 3.0," "must accommodate operator of height at least 5 ft 3 in but no more than 6 ft 2 in." *Alternative* objectives are not uncommon: "if the two (aircraft) crew positions are side by side, only the left-hand position is required to have flight controls; if seating is tandem, both positions must have flight controls." (Such an objective can significantly influence the weight of an aircraft.)

There may be added objectives or constraints, which take form only when some variant system configuration raises a new issue. For example, traditional aircraft ejection seats propel crew members upward. But in the B-52 bomber design, two crew members were placed on a separate deck almost directly below the pilot and copilot. This caused need for downward ejection, and accordingly changed criteria for ejection speed and altitude. Discovery of untraditional alternatives while in the process of trade-off analysis is common. It may require a visit to the chief system designer to determine if such an alternative is in line with other current design considerations.

It would thus be incorrect to suggest that trade-off studies require mere computation of system performance parameters for defined configurations and environmental parameters. More often, a trade-off study will require innovative rearrangement or redesign of parts of a system. For example, an early trade-off study for a fighter aircraft might consider alternatives including a very low altitude, subsonic aircraft or a very high altitude supersonic one. Clearly the early trade-offs will be of different character from those later in development, which tend to be at finer levels of detail, such as the width of a pilot's seat. For the most part, trade-offs early in design will require invention of different system configurations, with relatively simple analysis. Later, more detailed trades may require less invention but more detailed analysis.

7.4.2 Structure of trade-offs: *n*-space approach

In the most typical sort of trade-off analysis, which occurs during most of the design phases, an optimal set of design parameters is to be found for a more or less well-defined system configuration. Optimum is determined by conditions on a set of performance parameters. An analyst may thus envision the problem as exploration of a *multidimensional parameter space.*

Assume that a client's documented requirements, or perhaps some corporately divined market needs, have evolved a design (perhaps with major alternatives also defined) and a set of performance criteria. These represent what the requirements analysts and architects believe to be a desirable system. In some instances there will be a large number of system design parameters. The overall trade-off/optimization problem may be envisioned as an "*n*-space" defined by *n* system *design* parameters (Figure 7.2).

Only certain portions of this space, however, correspond to *realizable* systems. (Many design parameters must have positive values, for one thing, and most are limited in magnitude.) Related to each realizable point in the space is a value of each performance parameter. (You may think of cost as a performance parameter, in this context.) Although it is possible to imagine each point in the *n*-space as defining a *vector* of performance parameters, this is unlikely to assist you in visualizing options.

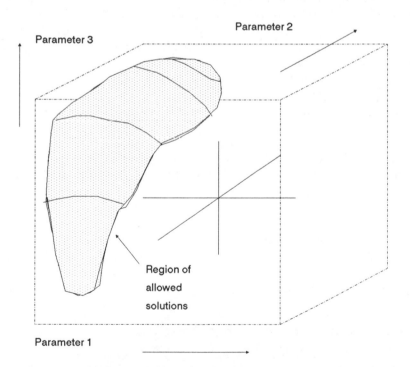

Figure 7.2. Envisioning system constraints in a multidimensional space is sometimes helpful for understanding. A designer may sketch such a space, for instance, to locate parametric regions where solutions are possible.

Rather than attempting to formulate so complex a problem mathematically (which we can't do, in any overall sense, anyway), the reasons for envisioning this n-space are:

- To grasp more realistically how many system defining parameters are *actually* available in design
- To "map out" conceptually the available and excluded ranges of these parameters, alone and in combination
- To determine how to separate the space into those design parameter subspaces which should be further examined

7.4.3 Simplifying the problem

Usually, the *known* combinations of system design parameters (those defined by existing systems, subsystems, and established standards, for example) are limited to a modest number of combinations representing capabilities of available subsystems or technologies. Certain portions of the parameter space must be eliminated because they lie beyond outside physical realizability.

If the realizable set of *subsystem selections* is small, one can sometimes optimize a system simply by enumerating the objective performance parameters *for each combination of available subsystems*. For example, the numbers of different available mainframe computers whose computing performance economically meets some given computation requirement is usually not more than a handful. It would do little good to define, say, a cost versus performance relationship, look for the desired value, and determine that the ideal solution is one for which a real computer does not exist.

If parameters are more numerous and contain both discrete and continuous parameters, you can separate the design parameters into two subsets, one for which only discrete values are meaningful (e.g., automobile body type), and the other for which virtually continuous parameter values are realizable (e.g., passenger capacity of an airliner, Figure 7.3). Now examine the supposedly continuous variables to determine how many differentiable (sensible) combinations there are. That is, even though airliner maximum range *could* be anything from 200 to, say, 12,000 mi, it would be reasonable to limit classes into "short-range," "transcontinental," and "transoceanic" subsets. These might be defined, for example, by examining the ranges of existing aircraft. Even though passenger count could be any integer value, *existing* aircraft could again be used as a basis for defining a limited number of capacity categories.

By this process, what appeared to be a large number of continuous design variables can be reduced to an enumerable set. Traditional optimization theory would suggest that each performance parameter now be expressed as a mathematical function of the design parameters. This

Parameter	Treated as continuous	Treated as discrete (examples)
Age	Years+fraction	0-1,1-3,3-6, etc.
Aircraft range	Nautical miles	Short-haul, intermediate, transcontinental, transoceanic
Passenger capacity, auto	Pounds	Nominal no. of passengers, (or range: 1-2,1-5,1-66,1-9)
Fighter A/C weapon capacity	Pounds total or pounds for each station	Specific missiles carried and on what stations
Voltage of power supply	Volts	110 V ac, 220 V ac (RMS) 440 V ac, 3-phase

Example: design of stereo equipment

Parameter	Continuous (C) Discrete (D), Standard (S), or Range (R)
No. of loudspeakers	D
Power output/channel	C
No. of auxiliary inputs	D
Power supply voltage	S (nation where used, R)
Level of distortion	C (below some upper limit)
Frequencies tuned	S,R
Volume selection	C or D
Max. audio frequency	C
Cassette tape speed	S

Figure 7.3. Many, in some instances all, system parameters in tradeoff analysis may be considered to assume only discretely many significant values. This can simplify analysis greatly.

is seldom reasonable, and we don't need to do it. Rather, we can use piecewise definitions based on analogies with known systems. We may if we choose retain some familiar continuous-variable relationships, such as aerodynamic drag force on an aircraft being proportional to the square of its velocity times the air density—or communications signals decreasing in power by 6 dB for every doubling in range.

Having spent considerable time attempting to understand the true nature of our optimization problem, actual computation required is much reduced. More importantly, we should not be faced with attempting to interpret reams of meaningless computer output. The first step, then, was to determine what trade-offs were actually available. The second step was to look for ways to consolidate continuous design parameters into a few meaningfully different values. And finally, we computed and examined performance parameters for all meaningful combinations.

Trade-off analyses at almost any given level in design should be reduced to a few handfuls, or fewer, of significant alternatives. Thus in defining a new automobile design, one might be dealing with body types, drive train configurations, overall carrying capacity, and vehicle weight. A second benefit of working from a limited number of specific and understandable alternatives is that the analyst (and others) usually find it easier to *understand* the conclusions than if they were merely stated in terms of a set of optimal design parameters. It will be possible later to explore the design parameter spaces around and between those cases found most favorable. Although optimization theories frequently locate narrow optima, which are very sensitive to the selection of design parameters, these should be avoided for robust systems.

Trade-offs can be further simplified by recognizing that design parameters can be separated into subgroups which interact little or not at all with those in other subgroups. For example, luxury of interior accommodations in an autoor aircraft—leather versus vinyl—should have negligible influence on vehicle weight. Design parameters should be considered only in connection with performance parameters which they influence.

7.4.4 Performing needed trade-off analysis

Careful setup and structuring of a trade-off problem should result in a limited set of options and of analytical computations which together will permit objective and quantitative selection of system design parameters.

Analyses which may be needed include the following:

1. *Static or dynamic* engineering computations use defined system inputs or conditions, with variable system parameters and may involve the computation of weights and moments, analysis of dynamic responses to transient disturbances, or acceleration and maneuvering calculations.
2. *Statistical analysis* of operation is based on *expected* (average) values of inputs, fixed or variable arrival rates at queues, probabilities for random events, and conditional probability estimates. These analyses may be used to estimate system capabilities assuming a probabilistic environment.
3. *Simulations* which use predetermined driving functions and variable system response parameters can deal with overall operational characteristics of complex systems They can be used to determine expected system accuracy, correctness of response, and other static or time-dependent characteristics.

4. *Monte Carlo simulations* emulate activities in actual system operation and can be used to describe complex sequences of operations and the system's responses, with probabilistic variations.

To illustrate these four different analytical approaches in attacking a single problem, the design of a wheeled military vehicle suspension system, one might:

1. Compute the response of a suspension when the wheel rises over a single step-obstacle over some reasonable range of step heights.
2. Represent anticipated terrain roughness by a spatial-frequency distribution and estimate the expected excursions of the suspension by frequency-domain analysis.
3. Predefine two-dimensional terrain height versus distance representative of actual terrain to be crossed, and analyze the performance of alternative suspension designs in isolating the vehicle from terrain undulations.
4. Generate a three-dimensional map of a representative battlefield, including buildings, trees, and shellholes, then simulate a high-speed trip across it; the driver would be able to steer the vehicle, but might have limited ability to notice shellholes from a distance.

Clearly these four analyses lead to very different results, any or all of which might be important. The first would reveal the nonlinearities ("bottoming") in the suspension; the second could reveal *resonances* which might represent unpleasant riding qualities. Although the third could also point out these same properties, it would better serve to verify the design selected using the first two analyses. The fourth analysis is more of a "final exam" prior to system construction, to evaluate the suspension while being steered through a realistic three-dimensional world.

The concerns above were for system *dynamic behavior*. Nondynamic system characteristics also lend themselves to several levels of analysis. Design or selection of an *operator's seat* for a new system might simply be based on prior practice, without analysis. Alternatively, anthropological measurements of a group of typical users might be made, with their average and extreme dimensions used as a basis for seat design. Or, a seat design incorporating several adjustments might be simulated to determine the range of contours to which it would be able to conform.

System trade-off analyses should be consistent with the importance of the system design and performance parameters being evaluated. Less important parameters may be treated in less detail or more approximately. The entire set of trade-off analyses, on completion, should make a clear case for design parameter selection. If it does not do so, there may be two reasons: (a) system operation or effectiveness *is* insensitive to the value of the parameter, or (b) the analysis was incorrect, or insufficiently precise to reveal significant differences. The analyst must be sensitive to either possibility.

It is not at all rare to find, after careful checking, that variations in what was thought to be a major design parameter did *not* produce *significant* differences in major *performance parameter* values. This would imply that its value is "don't care," *unless* there are secondary performance objectives for which differences *are* significant. In that case, parameters selection should be made on whatever basis makes most sense. If no alternative shows superiority over others, one usually dips deeper into the bag of objectives to find a selection rationale.

It is characteristic of many engineers to place high confidence in analysis. They may accept very small *calculated* differences in system performance or cost as being significant. Experienced business managers, however, are skeptical of predicted performance values closer than 10 percent, though they may consider departure of 1 percent from a *fiscal* estimate as unacceptable. Many engineers have the opposite problem: they have no fear of the most exacting technical requirements, but find it difficult to hold within 10 percent on budgeted expense, usually erring on the high side.

7.4.5 Pacing requirements

System design is often dominated by "pacing" requirements, ones believed essential to system success or known to be difficult to satisfy. This is most common when a system design must compete directly with others, i.e., with commercial aircraft of other firms or military equipment from potentially hostile nations.

Difficulty may be compounded by the *reduced* attention given other requirements normally important in this class of system. The client often assumes that all of the common criteria will also be met handily. For example, a system designer may be asked to develop a new automobile which, without sacrificing traditional features and comfort, operated using only 2/3 as much fuel as traditional designs. A military aircraft designer might be asked to design a fighter aircraft which, in addition to the high maneuverability typical of the class, had ability to travel at speeds over Mach 3.0 while enroute to or returning from missions. Either of these is a large order. If there is *more than one* pacing requirement, for example, coupling greatly improved performance with significant cost reduction, the unwary system contractor may have a potential business disaster on its hands.

The challenge of pacing requirements occurs fairly often in a competitive environment. In the commercial sector, it may take the form of matching performance or price levels established by a competitor, while continuing to meet one's traditional quality or customer-service standards. In the computer industry, long-term trends in price/performance has been steadily downward. A product whose design is *begun* only at the time a competitor's product *reaches the market* has an uphill struggle. It must fight both the visible competition and the expectation that competitors may soon offer similar equipment at even lower prices.

A pacing system requirement affecting only limited portions of the system can be handled far more readily than one which affects the complete system. For example, a requirement for a reduction in some spaceborne system's volume from 5 ft^3 to 2 ft^3 will likely affect most elements of the system. A requirement to increase the sensitivity of an infrared sensor system by a factor of 2, however, might be accomplished via an improved detector, with little or no change elsewhere.

Compromise may be required to improve chances of success for meeting pacing requirements. One possibility is to increase the funding. The "money is no object" paradigm has seen only limited use, in development of the atomic bomb, for example. Where it occurs, it is most often coupled with time urgency. Only a small fraction of such efforts ever succeed.

Where there is a pacing system *performance* or *schedule* requirement, a common technique is to approach it simultaneously from as many independent promising directions as funding will allow. In 1955, when *RCA* established a crash program to develop an alternative to its shadow mask color TV tube, two major alternatives and many variants were pursued concurrently. Urgency derived from the belief that competitor *Philco Corporation* would announce products using an alternative design of display tube. When this did not happen, the competing developments were abandoned by *RCA*. At another time (late 1960s), *IBM* had under way developments for competing magnetic-core and semiconductor memory technology. In each development the pacing requirement was *cost*. Semiconductors won. Since the competition was internal, selection could be postponed until 1969, when semiconductor memory cost dropped below that of magnetic memory cores.

7.5 Balancing design objectives or requirements: Design compromise

The skilled system designer is thus much like a juggler, keeping continual watch on many different objectives. It is seldom possible to isolate one objective from the others in this balancing act. Improving any subsystem's efficiency or performance almost always implies a heavier or more expensive subsystem, perhaps both. Reducing weight usually means use of more expensive materials, which also take more time and special facilities to form or join. The adage "there's no such thing as a free lunch" is very pertinent, in this activity.

The compromises required in balancing system objectives should be fully understood by the system designer. The designer must be able to explain them in terms meaningful to clients, subsystem contractors, or financiers. One might, for example, be required to estimate the additional acquisition and operating costs required, in a commercial jet aircraft under design, to carry one additional passenger. The typical approach would start from some familiar or attractive initial design, plan the direct changes needed to accommodate one more passenger, then determine what other parameters must also be changed (such as overall weight, fuel required, or cost of operation).

Unfortunately this is seldom a workable approach. One could not, for example, derive the design of a 747 jet via incremental growth from a Piper Cub. In most cases trade-offs should be made by examining initial system configurations which are *significantly different*, each being designed as a good solution for some particular set of requirements. By examining *if and how* one of these solutions evolves toward another as the parameters are changed, it will be easier to imagine the underlying problems and to understand the necessary compromises.

7.5.1 Parameter budgeting in system design

Where an extensive system parameter must be controlled, an intelligent system design approach is to *budget* it across the system. (Extensive parameters are ones for which the system value is the sum of the values for its parts. Weight, power, volume, or aerodynamic drag are all extensive.) Let's assume we are dealing with a *weight budget* for an air vehicle. A physical partitioning of the system at an appropriate level of detail must first be developed. A budget of weights for each subsystem is devised on the basis of experience and actual weights of any *specified* subsystems. Initial allocations will often be low, to provide some challenge; the chief designer may "reserve" the remainder of the allowable system weight as a "contingency" for later assignment if necessary. Each subsystem designer is made responsible for controlling his or her subsystems' weight, and reporting it.

As more design data is obtained and refined, individual subsystem weight estimates may prove pessimistic, more likely optimistic. Certain subsystems will threaten to break through the allocated weight budgets. The system designer may allocate more from an unassigned fraction, or negotiate with subsystem designers to find other ways to reduce subsystem weight.

Why not simply allow the weight to grow as needed? Because many aircraft design parameters are sensitive to the gross weight (airframe and landing gear strength, engine thrust, fuel required). An orderly, continually updated budgeting approach to control of weight, power, or other extensive parameters is an essential part of configuration control, which for performance-critical systems must be maintained *throughout the useful life of the system*. Weight added by modifications after an aircraft is operational must be at least as great a concern to the configuration manager. It is impractical to compensate weight addition in one subsystem by weight reduction in other subsystems. Furthermore, deterioration during operational life reduces the ultimate strength of the aircraft's weight-carrying subsystems.

It may be learned during budgeting of some important extensive parameter that the system decomposition is not well suited for controlling that parameter. This may be because the parameter, though additive, is not simply summed for all subsystems. While aircraft subsystem weights are added to obtain total aircraft weight, another important parameter, *pitching moment about the center of lift* is the product of weight of each subsystem times its average fore-aft distance from the *center of lift* (which is typically about 35 percent of the wing chord from the wing's leading edge; see Figure 7.4).

Similar considerations apply in determining the weight distribution between front and rear wheels of an auto. Not only is subsystem weight important, but (particularly if the subsystem is physically large or in several physical parts) so is the *distribution* of that weight. Likewise, electric power required by a vehicle is the sum of the power required by each subsystem. But the cost, weight, and power loss associated with distribution of power depends on the distances between generator and electrically powered subsystem.

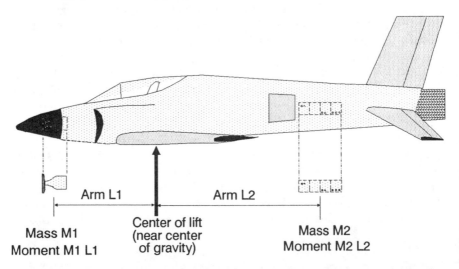

Figure 7.4. In an aircraft, not only weight (mass) but also its *distribution* about the center of lift must be budgeted. This is accounted for by computing pitching moment (mass times distance to center of lift) for each added subsystem. A few pounds at the nose or tail counts far more, in this respect, than much more if directly over the wing.

Wherever budgeted parameters are interrelated, the budgeting problem is more complex. Water supply and drainage systems, or hydraulic fluid source and exhaust, must go hand in hand. It may be better to budget both as a single resource. Large heat dissipation in system spaces lacking access to cooling air or water is also a problem. Though the system may be within total power-supply and heat-removal budgets, cooling must be redirected. Relationships between budgeted parameters may complicate, but should in no way diminish, the emphasis on budgeting these parameters during system design.

Parameter budgeting may in fact suggest major changes in system architecture. Most current aircraft contain many different electronic subsystems, each within its own housing and with its own cooling and power supply. Advances in electronic circuit miniaturization, principally as a result of integrated circuit development, have greatly reduced the size of traditional subsystems such as radios and navigation apparatus. Combined weight of these subsystems would be considerably smaller if they could be combined to occupy a smaller number of housings. However, a system design philosophy which packs electronic subsystems acquired from many suppliers within a common electronics enclosure will require many architectural, interface, and design changes and the standards to support them:

- Modification of electronic packaging standards
- Better definition and control of undesirable subsystem interactions such as signal interference, or heating of one subsystem by adjacent subsystems
- Test standards and criteria reflecting detection of interactions
- Redesign or repackaging of existing subsystems

An additional problem is that subsystem designers may not be certain at design time what subsystems will be placed adjacent to their own or be able to predict interactions with those subsystems. At least for digital circuitry, the problem is not insuperable, as evidenced by the great success of the IBM Personal Computer or any of its many clones. An enormous variety of independently designed and fabricated "adaptor" boards may be placed, in any arrangement, in the narrowly spaced "slots." Systems of more critical performance such as avionics, however, demand greater assurance that subsystem failure will not be induced because of interactions.

7.6 Design iteration and documentation

Design iteration during system development is a natural consequence of system complexity and of coupling between subsystems. Changes in one subsystem will precipitate changes elsewhere. One subsystem requiring slightly more electric power may demand a redesign of a central power supply, perhaps with extra capacity to allow for future changes. If uncontrolled, design iteration may result in repeated *major* redesigns. The designers' objective should be to see that each change does not cause need for even greater or more changes.

In design organizations lacking a continual supply of new projects, design iteration may be pursued for the lack of any more productive requirements. This is common where software is maintained, within an organization, for its own use. Availability of programmer time leads to continual changes. Design iteration expands to fill the available design-hours.

Although detail (e.g., component or subsystem) designers may emphasize the value of the improvements which their design iteration can yield, *at system level* these potential improvements may not be usable without other revisions. Increasing the resolution of a TV camera may require replacement of display devices used by all of its viewers and increase the bandwidths required in communicating signals, thus perhaps requiring new transmitters, receivers, and amplifiers. To accommodate the "improvements" in one subsystem may even demand *unrealizable* performance increases in others. Attention devoted to architectural design and system partitioning can pay off in this regard by allowing subsystems to evolve more independently with little effect on others. Viewing design iteration as a "loop" involving initial change and collateral changes and repairs (Figure 7.5), one management objective should be to minimize the number of active loops, another to keep each *necessary* loop as short and local as possible. This should result in minimizing cost, wasted time, and system errors.

Design iteration may unfortunately be *required* if specific requirements were purposely ignored and left unresolved during architecture and early design phases. This may occur because designers become engrossed in the portions of the work they consider most challenging or which they most enjoy. If this occurs, the system design manager must be held responsible.

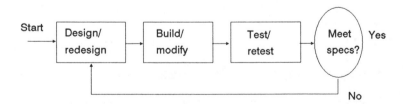

Figure 7.5. Design may be iterative if there is large uncertainty (or if it is managed poorly). Keeping the iteration loops short and local to a small part of the system, where feasible, helps minimize wasted time during development.

7.6.1 Design documentation

Most system designers and design engineers will acknowledge a need for user operating and maintenance manuals for the systems they develop. Such documents are most often prepared by specialists not involved in the system design itself. Designers seldom place importance on their own preparation of design documentation, that intended to be read only by other specialists or used by them in system integration or later enhancements.

Design documentation is an essential product from good design practice. It should permit designers and others as well to understand the features and limitations of a system design. It should serve as a basic resource for preparation of operating and maintenance manuals, rather than expecting technical writers to teach themselves how to operate or maintain the system from assembly diagrams, pore through obscure, incomplete design notebooks, and try to pry answers from the designers during brief interviews. A requirement to maintain redundant system descriptions: both written documentation, detailed system diagrams and computer-aided design data provides additional checks on correctness of the design and its implementation.

Using microcomputer or minicomputer technology and local area networks, with database management, text-editing, and CAD/CAM (two- or three-dimensional line graphics) software, all the facilities needed for easy design documentation are available and affordable. What is often missing is a means to relate these facilities so that needed information can be readily accessed, and design revisions quickly and accurately recorded. There are a number of "computer aided systems engineering" (CASE) software packages available which combine some of these functions. Most were created for development of software, however, rather than for interdisciplinary systems. No software of this sort is universal in applicability to system design. Inevitably, certain functions are inferior to those in popular general-purpose editing, graphics, or spreadsheet programs.

Quality of system design documentation varies with the personality of the designer and the demands made on the designer by management. Traditionally, industry has required only minimal design documentation of the sort recorded in "engineering notebooks." The result of casual attitudes toward engineering records is that almost no "lessons learned" information is recorded, and far less is published. Management and designers often feel that preparing documentation not actually required for delivery is wasted effort having nothing to do with success of the system in a competitive environment. By the time the system becomes a financial success, it is too late to capture more than a brief abstract of design information. Where technology is evolving rapidly, there is often a belief that little learned in one development will be of value in the next.

This just isn't so. Valuable analogies can be carried forward across vast changes in technology.

Ideally, design-level documentation should be captured without overt effort as a byproduct of other activity. Perhaps when computer-aided design technology has advanced to aid many additional areas of design activity, this objective can be realized. Meanwhile, a best alternative is to capture as much as possible of the design process in computer-readable form. The availability of a reasonably detailed and well-annotated chronological design records with text and associated computer-based graphics prove highly valuable. If one uses high-speed searching of this data for key words or phrases, even at present it provides a design organization with valuable leverage. More powerful abstracting and searching capabilities will make this information more valuable in the future.

EXERCISES

1. Follow the synthesis procedure laid out in Section 7.1, to devise a novel *personal, pocketable one-piece writing tool* which could draw parallel straight lines on plain paper *without aid from some other device*. (Don't worry about marketability, just have fun! Make sure you take full advantage of the freedom granted in the wording of the problem.)

2. (a) If a requirement for some portion of a system is to transport pedestrians *between floors* in a structure, list *known solutions* as completely as you can. Also add, at the bottom of your list, nontraditional solution *elements* which you envision as possibly addressing the requirement.
(b) Prepare a similar list, if the requirement is to transport *people moving large or heavy loads for which they are responsible*. Again append possible new elements to the list. (Don't get trapped into too narrow a definition, therefore a too limited perception of the possible options.)
(c) Repeat exercise (b), assuming that the people do not need to accompany the items, merely to define destination correctly.

3. A study of possibilities for a new restaurant food warmer system, to maintain each item of food, on each plate, at different temperatures has identified new elements including: (1) special platters having built-in warming electrodes, one in each quadrant; (2) variable-density masks designed with patterns to selectively reflect infrared energy from different quadrants, (3) an electrically powered cover with heaters in each of the four quadrants, and a single switch which can be set to a separate lettered position which adapts it to each combination of foods on the plate, and (4) hot/cold pattern locations on a counter surface on which a platter can be placed.
(a) List suitable evaluation *criteria* for assessing these options. (It will be considered unfair to devise a critical criterion with the sole objective of scuttling one of the proposed elements!)
(b) Using these criteria, evaluate the capabilities and limitations of each alternative element, adding even more elements if you can. (At this point you

are not examining combinations, but *elements*. Despite enthusiasm you may build for a new principle, phenomenon, or device idea, it usually requires additional system elements to produce a result. The selected system combination will ultimately determine system marketability.)

4. Try your hand at the better mousetrap problem discussed in the text. Carry it through to a proposed trap system concept, providing a description and sketches showing principles of operation.

5. Assume that you operate a company which designs, integrates, and installs central home entertainment systems. In their most elaborate (and expensive) forms, they can be controlled from anywhere in the home and will deliver audio or video anywhere, i.e., to audio and video output units, with any combination of programs and rooms. (The purpose of describing so broad a capability is to make your problem less system dependant.)
(a) Develop a list of questions which you might use in an interview to understand the requirements of a particular customer.
(b) Certain generic diagram formats, laying out desired system capabilities, should be helpful for formulating formal written requirements. Other diagrams may be helpful in making approximate cost estimates. Sketch diagrams which could be used as survey aids, in combination with your survey questions.

6. (a) With your present improved understanding of system design, establish a set of no more than 10 "ideal" personal requirements for a powered personal ground transportation system, being careful not to bias it as to medium or mode. Include as many wild ideas as you wish (a favorite of mine is that the vehicle should collapse into an attache case to avoid parking problems).
(b) *After* completing your requirements list, go back through it. For each requirement, *separately* identify what *available* systems, if any, could satisfy your requirement. Identify those requirements not satisfiable by current systems either as pacing requirements or desirable options. Then discuss each of these requirements in terms of the breakthrough(s) required to accomplish them.

7. Though some system users will demand specific quantitative system parameters, most requirements are either qualitative or approximate. Review *laws* within your knowledge or easy access: physical laws, statutes, or even perhaps unwritten codes. Look for quantitative legal requirements which might control system requirements, and observe whether they have actually done so. (An example: In the United States there was for more than a decade a 55 mi/h maximum speed on highways. How, if at all, did this influence automotive system requirements? You may wish to consider as a counterexample Germany, which has no speed limits on major highways.)

8. Discuss the *feasibility issues* in the proposal below for which you understand the underlying science or technology, or for which you can seek advice from a specialist:
(a) A single powerful nuclear attack warning siren, audible throughout Manhattan and certain portions of the other boroughs of New York City, to be mounted conveniently atop one of the World Trade Center buildings.

(b) A food packaging system by which foods such as bacon, lettuce, and tomato sandwiches, or poached trout could stay edible and tasty indefinitely.

(c) A computer-output sheet-fed printer capable of output rate as high as 10,000 pages per minute.

(d) An 8-oz transformer, in a 2-in cube package, capable of operating a 1750-W electric heater, designed for 110-V operation, on a 220-V power line.

9. It sometimes happens that the accuracy with which independent variables *in a system environment* can be estimated is very rough, but a collective variable, determined by combining the others, is more accurately estimable. This may be due to (a) statistical averaging (in probability the *law of large numbers* makes the average of a number of independent random numbers vary less than its components), (b) group behavior characteristics, or (c) time-periodic phenomena. Think of system-related examples of each of these three factors: a case where it complicates system analysis and another where it simplifies analysis.

10. Devise a set of *general rules* you might suggest to determine the most appropriate number of significant figures to use in elaborate computations involving large amounts of experimental data. The objectives, both of equal rank, are accuracy and economy.

11. Develop, for your home-town locale, a simple earth-temperature model based on the amount of sunlight falling on a given area of ground during each day of the year. Develop another simple model for the escape of ground heat. Use these to model local long-term-average *maximum daily temperature*, over a year. Explain, and attempt to justify, the simplifying assumptions you used at each point, and how they were applied, beginning with your handling of diurnal averaging. Also explain any approximations used in the spherical geometry part of the problem.

12. Varying the model parameters, fit results obtained using the model you developed in Exercise 11, with actual long-range maximum daily temperature data recorded in your locale. Rather than attempting to verify that your model can *predict* the actual variation, you may simply demonstrate that the *shape* of the annual maximum daily temperature curve can be closely approximated.

13. Devise a detailed hypothetical trade-off analysis problem in which requirements might be met by either a bicycle, moped, or small automobile (though not by walking). Include qualitative and quantitative requirements which suggest trade-offs between the three modes, based on a model of the problem. (Suggestion: One possibility is the problem of a delivery service on a small island with few roads and more bike paths. See how subtle you can make the trade-off problem without becoming ridiculous.)

14. (a) *Without* measuring rooms in your home, devise a *budget allocation for floor area* (nonoverlapping) to the following residential functions: food preparation and dining, casual activities, sleeping, dressing, bathing/toilet, access areas/halls, and storage. For purposes of producing the budget, you may want to determine the fraction of an average week during which each activity is pursued, and the number of family members who pursue it. You should also

include a factor which reflects the average amount of space per person required to carry out each activity.

(b) Measure the living areas of your home and assign each area to one of the functions in the budget. Where a room is used for more than one function, perhaps at different times, assign its area fractionally to the appropriate budget items.

(c) Analyze the results of (a) and (b), critique any significant differences, and suggest what sort of adjustments you might make—either to your area budget or your use of space.

15. Find a limited-size set of design-level documentation for some implemented design project. This could consist of design drawings for a home or office building, or design documentation for a small product produced by a company. Examine it carefully (that's why a small set is suggested!).

(a) What other documents or standards are referenced?

(b) In what ways does the documentation fall short of what you believe a builder would need to complete the system, considering both the documentation and formal or informal but widely known design rules or standards? (Cite applicable design rules or standards.)

(c) What techniques are used, in these documents, to describe the design (e.g., assembly drawings; plan drawings; "three-views" showing views from three orthogonal directions; design-detail drawings; lists of materials; or assembly sequences). Suggest other techniquess you believe would have been worth additional effort. Indicate which of those techniques included which in your belief were poorly done or represented a wasted effort.

BIBLIOGRAPHY

Maurice Daumice (ed.), *A History of Technology and Invention*, Crown Publishers, New York, 1979 (English edition; original was in French). Describes every generation's better mousetrap.

Howard Eisner, *Computer Aided Systems Engineering*, M. Dekker, New York, 1988. Eisner describes a wide variety of analytical techniques applicable to systems analysis.

Bertrand Gille (ed.), *The History of Techniques*, Gordon and Breach, New York, 1986 (2 volumes). Technologies and systems down through the ages, evolving with technology. This should give you ideas.

F. P. Preparata and R. T. Yeh, *Introduction to Discrete Structures (for computer science and engineering)*, Addison-Wesley, Reading, MA, 1973. This textbook describes use of many mathematical representations important in systems analysis.

A. P. Sage (ed.), *Systems Engineering: Methodology & Applications*, IEEE Press, New York, 1977. A collection of reprints which emphasize policy development and decision analysis trade-offs.

T.I. Williams (ed.), A History of Technology, Clarendon Press, Oxford, (seven volumes through 1978). Covers technology "from early times," with two volumes covering 1900 to 1950. A charming history plus how-to information.

Design documentation is seldom addressed generically except via government standards. In the case of software, DOD-STD-2167A (Feb., 1988) prescribes the formats and contents of a set of "data items" which can be required for developments including software.

Other Design Considerations

Earlier chapters on system design dealt with product and process factors which influence design goals, with synthesis and the various types of analysis system designers need to carry out. This chapter treats a collection of other topics, to complete our treatment of system design.

We first examine the need for architecture and design support tools, and some of the existing tools which have value in architecture and design processes. Because computer-aided design technology is progressing rapidly, our approach in this area is to point out functional needs, i.e., what to look for when selecting design aids. A view is presented of features desirable in future design tools for conceptual levels of design.

The author has long been convinced that highly valuable design guidance can be obtained through considering the *tests* needed to verify correctness of design and continued correctness of the completed system. Though traditionally testing is most often an afterthought, its design contribution is that it is a different point of view and cuts across technology and applications. Test strategies applicable in each major phase of the life cycle are discussed.

Design rules are devised by, and put to use in, design organizations as means to ensure design quality and to reduce likelihood of failures with far less effort than would be needed to discover every possible cause of failure in a system design. While they complement design *standards*, most design rules are applied more locally, i.e., *within* subsystems rather than at interfaces between them.

This chapter also includes discussions of several important economic issues, related to system design but distinct from management organization, project cost, and scheduling issues taken up later in Chapter 9. One is the question "Make, or buy?"(as applied, for example, to parts or subsystems) which must be repeatedly answered throughout system design. Others deal with the type of competitive market in which design is pursued (i.e., commercial or government markets) and how this influences design activity.

8.1 Design aids and system data

Designers of all kinds have, for the past 3 decades, been drawn increasingly to the use of computers as design tools. Large mainframe computers from the first were called upon to lay out and check the wiring of their successors. The microprocessor has in the past 10 years led a new revolution in design tools. Computers have proven their utility in complex but routine aspects of system design. Their contributions to the less complex but more subtle architectural and early design phases are much more immature.

Computers are valuable for several reasons, and in several ways. It is possible to store, on computer-readable storage, design details (wiring tables, instructions for a numerically controlled machine tool, etc.) whose control would be virtually impossible using traditional paper-document storage. This data can be copied and augmented without damaging the earlier data. Certain kinds of consistency checks can be made; for example, a circuit designer can verify that wiring connections on a printed circuit board design are free of common-variety errors.

Many tasks which are *technologically feasible* using computer automation are as yet *economically* impractical, however. This usually arises from the length of computation required. This barrier will fall as the future continues to bring lower computational costs. The more serious problems are those where major software developments would be required and economic incentive (market for the software) is not large enough to warrant speculative risk. Software costs appear to be rising with time, rather than falling.

Many powerful, widely used personal computer programs can be valuable for certain aspects of system design. *Computer spreadsheets*, of which Lotus Corporation's *1-2-3* is the most familiar, can be used to verify feasibility in many aspects of system design, particularly after analysis is divided into problems of manageable size. Later we will discuss both current spreadsheets and their possible future evolution to more capable system design tools.

Likewise, *word processing* programs, or separate *outline manipulation* programs are helpful during those phases of system design when requirements or design alternatives are represented mainly by prose descriptions. Such activity recurs over the entire life cycle. In later parts of the life cycle, it is more useful to append added design information directly to that stored in the form of graphics files or as database records.

Word processing is required for preparation of training curricula and operating or maintenance documentation. It can also be used in handling of subcontracts, manufacturing instructions, and numerous other documents. In most of these activities, it is valuable to be able to search through large quantities of information seeking a particular identifier or phrase. Though *any* word processor can search one document currently being edited, some word processors or special programs can search rapidly through wide areas of computer storage. If data is formally organized into records, *text searching* is seldom a worthy substitute for a database manager (see below). However, it can be of great value for examining text material or other data files which contain identifying character sequences.

Database management software is useful if large numbers of similar items of information must be arranged and listed in various arrangements. Such records are characteristic of later phases of systems engineering, when data mass is large and its organization more critical. Listing of system subsystems, components and hardware, and their sources needs to be accessed in a structured way. Text content is seldom a good way to search through this kind of data. Many searches must examine numerous similar data entries to find those matching user queries. There may be from two to a dozen separate points of similarity or dissimilarity which must be checked. Shrewd design organizations having access to and knowledge of database management will develop permanent databases listing suppliers, available equipment characteristics and other information which must be searched repeatedly. A *project database* should be set up for each project as soon as its needs can be understood. Among the formal records maintained should be ones describing documentation and graphic records, with brief descriptions, identifying data, source, date, and storage location.

Database-management programs are best for handling collections of data having *uniform formats* and relatively *simple connectivity*. To illustrate this, imagine using a database record of defined format, to capture properties of any of a collection of electrical *and* mechanical parts. The technical parameters of interest for a *mechanical bolt* include diameter, thread pitch and shape, head shape, slot type, dimensional tolerances, material, surface treatment or coating, and special heat treatment or other processing. Those for an electronic *capacitor* include nominal value in microfarads, tolerance, type (the phenomenology and materials inside), case and connection dimensions, and storage and operational temperature limits.

It would make little sense to include, in each part record, space to record the combination of these parameters (and perhaps many others as well). One must either design a separate database record format for each class of objects or fall back on textual descriptions to distinguish them. Putting technical data in verbal descriptions within database records loses many of the benefits of using a database manager.

Data base limitations in handling *connectivity* can be illustrated by imagining that we stored *system block diagrams* as sets of data base records. If different blocks are described by different measures, we would have the same problem as in the bolt-capacitor example above. We would also need a scheme to represent block inputs and outputs, in a record representing a block. Though this would not in itself be difficult, maintenance of correct data *after altering the diagram* would be a difficult task unless special software was used with graphic presentation of block diagrams. In any computer programs used for design, it is essential that parameter changes be as simple and natural as possible. If record-keeping requires detailed follow-up by designers or a data-keeper, it will usually *never* catch up.

8.1.1 Computer graphic methods

Graphics has been an essential design tool ever since a prehistoric ancestor traced a finger in the sand to describe a new concept for an arrowhead. Its use has continued and grown throughout civilization. It was and is an essential tool of design.

When computers first came into use for design, they offered no interactive capability, graphic or otherwise. Interactive graphics has grown rapidly, in its utility for design, over the past 15 to 20 years. Early computer graphics merely allowed automated *plotting* of designs which had to be stored as large numbers of numerical constants. Interactive computer graphics now allows designs to be created and changed graphically. Directly interactive workstations and personal computers have restored some of the finger-in-sand interactive nature of graphics. However, display dimensions are still equivalent only to the smallest sizes of traditional paper design drawings.

A central feature of design is frequent need to examine some detail, then with the flick of an eye and hand, shift to some other detail or review the entire design. Thus far, this has been denied computer-based designers except in a few detail design areas. The ability of a designer to convey different meanings by different strokes of chalk or drafting pencil is only now beginning to be recognized by interactive graphics system designers as an essential part of interactive graphics.

Availability of graphic design aids suitable for system architecture or design depends on the type of system and level of detail. As mentioned in Chapter 4, capable CAD (computer aided design) software has for some time been available for design of integrated circuit chips or electronic printed circuit boards. It can deal with the most complex of electronic circuits because they are highly *modular*, with frequent repetition of a few building blocks. Their form also depends little on the functions the circuits perform. Analog or digital, audio or microwave, these circuits differ little, save for size and shape of planar connecting conductors. By contrast, graphic design software for *mechanical* devices must deal with much greater structural variety.

Computer graphics serves as a vehicle by which designers can visualize spatial relationships. Its utility spans systems or details which may be very small, or very large, on human scale. One of the most valuable applications of computer graphics is to permit designers to "see" two- or three-dimensional relationships or clearances between system parts. A new automobile suspension design may be flexed—computationally— to determine if suspension parts will hit other parts during their normal motion. Similar programs allow designers to determine, using only the computer, whether oddly-shaped subsystems can be introduced into or removed from a system via planned access doors, without disturbing other parts. (Anyone who has built a boat or airplane in the basement, as the author did, could benefit from such a program!)

To deal with continuous (rather than discrete) phenomena, such as flexure of structural members, design programs must carry out *finite element analysis*. Continuous structural members may, for analysis, be described as models composed of tiny girders, as depicted in Figure 8.1. This model may be intended to examine the nature of mechanical flexure of the part when certain forces are imposed. Modeling of this type is usually accurate enough to permit dimensioning of system parts. It can be followed up by more precise models or by experiment. A designer must always verify correctness of computer models used in design.

Computers are as yet of limited help in architecture or conceptual design of systems. Ideally, computer support here would help designers sketch ideas roughly, analyze them, then aid the designer in making changes and iterating the process. Computer graphics and its underlying software do not yet help much with the fragmentary and incomplete problems of early design. The needed computer representations and related program algorithms have not yet been devised. They should be capable of dealing with systems and concepts described using what engineers refer to as *block diagrams*. Such diagrams may be used for a variety of representations and in a variety of ways, and are central to conceptual design.

8.1.2 Integrated computer aids: System design environments

Most available design software suffers from a familiar system shortcoming, the lack of integration or of compatibility of programs with complementary software developed by competitors. Some word processors can incorporate text or graphs produced by the most popular spreadsheet program. Some spreadsheets can incorporate data prepared by a particular database manager program. However, word processor output files are almost worthless as sources of input *to* spreadsheets or database-management software. It is reasonable to say that we don't yet know how the ideal *system design environment* should be put together for most design applications. There is too little experience, existing products are

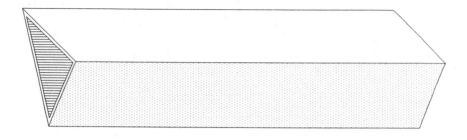

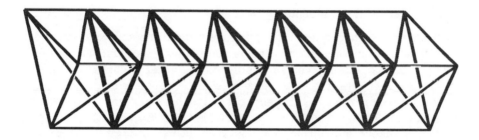

Figure 8.1. Computers cannot easily solve problems involving continuous variables or structures. One method which has been used extensively to represent mechanical structures is termed *finite element analysis*. (This distinguishes it from the use of infinitesimals, as in calculus). A continuous structure may, for example, be represented by a structure built up of miniature girders. Even inexact solutions can often be used to understand vibration modes or other features of interest.

limited in quality and quantity, and the cost discourages users from attempting to build their own system design environments.

A recognized key to system design is an ability to *describe* the system in some *computer-understandable* form. There have been numerous unsuccessful attempts to define broadly useful, machine-interpretable *system descriptive languages*. If such a language is to have high utility, description of a system in this medium should permit computer-based design evaluation, correctness testing, and requirement tracing. The languages thus far proposed for this purpose have demanded that designers master complex and unnatural representations or rely on specialists to interpret their ideas via the language. Neither has been acceptable to most designers. Eventually it may be possible for a computer program to adapt to the *context* in which a designer expresses system characteristics. Then natural-language (e.g., English) system descriptions might be translated into machine-usable form without semantic alteration, loss of content, or embellishment. Such capabilities are probably years away.

Powerful support systems based on prose descriptions would still fall short of what most designers need. Manipulation of concepts graphically has far greater power to portray systems, but here also the computer cannot yet capture *meaning*: relationships, constraints, and other design implications. Earlier system description languages were not accompanied by interactive building and or modification capability. We see future design support tools *combining* graphic, numeric and prose descriptions, and communicating with the designer to verify any unclear entry. When this sort of support power can be combined with on-line equivalent of handbook information and histories of previous design experience, the designer's intellectual capabilities should be truly multiplied.

8.1.3 Near-term possibilities

Word processing, computer graphics, and database software are more useful to systems designers than to detail designers, in part because system design often involves dealing with larger volumes of textual information. Certain highly specific system architecture and design areas, however, receive little support from these general-purpose computer capabilities. One such support area is *test definition*. While testing of homogeneous monolithic designs (such as for integrated circuits or printed circuit boards) has been dealt with extensively, testing of inhomogeneous or distributed systems has received far attention. Likewise, only scattered problems in system partition and integration have been dealt with. The best known example of the 1980s was probably Digital Equipment Corp.'s expert system originally known as *R1*. This program determined cabling and hardware requirements for connecting pieces of computer systems ordered by DEC's customers.

Tools for early system design In early phases of design, when system concepts and architectural decisions are being formulated, problems are neither highly specific nor usually so mathematically complex as in more detailed design phases. Aside from programming languages such as Pascal, C or Ada, the type of broad-spectrum computer tool in widest use today by decision makers and those who support them are *spreadsheets*. Lotus *1-2-3*, though not the first, remains the most popular. The key features of a conventional spreadsheet are:

- **Cell array** A rectangular array of *cells* is arranged in rows and columns resembling an accounting sheet (for which it often substitutes). Any portion of the entire array, which can be hundreds of columns wide and thousands of rows deep, can be displayed on the computer's screen.
- **Cell contents** Each cell contains either a number, a text string referred to as a *label*, or some user-defined formula which can mathematically or logically combine the values of the contents of other cells. For this

purpose, cells are numbered: A1 occupies the first row and column, C5 is in the fifth row down and third column from the left.

- **Interactive change of contents** The user can change contents (number, label, or formula) of a cell freely. Values of other cells which depend on that of the changed cell are also changed almost immediately.
- **Flexible array operations** There are a number of user-friendly interactive operations to perform such functions as copying formulas from one cell or a block of cells, into one or many cells (or blocks) elsewhere in the array. Unless modified, on every replication of a formula, cell addresses used are based on the original formula but with variable addresses shifted by as many rows and columns as the location of each replica cell is displaced from the cell copied. (In practice this is simpler than its description!)
- **Graphs and outputs** Other features include output, to a printer, of the values or formulas in a selected array of cells. Graphs corresponding to numeric data in selected rows or columns of cells can be viewed immediately after any variable parameter is changed. "Canned" sequences of interactive steps, known as *macros*, can be stored and played back repeatedly.

After brief exposure to a computer spreadsheet, nonprogrammers grasp its operation and utility. There is nothing else quite like it, combining visual presentation with instant updating and without programming as such. If, for example, one wishes to compute time-dependent movement of a vehicle using a spreadsheet, one might use a single row to represent a single time. The vehicle's location, velocity, and accelerational forces at that time could be placed in different cells in the row. An updated velocity component could be determined from velocity and acceleration components in the row above, and an updated *position* from position and velocity components in the row above (Table 8-1). After the formulas are defined for one row, they can be copied to all other rows (times) in one spreadsheet operation. Figure 8.2 illustrates a result plotted by a *1-2-3* spreadsheet. The author used the same process to model a trajectory of a hypothetical multistage ballistic missile. Though not adequate for detailed design, it sufficed to illustrate dependence of the maximum range on downrange steering.

In their available forms, computer spreadsheets have proved nearly ideal for handling "what-if" problems in business enterprises of all sizes. These problems are similar to early system engineering problems in that computations are simple but a number of variations must be examined in a short time. A dozen years ago, most such problems were solved by hand on paper and were frequently in error.

Despite sophisticated data-keeping and computational abilities, however, the spreadsheet structure falls far short of meeting many needs in system architecture and top-level design. A hypothetical future

Ballistic trajectory calculation using a spreadsheet

Formulas, as entered in row 5:
B5: 700   **C5:** +C4 + (A5–A4) × B4   **D5:** –32.2
E5: +E4 + (A5–A4) × D5   **F5:** +F4 + **(A5-A4)** × (E5 + E4)/2.

(These are also copied to all rows below.) Row 4 contains only constant (initial) values.

Row ↓	Col→ A	B	C	D	E	F
1	Elapsed	Horiz.	Horiz.	Vert.	Vert.	Vert.
2	time	vel.	posn.	accel.	vel.	posn.
3	(s)	(f/s)	(f)	(f/s^2)	(f/s)	(f)
4	0	700	0	-32.2	700.0	0.0
5	1	700	700	-32.2	667.8	683.9
6	2	700	1400	-32.2	635.6	1335.6
7	3	700	2100	-32.2	603.4	1955.1
—	—	—	—	—	—	—
—	—	—	—	—	—	—
44	40	700	28000	-32.2	-588.0	2240.0
45	41	700	28700	-32.2	-620.2	1635.9
46	42	700	29400	-32.2	-652.4	999.6
47	43	700	30100	-32.2	-684.6	331.1
48	44	700	30800	-32.2	-716.8	-369.6

Table 8.1. Parts of a spreadsheet computation describing the trajectory of a missile in airless space. Rows 8 to 43 are omitted to save space.

capability, closely related to spreadsheets, might permit the user to describe "capability blocks" like those portrayed in conventional *block diagrams*. Each block would be handled internally as an independent entity, having possible "input" and "output" connections with any other block. Block characteristics and connections would be defined by entering descriptive data, perhaps into small spreadsheet-like arrays.

A central feature would be a capability to expand any block revealing an internal structure including a combination of cells containing block attributes and *lower-level blocks*. These lower-level blocks could, in turn, be expanded, down to some lowest level. Clearly, a very desirable feature would be the ability to "zoom" for lower- or higher-level examination of the system (Figure 8.3). When dealing with blocks at any level it should be possible to suppress lower-level structure, for purposes for which it did not matter.

While this might seem a relatively simple and straightforward capability, it raises many subtle realizability questions. A major one is

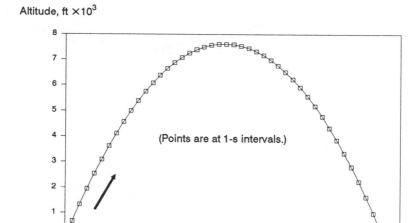

Figure 8.2. Plot of the data of Table 8.1 describing the approximate trajectory of a ballistic object aimed at 45° to the horizontal with a velocity of 700 × √2 f/s. Vertical acceleration is assumed constant at 32.2 ft/s/s. Wind resistance is ignored.

the representation of connectivity between or within blocks. As the level of representation is shifted, how will connectivity be shown? Should "minor" connections (whatever they might be) not be shown? How should the inhomogeneous internal block structure, comprising both cells and smaller blocks, be displayed? How, and where, will the set of *external* attributes of the thing a block represents, such as dimensions, outputs, or color be described?

If these knotty problems can be resolved, the new design tool should also provide convenient ways for the *data and the system represented* to be decomposed by designers. One designer might wish to take data on one partition home, work on it, bring results back, and mate them to partitions studied separately by colleagues. To handle this requirement, the design support tool must enforce interface definitions as defined at the time the system is partitioned. Additional program design decisions now arise: Will blocks at different levels be allowed to communicate (i.e., be connected) directly?

There are many possible applications which should be examined, including comparative system evaluations, feasibility analyses, or computing dynamic system characteristics. The user interface to this tool would have conflicting requirements: relatively simple to use, yet capable for building complex models rapidly.

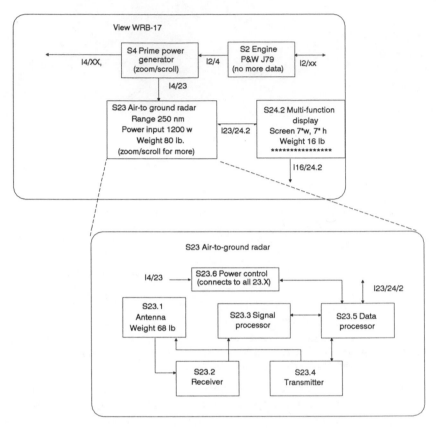

Figure 8.3. This illustrates a "zoom" capability of a possible future system design aid which might store system and subsystem parameters, providing graphics in block diagram form, and enabling computation of system performance and other parameters.

No such capability exists at time of writing, nor can we even present a proof that one is feasible. Yet it seems clear that its utility would be great in systems engineering as well as many other applications.

8.1.4 Control of system configuration data

We turn next to a system design and system management application of computers which is certainly feasible today although not in fully integrated fashion. *System configuration data,* taken as a whole, comprises the combined and refined output of system design, production, maintenance, and life cycle management activities. It includes such items as mechanical drawings, interconnection layouts, parts lists, parts descriptions, specifications and tolerances, part source and ordering information, assembly sequences, calibration data, and test and repair procedures. It should include all data needed to maintain and enhance a system design, its embodiments and evolutions, over a long useful life.

If all systems of a certain design, model, etc., were identical, and they were never altered in any way, a single set of configuration data could describe all. More often, a system design is produced over a period of months or years. Especially with large, complex systems produced in quantities of tens, hundreds, or more over several years, there will be many *different* configurations in the field at one time. Systems are often built separately, installed in different locations at different times, used for slightly different purposes, and maintained by different organizations. In this case, each configuration may be significantly different. Even where there is need to maintain all versions of a system in the same configuration, changes or enhancements seldom take place simultaneously. Here also many different configurations may exist.

Configuration management is a term applied to the work of: managing changes of system design, maintaining records of the current configuration of each system as part of overall configuration data.

Existence of only a single copy of some complex system would seem to simplify its configuration management. But the per-system cost of maintaining system data to proper standards is higher if not shared between many systems. Comprehensive configuration management is so costly and time-consuming that it is often simply ignored by developer and users. However, owners of complex, costly systems usually anticipate long useful lifetime punctuated by occasional updating. Orderly, coordinated, configuration management, in the long term, will be the most economical approach but must be planned for early and put in place *before* the first system is fielded. Its costs should be incorporated into system acquisition and/or operating costs.

Far too often, configuration management, whether done by developer or user, is handled carelessly. This is easy to understand. After a system is turned over to the user, and its initial shakedown completed, it is in its most reliable and robust period. At this point it will be working less intensely than later, when users' needs have grown and the system is showing signs of age. At the time most favorable to practice configuration management, it thus appears much less important than later, when the system suffers from age or obsolescence and can obtain the benefits of having had its initial configuration and repair/modification history carefully recorded.

If the system is built for open sale by a commercial system developer, that developer is best equipped to handle configuration management (presuming it will remain in business). If there is but one customer for the system, after system installation(s) the developer may see no promise of additional system sales, or even maintenance business. The user is the more appropriate configuration manager in this case. If responsibility is not assumed until too late in the life cycle, neither developer nor user will have the data required to assume and maintain configuration control.

Since configuration data assumes a variety of physical forms, separate facilities are required for updating and storing each form of the data. These may include computer databases, engineering drawing files, and hand-written installation, maintenance, or modification records. Not only must these various records be stored, but configuration management must include means to locate or access each record and means to record changes in an efficient and timely manner.

Configuration management for multicontractor systems The formidable problems of configuration management are made even more difficult by the involvement of many subcontractors, each generating configuration changes. New problems arise. A subsystem may have been selected "off the shelf" in the belief that it would be of higher quality and longer life than a specially developed unit. If its developer replaces that product with a new model, and perhaps suspends maintenance or spare parts availability for the original product, reliance on that supplier for maintenance parts will, of necessity, cease. Though the supplier may have performed good configuration management, it is unlikely that the database and other resources will be turned over to a particular user.

A supplier may make internal design changes to a popular product, without altering its model identification. The first warning the configuration manager may receive is when an item in spare parts inventory several years was found not to work when installed in a system. There is no certain, economical way to eliminate this possibility. Suppliers familiar with configuration management, however, are more sensitive to this kind of problem.

If many contractors are responsible for different parts of a complex system, the system developer, system owner (if one organization purchased all copies of the system), or some other party (under contract) may be assigned configuration management responsibility. The task may be a complex one, as evidenced by this scenario: (1) A system fault occurs after some subsystem is repaired or replaced. (2) The subsystem supplier, on request or by threat, sends a representative to study the problem. (3) This individual makes tests, then declares that the problem lies with a second supplier's subsystem. (4) The second supplier, responding, denies the allegation that its equipment caused the fault.

It should be the responsibility of the configuration manager to respond, negotiate, recommend a set of actions to resolve the problem, and if required expedite the action required. To get the cooperation of subcontractors, a configuration manager may need to threaten to enforce contractual obligations or persuade them with reminders of future business opportunities.

8.2 Subsystem and system testing

Little testing, as such, is *performed* by system designers. However, they must plan it, providing not only the tests but their expected results as part of system design. Later in development, system designers should be assigned to oversee, or provide counsel for, the integration and testing steps. The most complex activities associated with integration and testing are setting up the tests and interpreting test results.

Since integration on occasion yields *unpredicted* results, it is in practice an iterative process, not complete until the system has passed all specified developmental tests. After system installation, testing continues to ensure system availability, perhaps including both *on-line* testing and (scheduled and nonscheduled) *off-line* testing.

System designers should be involved in design of or at least verification of design of testing facilities built into the system. Users should not be expected to develop test procedures and equipment for complex systems. Many maintenance tests are local to subsystems. Their design must therefore be the responsibility of the subsystem designer. System designers, however, should be responsible for the overall concept and strategy of maintenance testing. Only they can ensure that the system has balanced maintenance concepts consistent with its expected reliability.

8.2.1 Testing objectives

Testing usually has, as a top-level objective, verification that the system operates *as required,* under the specified conditions. A complex system including software may have thousands, even millions, of potentially testable operating modes and conditions to be tested. Even a simple one may have dozens.

Consider, as a simple example, a pocket-sized portable radio receiver covering only the AM-radio band of 550 to 1600 kHz. Those receiver performance characteristics related to user satisfaction include:

- **Sensitivity** (the smallest signal level which will produce normal output with less than some specified amount of noise in the output, nominally perhaps 50 μV)
- **Tuning range** (the band of frequencies over which signals can be tuned and received, nominally 550-1600 kHz)
- **Signal output** (the maximum audio power which can be delivered to the speaker, while the volume control is turned up high, before distortion in the output reaches a specified percent, nominally perhaps 20 milliwatts for headphones, perhaps one watt for a speaker)
- **Selectivity** (the ratio of signals reaching the output from a transmitter at the selected frequency, divided by that from a transmitter of the same

power level and distance, but at a nearby frequency, nominally at least 100:1 for a station 30 kHz distant)
- **Thermal drift** (the maximum rate, say in kilohertz per second, at which the actual tuning frequency will vary because of internal heating while in use, nominally no more than say 0.5 kHz)
- **Battery life** (the number of hours of playing time for a set of batteries, at average volume and with an intermittent schedule)

To test for all these characteristics requires, *at the minimum* (a) a radio-frequency signal generator covering at least the tuning band of the receiver, modulatable by an audio tone, finely adjustable and accurately calibrated for both output level and frequency; (b) a second similar signal generator to simulate the second station for the selectivity test; (c) an accurate, multirange audio power output and distortion meter having sufficient sensitivity to measure the receiver output; and (d) a clock. With this (non-automated) test equipment, production tests would be slow and costly, however.

The portable radio receiver is trivial to test, by comparison with a large aircraft containing several two-way radios, several large engines, a radar, hydraulic equipment, cabin pressurization equipment, navigation equipment, etc., etc. It might appear that the opportunity for failure would be so great that satisfactory test results might never be obtained. This catastrophe is avoided because: (1) Components and subsystems *are* of high reliability, are tested before being installed in the system, and are given periodic overhauls including stringent testing; (2) system designers anticipate variations in parameters and take positive steps to reduce or eliminate their consequences; and (3) many system parameters exceed requirements for normal, safe operation.

In any system, as many of its subsystems as possible should be fully testable even when *disconnected* from the rest of the system. This isolates any problems within the subsystem, which is desirable, but does not test the part within the system context, which is undesirable. Commonly, test procedures addressing a subsystem believed faulty from system observation involve disconnecting the subsystem and testing it independently of inputs from the system. If a problem is found in the subsystem, it is repaired or replaced. Ability to rapidly isolate each subsystem for independent test is a valuable architectural feature.

It sometimes happens that a fault which appeared during system testing *cannot* be found by isolated testing of any subsystem. In this case, it is most likely a result of interactions between two or more subsystems whose parameters are faulty or out of adjustment. However, the problem may still be observable only when the responsible subsystems are tested *together*. Such problems are difficult to isolate. Once identified, if such a problem appears likely to recur, design changes may be in order.

Stressed testing If, during normal operation, systems are subject to possible damage or destruction because of physical stress, representative stresses should be incorporated into the proper phase of system testing. Conditions of operation such as high peak system workloads can cause *temporary* system malperformance. These must also be understood and represented during both developmental and operational testing.

Physical stress may take a variety of forms: heavy mechanical or electric power loading or dissipation, rapid acceleration of change of timing, abnormally high or low temperatures, pressures or humidity levels, or the occasional presence of explosive fumes.

Test stresses which could destroy the system or damage more than merely a weak part should be avoided, except when destructive testing is intended (see the next section). Other stresses may be introduced routinely during integration testing and in routine operational testing. *Every predictable* stressed operating condition likely during *normal* system operation must be thoroughly explored, at appropriate times during development and integration.

Imposing physical stresses which occur naturally in system operation may require complex test apparatus, such as "shake tables," to simulate the working environments of mobile systems. Test operation at extremes of atmospheric temperature, pressure, and humidity likewise requires special facilities. Ideally, *each combination* of stress to which the system will be subjected in operation should be simulated during testing. This is seldom fully practicable. When it cannot be done, the best alternative is to identify, and act to reduce, critical *stress sensitivities* of the system. Parts which could vibrate loose should be redesigned, the cooling improved in parts of the system most likely to overheat.

Workload stressing is also challenging to simulate in system tests. Often, very complex test conditions must be generated. An example is a military aircraft's *radar warning system.* This system may operate by examining each radar pulse it detects, sorting them by frequency, length, approximate direction of the source, or other attributes. In some regions of the world there are hundreds of radars, near enough that their emissions can be detected simultaneously. If each produced on average 1000 pulses per second, there could be hundreds of thousands, possibly millions of pulses per second to be analyzed. This sort of environment is difficult to simulate for testing. Test apparatus must simulate hundreds of independent pulse sources, without introducing correlations absent from the signals from independent radars.

A similar problem would be encountered in testing a computer system on which hundreds of users might request data within a few seconds. Unlike physical stress, workload stress should produce no *permanent* damage, *to a correctly designed system*. Information systems, however, sometimes "crash" in ways which destroy data and programs as a result of stressed workloads. Such behavior should be studied, by representative workload simulations during developmental testing, to identify sources of irreversible changes which would produce catastrophe in system operation.

Destructive testing System damage or destruction are the expected results of certain tests. It is common, for example, to test new aircraft structural designs "until something breaks." This may be done to verify the theory of structural strength on which the design was based. Or, it may serve to verify that production processes are understood and controlled.

Certain systems are intended to be *expendable* after one use. Most of these are developed for the military: missiles, bombs, and other ammunition, flares, etc. Where testing results in system destruction, special testing processes and equipments are often required.

Systems such as space rockets or military guided missiles cannot be examined well enough after operational testing to locate subtle sources of failure. Test facilities must examine these items *during* the test. If events happen rapidly, test equipment must be able to record the dynamics. Systems are often examined during tests by installing sensors and telemetering transmitters in system volumes normally occupied by unneeded subsystems, fuel, explosives, etc. All *measurable* variables of interest are sampled, digitized, and transmitted to ground-based or airborne instrumentation receivers, while the test object seeks a target or performs maneuvers. (Figure 8.4 illustrates representative elements in this sort of test.) Expendable test items must often be built in quantities of 10, 20, or more, enough "shots" to verify correct system operation. Because of instrumentation complexity, tests and their planning require exceptional efforts. Their costs are correspondingly high.

Destructive *load tests* of mechanical structures are carried beyond the elastic limits of structural parts. (They of course make the structure unsatisfactory for further use.) These tests are usually carried out in permanent instrumented test facilities. The cost of the test item(s) and weeks or months of time of test personnel, as, for example, for wing or tail structures of a large aircraft, make major load tests expensive.

Structural designers have a difficult objective in designing three-dimensional mechanical structures: to make them fail above, but not *far* above, the calculated load limits. This capability has improved with use of computer-aided design. The accuracy of predictions depends on the precision of the fabrication technique and how accurately it can be computer-modeled. Structures assembled from standard structural shapes (I beams, etc.) and fastened with bolts or rivets of known strength can be sized more precisely than those involving more complex fabrication techniques, such as adhesives or plastic/fiber composites.

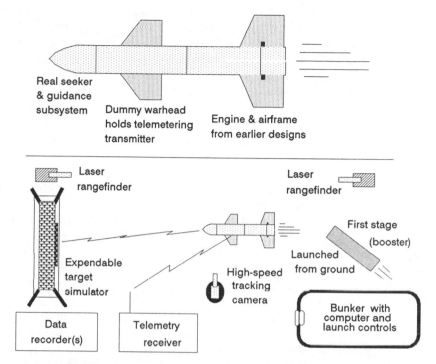

Figure 8.4. Testing a system normally intended to destroy itself, such as a guided missile, requires elaborate real-time sensing, telemetry, and safety provisions. If a missile normally fired from a high-speed aircraft is to be ground tested, it must usually be fitted with a booster rocket to bring it up to more than an aircraft's speed.

During operational lifetimes of expendable systems such as missiles, system-level testing should continue periodically to verify that deterioration has not disabled them. (Expendable systems often cannot be disassembled for testing.) This may take the form of periodic test firings of weapons equipped with nonexplosive warheads. Since modern military weapons might be retired before being fired in war, test firing is usefully combined with operator training. Missiles replaced by improved designs may be expended for training, or serve test vehicles for new missile components. Frequent failures of expendable systems during use or testing suggest poor design, improper storage or bad maintenance.

Establishing test environment   In testing of installed systems, systems being integrated, or of subsystems, certain environmental features normally present during system operation will often be absent. Presence of numerous human operators or of system inputs representative of actual operation, may have to be *simulated* for testing. Ignoring them makes realistic testing impossible.

In a testing context, simulation may be simple or extremely complex, depending on what must be simulated, and with what fidelity. As a simple example, imagine that some system input should receive alphanumeric character strings, at rates not important to correct system operation. This input signal can be simulated by a terminal keyboard capable of producing the proper signal format. Test sequences may be keyed one character at a time, while their effects are observed. This should permit verification that the system under test responds correctly to each keyed test sequence. Where timing is not important, manual inputs are often an effective way to examine a few responses of a system to key input, switch closings, light impulses, and other *single-source* stimuli. If the number of test sequences is large, it may be worth the cost of preparing to both input them and record the results with computer assistance.

If several simulated inputs must be synchronized, or a single input requires rapid or precision timing of successive inputs, manual simulation sequences will not be feasible. Don't give up yet, however. . . If the system described above operates at high data input rates, it will probably be designed to abort when unexpected delays occur between successive character inputs. This could occur if inputs were manual as suggested above. In the interest of simplifying initial testing, it may be desirable (if practical) to *disable the system timeout feature* and test manually at a slow character rate. Once the system's *logical correctness* is proven—or corrected—a similar test at higher speed will often not be required.

This is one example of a powerful test principle which is often applicable: *change the system under test temporarily, if it will simplify early testing.* After you have established correctness without the complication added by the disabled features, they can be reenabled. A relatively simple additional test can usually demonstrate that operation is still correct. There are many ways involving hardware or software to *disable or isolate* features which inhibit easy testing.

Test input sources   In years long past when electronic and mechanical systems received principally continuous (analog) inputs, the conventional input simulator was called a *signal generator.* This device usually produced a sine-wave signal of controllable voltage amplitude, or perhaps periodic sawtooth or square-wave signals. The "shake table" is a mechanical analogy. This type of apparatus, in more elaborate forms, remains in use for dynamic tests of analog systems and subsystems. Control and signal inputs to modern systems, even mechanical ones such as jet engines, are now mostly *digital,* however. Most electronic or electronically controlled *analog* test equipment used today in system-response testing is also digitally *controlled.*

This revolution has been spurred by availability of compact high-speed microprocessors. The special "tools" for this sort of test simulation are often programs which may be executed in response to a triggering input, or which operate continuously in a program loop. Some test instruments are "programmed" by setting control switches or selectors. Others require a computer program to be written. Simple documentation should be incorporated with even such simple *ad hoc* test programs, so they can be understood for independent verification or modified for slightly different applications.

A modern system test environment should make it easy for test personnel to create simple simulations, or modify more complex ones, during the testing process. *Testing must be responsive to what it reveals.* Ability to modify tests quickly and accurately is one key to rapid, efficient system integration. Certain system environmental factors require elaborate simulation and involve major capital investments. These installations may include indoor or outdoor *test ranges*, the generation of large numbers of synthetic electromagnetic signals, visual scenes which can be displayed to a system operator through a display simulating a port or window, and control/display consoles manned by test personnel who simulate operators of the system under test, or of external systems with which it interacts.

As testing moves up from small subsystems to larger complexes and finally to the complete system, the nature of the tests usually changes. Early testing is devoted largely to technical measurements (image resolution, processing rate, cruising speed) and *robustness testing*. In this latter, a test sequence may be repeated for millions of cycles or until the unit under test wears out or breaks. An example is a door-slamming tester for automobiles which repeatedly opens and closes a developmental door, until something breaks. As testing moves to higher levels of integration, focus shifts to interaction of the system's parts. Finally, in operational testing, emphasis shifts to interaction of the system with users and environment. In automotive testing this corresponds to test tracks, and to employees driving new models long prior to full-scale production.

Tests involving human subjects Systems which require presence of human operators in normal operation usually require them for certain tests. In particular, human subjects are needed in tests where they operate system controls and displays to test ease of use, or if humans are "in the loop" during normal system operation and the system does not operate correctly without them.

Human-subject testing may be separated into two categories: (1) tests in which skill and speed of the operators are unimportant to test results, and (2) tests where skill or speed influence the test results. Tests of the first type will usually be carried out with system development personnel performing as user surrogates. This is convenient and minimizes confusion. Where operator skill and timing are important, subjects should be representative of actual system operators.

Humans have much higher error rates than machines, and usually lower precision. The system under test usually has no way to determine if a human subject correctly performed a test action such as pressing a given key sequence. Correct operator actions must in such instances be verified by human observation or video recording. Where several operators and test coordinators are at various locations around a system under test, an *audio intercommunication system* ("intercom") is essential for coordination and for system and personnel safety. Its activity may also be recorded, preferably including time information so that event sequences can be determined accurately.

In this kind of intricate test, there will often be instrumentation recorders as well, to capture sequences and values of signals, switch-settings, or other features of system operation. All records should be synchronized by signals from a single timing clock.

The most subtle involvement of human test subjects is required when humans are in the loop and system performance is strongly influenced by their speed or accuracy. A system developer, given opportunity to improve the system's score, might elect to use intelligent, well-trained subjects with long experience in operating the system, i.e., the firm's own engineers or technicians. The system purchaser is most interested in how well the system performs when operated by the personnel who will actually operate it. Comprehensive testing requires use of operators whose intelligence and skill profiles match those of expected system operators. Each should have been subjected to a training regimen representative of that expected to be provided to actual system operators. There should thus be a strong incentive for the developer to put together an effective operator training course or to determine that an appropriate course is available in time for operational testing.

A complex system test involving planned sequences of inputs and time-critical operations by human subjects demands well-laid-out choreography. A master clock should be used to control activity. Test director(s) can be called upon, by cues driven by the clock, to initiate stimuli at particular times, to read informational messages, or to respond to representative operator inquiries. Some test scenarios may last several hours, as, for example, in simulating a long military mission. When sequences are very long and involve many human activities, it may be necessary to separate events into those (if any) which must occur at a particular time, and those which may be delayed because of variation in human (and sometimes, system) response times.

The *direct* record of a system test scenario involving many human subjects may encompass hours of recordings in various formats and hundreds of pages of time-annotated notes of observers. Comprehensive analysis of such a record involves transcribing audio and video information into written form, editing, and integrating it, identifying key steps and results and adding descriptive commentary. Carrying out these involved and costly tests without good follow-up is wasteful. Synchronized timing permits tentative removal of uninteresting segments prior to major analysis work.

8.2.2 Test design

Chapter 5 deals with design *for* testing, the selection of design features which aid in testing or reduce necessary tests. Here, we examine the design *of* tests for use during system development, operational assessment or operation. What constitutes adequate testing, for a given system? This question has no single answer, since different test activities have different objectives.

Predevelopment testing is part of research or exploratory development projects, intended to confirm system applicability of a technology or realizability of a system capability.

Development testing serves principally to validate correctness of the design and of the system's integration, and the meeting of system specifications.

Operational testing attempts to verify the system's usefulness for intended applications in a test environment identical to the intended operating environment, or nearly so.

Maintenance testing verifies correct operation, continued adherence to specifications, locates system faults which occur during operation, and may also be employed to recalibrate the system.

Each of these four testing situations requires different test design approaches.

Exploratory or prototype developments prior to system development are frequently based on one-of-a-kind, short-lifetime test articles. More cost-effective demonstrations (more "bang per buck") are possible if selected test parameters permit simple, convincing demonstrations rather than also attempting to meet system requirements unrelated to the demonstration objective. These demonstrations inherently have great uncertainties, and there is no need to add to them unnecessarily.

Consider, for example, a prototype fighter airplane, without mission equipment or weapons but with major structural innovations. This may be put together for testing, for 20 percent or less of the cost of developing a producible, mission-capable system.

Approaches to integration testing carried out as part of design also must deal with uncertainty, especially if the subsystems, system concepts, or applications lack precedents. A step-by-step process of integration ("build a little, test a little") has consistently worked better in practice than assembling the system completely before serious tests. Late arriving subsystems or parts often complicate matters, however. Simulation of these missing elements is feasible only if their characteristics and performance are *fully* understood. If that is not so, it could be more prudent to delay integration activities until a more orderly process is possible.

Operational testing's approach is straightforward: Test under the system conditions in which the system will be used. If the operating environment is well understood, along with the system functions, tests can be relatively simple. If either or both are without precedent, it may be necessary to carry out operational tests in a highly simulated, totally synthetic environment. (Consider the problems of early testing of a "Star Wars" anti-ballistic-missile system.)

Routine maintenance testing and system calibration continues, often for many years, for *each* duplicate system. If the system is built in large numbers, this component of testing accounts for more total resources than all other testing combined. It is pertinent to the user that this testing be overall cost-effective: Less testing may mean more system down-time. In general, there is an optimum. If able to anticipate long-term test needs, the system developer can often justify development of special test equipment of use both during development and later in system maintenance. The best approach to maintenance testing design is strongly dependent on the maintenance budgets available for personnel and equipment. If these are too low, the system will be continually in difficulty. If they are generous, special diagnostic equipment and well-filled spare parts inventories will allow diligent maintainers of limited skill to make a system perform well.

How much testing is enough? The most difficult task in test design is establishing *how many and what tests* to perform, and what to measure. If a system appears to be operating correctly after study of a few key parameters or collection of limited test data, how much further test is needed? How safe is an inference of system correctness, when made from limited test results? Developers prefer to limit tests because they are costly and time-consuming. Also, testing with exceptional vigor sometimes unleashes unpleasant system characteristics. Even if they would not influence normal system operation, once revealed their existence cannot be escaped and it may be necessary to eliminate them.

The answer to "How much is enough?" will differ with the conditions and objectives of particular testing. There *are* analytical processes available, based on probability plus a clear understanding of what *can* go wrong, to assess how much is enough. In testing the sole example of a large system for only a limited period, the test data alone will be deficient as a basis for conclusive assessment of system capability. Previous demonstrated performance of its subsystems and analogies based on tests of similar systems and environments must often be combined with test data, to form a credible overall conclusion.

Predevelopment testing (of prototypes, for example) is typically carried out to determine if a system *could be designed* to perform a certain function or achieve certain performance. Short of thorough testing, you have wasted the funds spent on test articles. After nondestructive tests are complete it may make good sense to carry some tests through to failure or destruction of the test article if this would improve knowledge of the system functions demonstrated.

Developmental testing, too, must be relatively thorough. How exhaustive these tests should be depends on the marginal value of obtaining additional data. In most cases, bottom-line economics determines how comprehensive developmental testing will be for a system. Not surprisingly, management often favors limiting testing, while engineers typically argue for more. Though developmental testing is considerably system-unique, we will suggest a rule. A first set of developmental testing should carry tests to a point at which there is positive evidence that every performance requirement has been met and that each function is performed correctly in isolation. Additional tests are designed expressly to test *interaction* of primary parameters or functions, while acting as a partial second check on individual parameters. For example, when testing a new aircraft, all features are first tested on the ground. The first flight exercises only simple take-offs with almost immediate landing. Later flights first exercise individual features, then test them in combination, to complete the test repertoire.

Throughout development, and even in maintenance testing during the system's operating lifetime, most tests are based on *technical performance* parameters. These are measures such as speed, power output, range, or capacity. In *operational testing*, however, emphasis is on *user-pertinent* parameters, measured in a complex real-life *user-pertinent* environment. Operational testing must be thorough enough to demonstrate that the system can perform required *functions* to required levels of *performance*. These functions and measures may be complex. For example, the function might be the production of artwork by human artists using a graphic design computer system. The performance measures, say, *production rate* and *accuracy*, must be defined by or for the user. Assuredly, they will be only indirectly related to the speed of the computer, and not at all to its accuracy. In the case of a weather radar, developmental testing may relate to power output, receiver sensitivity, and antenna pattern. Operational testing will require that it detect dangerous weather with certainty, far enough away that course changes do not greatly lengthen a typical flight in thunderstorm weather.

Plans and schedules for development testing Development tests accompany and follow system integration. These activities should be structured and tentatively scheduled after system partitioning, and as early as reasonable estimates of subsystem availability dates can be obtained. Schedules for availability of special test facilities and test personnel may be constraints on integration and testing schedules.

To save elapsed time during development, integration and testing activities are often pursued in parallel, for groupings of subsystems. Effectiveness of parallel integration/test depends greatly on how independent the subsystem groupings are during normal system operation. In the limit, one could test single subsystems by simulating *all* interfaces with other subsystems. Aside from the technical difficulties imposed, this would typically be too costly and risky a plan. Usually a workable parallelism in integration will be almost obvious from study of the partitioned system. Those subsystems which must operate most cooperatively should be assembled as a unit. If key subsystems in these groupings will certainly arrive late, they may need to be simulated for early integration activity.

Test activity during system integration is often punctuated by discovery of design or production errors. Some can be repaired quickly while others require longer efforts. The most serious may require significant redesign, followed by hardware or software changes and additional tests. Iteration *must* be allowed for in test schedules, and this admits that schedules will change. Different schedule-management techniques may be favored by different managers. These differ mainly in the way *schedule slack* (the time available, if everything proceeded without a problem) is allocated (Figure 8.5).

In the most aggressive schedule-management style, the slack time remaining is never allocated. Assume it is expected that integration and related testing can be completed in no less than 60 nor more than 90 (working) days. A 60-day schedule will initially be created. Rescheduling, with possible extension of estimated completion date, will occur if and when needed.

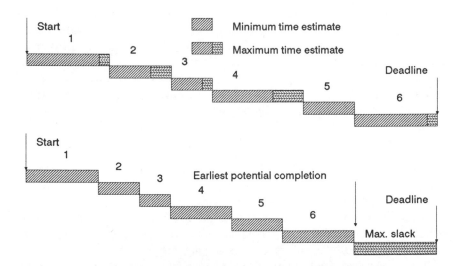

Figure 8.5. Two approaches to scheduling activities in the presence of uncertainty. In the upper, each activity is allowed its maximum time. In the lower, the total difference between minimum and maximum times for execution of the tasks is put in reserve and is allocated only if necessary. The second approach usually saves time over the first; however, the manager must be prepared to yield up some of the available "slack" time.

The least aggressive schedule management style schedules, from the outset, for 90 days, dividing the slack uniformly throughout the schedule. (Intermediate approaches may initially allocate only half of the slack, or 15 days.) The more relaxed management styles produce fewer crisis exercises ("fire drills"), and put less pressure on employees. However, they do not provide a way to capture savings available through exceptional performance, too-generous initial estimates, or plain good luck. If one integration team sees itself having slack, it will assume a more relaxed posture since it must otherwise wait for others to catch up.

Poorly planned or ineptly managed integration and testing activities may extend development completion by months or years. When a program gets months behind schedule because of integration/testing iteration, its chances of recovery are often slim. By this time, its original design team will have dispersed to newer tasks. Test personnel will lack the motivation and skills to propose or carry out major redesigns. Management must remain alert, and take timely and intelligent steps to avoid the project's losing forward momentum.

Operational testing: environments and scenarios Generic subsystems such as motors, amplifiers, and switching centers are well characterized by *performance specifications*. This is also true of some general-purpose systems, such as data processing systems or utility vehicles capable to carrying almost any kind of load.

Even most special-purpose systems usually have a few performance specifications against which they can be tested: speed, size, weight, range, and the like. Performance specification often fails to properly capture the needs of users, especially when applications are specific and objectives complex. This is often the case with military equipment, where both offensive and defensive capabilities are required in many weapon systems.

Application-unique requirements are difficult to describe in generic terms. Airframe and landing gear of Navy sea-based (carrier) aircraft are subject to landing stresses much more severe than those characteristic of ground landings. If the scenario (keep power up, drop it in, catch the cable with the tail hook, be ready to go around if it doesn't work) is not reflected adequately in system tests, those tests will not be representative. A poor way to express this requirement would be to specify "landing gear able to support twice the weight of the aircraft." The right way is to require test aircraft to execute representative numbers of actual carrier landings without damage. An even better way is to incorporate these landings in demonstrations where test aircraft carry out complete and representative missions. The *measures of performance* which are characteristic of generic systems or subsystems must, for operational testing, give way to *measures of effectiveness* based on performance in actual system operation. A useful measure of effectiveness of a giant strip-mining shovel, for example, would be how much earth it could remove from a pit of certain dimensions in one day. One for an assault rifle might measure how many shots it could place inside a 6 in circle in 3 s, in the hands of an average marksman.

Operational tests often demand sophisticated and expensive facilities. For testing in the Apollo lunar landing program, contractors built huge vacuum chambers in which full-scale spacecraft could be suspended for tests. Testing military systems against nuclear weapon radiation effects has for many years been carried out underground in excavated chambers. An adjacent chamber contains a real nuclear bomb. A long tunnel connecting the two is closed after the radiation event, before the blast effects reach the test area.

Except in rare cases such as those involving nuclear weapons, there is no better full-system test environment than that in which the system is to operate. Tactical weapon systems are tested operationally on huge *ranges* populated with representative targets and defenders. If such facilities are not affordable, partial or total *simulations* may be used instead. In a partial simulation, some of the elements of the real environment are missing. In total simulation, many parts of the system under test may even be simulated by computers.

Environment and operational scenario simulation often lacks details present in the actual situation, but it also has several advantages over real environments. Simulated environments can if desired be made more stressful (to equipment) than the current real environment. (This allows future evolution to be simulated.) Simulations not involving human subjects can be repeated in complete detail, as often as desired.

Maintenance testing and repair strategies Maintenance testing should not pose the same uncertainty as that for a system still in developmental tests. However, neither will most maintenance personnel be as skilled and experienced as developmental test personnel.

System maintenance environments differ in terms of: the importance of rapid system repair; availability of repair personnel and their skill levels; comprehensiveness of available test equipment; proximity of repair facilities; and the stocking of spare components and larger system elements. The selected maintenance strategy should match the limitations of the operational maintenance environment. Systems acquired should be compatible with that strategy.

Major maintenance strategy alternatives are associated with particular repair or replacement strategies. These (Figure 8.6) merit further discussion.

1. **System replacement strategies**: Non-essential equipment which is intended to be discarded when it fails, or returned to its manufacturer for repair, may have *no* prescribed test procedures or criteria. When

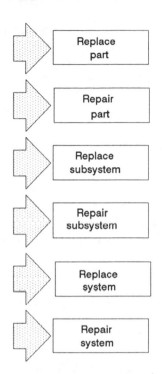

Low-cost part, subject to wear, or cannot be removed without high risk of destruction.

Replace part

High-cost part, worn portions replaceable, designed for removal, no damage problem during removal.

Repair part

Unrepairable, or high cost to repair subsystem, or subsystem to be repaired remotely from operating location.

Replace subsystem

Subsystem designed for fast repair by operational maintainers, or not easily removable from system.

Repair subsystem

System damaged beyond repair or must be sent to repair facility.

Replace system

System repairable, designed for easy repair in the field, or replacement of parts planned during scheduled maintenance.

Repair system

Figure 8.6. Field maintenance strategies may replace or repair at part, subsystem, or system level, depending on the capabilities available in the field. While replaced small parts may be thrown away, most parts and subsystems will be repaired or returned for repair. The cost of maintaining a full set of spare subsystems for almost immediate availability may double the cost of ownership.

users discern that the equipment does not perform correctly, it is replaced or returned for repair. Most small consumer products (kitchen appliances, cameras, watches) are, in fact, dealt with this way, since repair has become more costly than replacement.

More costly equipments may be replaced, the failed unit being returned to a repair depot or to the manufacturer for repair if feasible, or to be scrapped if unrepairable. After repair it may be returned to inventory. If the defective unit was "traded in" to the manufacturer or a rebuilder, it can be marketed as a "rebuilt" unit. This is probably the most common maintenance strategy in use today for using limited local maintenance resources (and, often, no repair resources) to handle complex systems.

2. **Subsystem or module replacement strategies:** Many complex subsystems are too costly to afford maintaining an inventory of complete units. Or, they may be too heavy or bulky to return conveniently to a remote repair point. Such equipment or its users may be supplied with means to locate failed *modules* within the equipment. Modules are expected to be easy to access and replace. Systems for which

module-replacement maintenance strategies may be employed should be provided with simple means to verify module failure. However, the most common test is to replace each suspect module in turn with a module of the same design which is *known to be operable*.

The module types constituting all failure-prone elements of all such subsystems *must be inventoried on-site* if rapid repair is essential. Otherwise, replacement modules may be obtained from suppliers only when needed. This is now a common maintenance strategy, if limited or no maintenance capability is available in the operational environment. It is widely applied to military systems and to computers of all sizes including personal computers. It usually assumes a single-failure situation. It is useless where the system has been suffered from massive physical damage or where there has been failure in apparatus which must operate correctly for valid testing.

3. **Parts-replacement strategies:** Although subsystem- or module-based maintenance strategies are the most common at present, part-replacement strategies may still be used in situations where: systems are mechanical in nature and huge in size, such as in metal-rolling mills, printing plants, or for farm machinery; older systems are employed, which are nonmodular; systems require parts not available from manufacturer or other suppliers but which can be manufactured by the system user; the failed parts are generic hardware (such as fasteners, wire, cable, or tubing), or *expendables* such as paper, sealants, lubricants, or fuels. Failure of a minor part may render some systems useless until it is replaced. Even the most modularized systems always contain some parts (usually long-lived, however) not contained within readily removable modules.

Certain relatively simple parts, for example, a pressure-seal between two subsystems, or connecting bolts which connect subsystems together, sometimes give rise to maintenance problems far out of proportion to their replacement costs. A common reason for this is shortage of design attention, during design and initial integration, to system elements other than major subsystems. For example, many system designs use large numbers of tiny machine screws of various types and lengths, often to close cabinets or retain access panels. This virtually guarantees that service personnel will frequently drop or misplace these parts. Designs intended for easy maintainability should employ rugged, easy to operate, captive connectors, which are less likely to be damaged and impossible to lose.

Built-in test and test equipment To support preventive maintenance and achieve high system availability, it is essential to include, *within the system*, devices which perform "built-in test." In automobiles, the familiar battery voltage, radiator temperature, and oil pressure gauges are built-in test devices. So are the so-called *idiot lights* which identify limiting thresholds of these same parameters or signal possible engine failure. These pieces of built-in test equipment perform *continuous* tests of certain parameters. They assume a vigilant operator and take no action beyond producing an alert signal. There are many critical automotive parameters for which there are no test indications: oil level, brake-fluid or engine coolant quantity, tire pressure, or operability of electric lamps.

There is scant evidence of attention to built-in test in the designs of most consumer products or systems. Many hot-water heating furnaces are provided with temperature gauges, though these are infrequently checked and less often understood by owners. Designers understandably resist installing test indicators which a typical user will ignore or which might confuse many owners. Safety-critical consumer products, though seldom equipped with built-in test indicators, often have safety devices which disable the units if critical thresholds (of operating temperature or electric current, for example) are exceeded during operation.

Commercial and military systems whose failure could result in large loss of life or other major loss usually have more capable built-in testing. Systems containing programmable controllers or other internal computers already contain most of the parts needed to analyze system problems, lacking only sensors and an appropriate software routine.

Users of large or complex systems whose malfunction could have costly consequences have a right to expect some form of built-in test capability throughout the system. This is difficult to implement if subsystems of varied origin and design ages are combined into a system. Newly developed systems should, wherever reasonable, have a specific *built-in test architecture* with test capabilities coordinated with repair or replacement strategies.

Maintenance support has occasionally employed as a sales gimmick when the system had poor capability. Built-in location of system failure only down to a subsystem containing dozens of replaceable parts is better than no clue at all. It still forces the user to carry out complex maintenance activities. Subsystems should be obtained from suppliers who will support system test concepts (Figure 8.7).

8.3 System standards and design rules

A large and complex system should derive a certain unity from its architecture. One feature of an architecture is a hierarchy of *design concepts* addressing various levels in the system's hierarchy.

Below the level of broad architectural features of a complex system, there should be a set of rules to be followed *by those actually designing the system*. If these did not exist, the designers of each part of the system would surely introduce many unique variations in design details. Most of

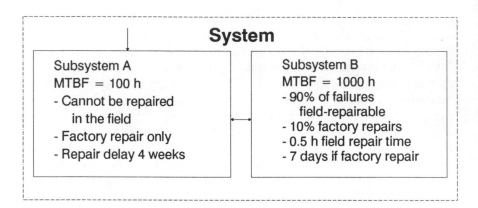

Figure 8.7. Though it is in some instances unavoidable, a system containing two subsystems as different as these in MTBF and repair attributes will complicate system ownership.

this would occur for no more important reason than that designers must make hundreds of detail decisions. Individual designers' preferences would be the basis for most such decisions.

We arbitrarily separate these details into two subsets, on a basis of complexity. The term *design rule* is often applied to a design practice *mandated within a corporation or design organization*. The term *standard* is usually applied to practices which are applied more widely, perhaps among firms in a particular industry, perhaps edicted by a national government, or less often agreed upon by representatives of several nations. Terminology here is not uniform. Though many industrial companies refer to their design rules as "corporate standards," we reserve the term standards for practices which transcend corporate or wider boundaries.

Both (1) standards and (2) design rules require *codification* in documented form. Adding (3) *patents*, these three categories represent the majority of non-product-specific, formal documented system design knowledge. A reader seeking design knowledge through professional-society publications will find it mainly in the organization's published standards. The technical journals are mainly the domain of academic researchers. General and survey publications provide little design information. Most industrial firms guard design information as proprietary matter, do not encourage its publication. One result is that design engineers get little practice writing for outside readers.

Useful snippets of current design information can be gleaned from trade magazines whose income is derived solely from advertising. These are available, free, to those active in the industry. Industrial standards, after their approval by the standardizing bodies which publish them, can be purchased from those organizations, which include the major engineering societies, some trade associations, and national or international standards organizations. Design-rule documents must be readily at hand to all designers within a firm. Copies of this material, however, are usually closely controlled to keep them out of the hands of competitors.

8.3.1 Industrial standards

Industrial standards are usually concise, formal documents ranging from a few to a few hundred pages. Some are framed in "legalese" terms, to avoid ambiguity. Typically, each is a stand-alone document describing a system interface, some set of definitions of product terminology, or perhaps a particular set of parameters (such as screw diameter or thread pitch) which will be used for certain types of products.

Pragmatically, there are few fixed bounds on what standards may treat. Though some represent long-absent codification of *ad hoc* standards in practical use for years, most are prepared with an eye on the future. Limitations which would quickly obsolete the standard due to advances in technology are avoided, where possible. Though there is always an attempt on the part of the standardizing committee to be thorough, differences of opinion between its members on minor points often leads to intentional omission of detail. This may leave designers leeway to develop different systems, all compatible with the standard though not with one another. Such experiences should be reported to the responsible standards organization, or to members of the responsible committee, if they are identified, and the committee is still active.

Though some design engineers will view a standard as a constraint on design freedom, many industrial standards are valuable sources of design guidance. They can save large amounts of design time. There may also be related handbooks and supporting materials available from standards organizations or book publishers. Published standards or *open architectures* invite commentary and explanation.

Many standards have been around for many years and are outdated. A copy of such a standard might yield a different impression since the enthusiasm of original developers still appears in its pages. Standards publishing organizations will if asked provide guidance as to what standards have been functionally replaced by later ones. Though this guidance might also be obtained from designers currently active in the same field, like other professionals not all system designers keep current with changes in their fields.

Care is required in selection and application of standards to make sure that your intended application is within the system context for which a particular standard was intended. It seems unlikely that a standard intended for aircraft application would be mistakenly applied to seagoing vessels (or the converse), but it could happen, quite possibly with poor results. A standards user should examine any standard for clues revealing whether it was intended for the proposed application, or would be in any way limiting.

Active and well-known free-lance designers, or design managers of a firm involved in certain areas of design may be solicited for work in standards activities. This work should be viewed as a responsibility of being a recognized authority in a design field. The sense of accomplishment is different from that which comes from well-received design work. It also may involve negotiation and compromise between proprietary views. Limited amounts of such activity add stimulation. Being unpaid, too much of such public-service work can get financially painful.

8.3.2 Standards used in government system acquisitions

Standards are also prepared by the government for use in its acquisition of supplies or equipment. The U.S. Department of Defense publishes elaborate standards, not only for generic items it purchases but also for almost every imaginable aspect of the special-purpose systems it acquires. A billion-dollar Defense system may, in its statement of work covering the system's development, reference 40 or more *military standards* as well as related handbooks and many other directives informing prospective bidders how to respond.

Military standards are intended to pose design constraints. In the main they are well founded, based on diligent work by knowledgeable engineers—government employees or contractors hired for the purpose. The standards which are cited most often are the most generic. For example, the military aviation electronics temperature-range standard specifies temperature range over which an electronic system must be able to operate and to equal or exceed the performance required for the particular system.

A chief difficulty with military standards is that they are frequently overemployed. The specifications for a system which will be used only in a large, air-conditioned room may reference the aviation electronics temperature standard. A typical reason for this is that the government engineer who cited the standard was an avionics specialist unfamiliar with any less stringent standard. Assuredly, the requirement should certainly meet these less stressful needs. Devices which satisfy the stricter criteria are usually selected from commercial products and may occur only once in every 1000 devices acceptable by commercial standards. The result is far higher costs to the government not only for the original acquisition but for each spare purchased over the lifetime of the system. Overspecification via standards, unfortunately, is not a rare occurrence, even in large industry. An engineer responsible for setting specifications may see no reward in practicing economies. System failures which might be traced to less stringent specifications will certainly exact some personal penalty. On those terms most people would opt for self-defense.

Ad hoc standards The term *ad hoc standard* usually applies to an interface design which originates in one organization and spreads to others. *Ad hoc* designs of broad utility and careful construction are often accepted widely without formal description or the critical reviews accorded industry-approved standards.

For a particular design or design element to become successful as an *ad hoc* standard, its originators must be willing for others to use it. Threats to take legal action against any who applied the design will cause others to turn toward a design promising no legal problems. A firm which was first to evolve an interface design may take satisfaction in it becoming an *ad hoc* standard. This is especially so, if the firm's name is used as an identifier and if it appears that its business will be improved as a result.

The great weakness of reliance on ad hoc standards is that, until (and unless) they are given official status, essential details may be left to individual users: a voltage level, the diameter of a bolt, number of circuits in a connection plug, etc. Full compatibility of a design with other versions of "the same" design may not occur without specific effort.

Ad hoc standards are often far less well articulated and complete than formal ones. In fact, there may not even *be* a written description. The initial developer of what becomes an *ad hoc* standard designs it for a particular purpose and may not be concerned about generality. The next users may have a slightly different requirement and see fit to "enhance" the standard.

One of the clearest examples has been in computer programming language development. Traditionally, most programming languages have originated with an individual or design group who also design(s) a language translator, i.e., a *compiler*. Initial versions of the compiler typically lack features thought important by other users. They change and extend the original language to suit their own needs or views. Eventually it may be given legitimacy, in the United States, typically through *American National Standards Institute* (ANSI, which is not a government agency). However, the official standard may define only a limited *common subset*, relatively primitive by comparison to some existing implementations. The organizers of the DoD programming language *Ada*, whose development was initiated in 1977, avoided that problem by registering *Ada* as a trademark which may only be used to describe *full implementations* verified by government-approved product tests. The long evolution typical of most successful programming (and other) languages was intentionally avoided. At time of writing, DoD had, however, not determined how to solve the problem of achieving efficient real-time computing with so complex a programming language.

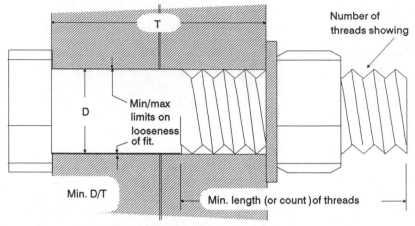

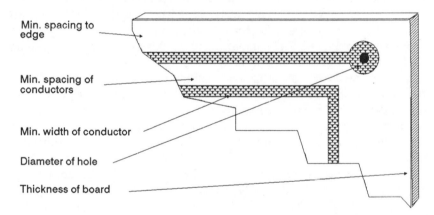

Figure 8.8. Design rules specify recommended or required practices. Those illustrated are all mechanical in nature, applied to a mechanical assembly (top) and electronic circuit board (below).

8.3.3 Design rules

Design rules (a few examples are depicted in Figure 8.8) are frequently established within a firm's design or manufacturing organization for use in both activities. They are of interest to test personnel as well. They may be afforded the same status as standards, if enforcement is feasible. Interface standards enjoy a natural form of enforcement; the design groups designing the two halves of an interface in essence enforce one another's adherence to standards. Design rules may sometimes be expressed either as *required* or *recommended design practices*, depending on their perceived importance. Deviation from a required design practice, for example, may require approval of a design board. Formal design rules are uncommon in small firms, or ones which have only recently entered a particular design field.

Design rules in traditional engineering design Design rules for computer circuits usually include limitation on the maximum length of connections between two digital circuits, based on the operating speed of the computer or "family" of hardware. These criteria are developed by extrapolating from past experience to new design requirements. Similarly, rules will usually be specified for minimum transverse spacing between integrated circuit devices on a printed-circuit board, and for minimum allowed spacing between boards lying parallel to one another. Design rules evolve from experience. In system design organizations where there is little similarity from one design to the next, design rules are less likely to evolve. Known problems which design rules are set up to *avoid* may pop up, to surprise firms operating without adequate design rules.

Design rules would encompass spacing of switches mounted on a panel, and the sizes and types of bolts to be used for certain purposes in a mechanical system or subsystem. In software, design rules might dictate that source programs be structured in certain ways, that comments be provided in particular locations, and that software name-labels be based on the data or program *functions*.

Design rules are often fragmentary and deal with lower levels of detail than interface standards. In a large design organization with good engineering discipline, a set of design rules may be organized and documented for application to a major development. New ones may be documented during or after some development experience in which the need for a rule became clear. In some design organizations, and for certain areas of design, a set of design rules will be evolved as an architectural supplement to *each* design effort. Selecting and refining from applicable past practices, adding new rules required to deal with new technology, new manufacturing processes or new record-keeping processes, the organization adopts a set of design practices which it will apply.

Where *computer-aided design or manufacturing* capabilities are routinely employed, some design rules can be imposed by locking software parameter settings. This could be applied to minimum line widths and spacings on circuit board designs, for example. It might also apply to *tolerances* of dimensions or of parameter values:

> *Design Rule 15.1 Resistor values shall be selected from Table 15-6, with nominal tolerance of ±5 percent unless the design requires a greater precision.*
> *Design Rule 12.4 Hole diameter tolerances for non-moving part installation shall be +1 percent, −0 percent or +.010 in, −0.0 in, whichever is smaller, for holes under 0.5-in nominal diameter.*

Some design rules may be too subtle, or perhaps too vague, to be enforceable via CAD/CAM program constraints. They will require attention by the designer and design reviewers:

Design Rule 16.8 Circuit functions shall be grouped on a board so as to minimize interconnection distances.

Design Rule 12A. Smoke detectors shall be installed in each internal equipment or personnel space which can be closed from others.

Design rules which conflict with the CAD or CAM capabilities in use obviously should not be imposed unless the design software can be changed. Certain design rules may be imposed on design of a particular system because of user requirements.

How many design rules can reasonably be enforced? If imbedded in design software, design rules should not be allowed to constrain the designer so that reasonable state-of-the-art designs are prohibited. If they must be invoked by designers, a designer should be able to learn and apply the rules in a small fraction of the available design time. The problem is complicated if rules are minor variations of other rules, perhaps applied under subtly different conditions. If there are many rules, but each is applicable only in very special circumstances, high intellectual concentration will be required.

Experienced and capable designers take care to apply design rules. They recognize the importance of consistency throughout a design, especially if personally involved during rule development and updating. Some organizations routinely carry out formal audits to enforce design rules. Others do so only under special conditions: to evaluate new designers, for spot-checking of adherence, or because of special user requirements.

Occasionally, circumstances will suggest need to override some design rule. There may be some which are considered unalterable because they bear heavily on system feasibility or reliability and cannot be excepted. Organizations may allow informal override by a designer, with a note to the design file giving a rationale. In other cases, review by other designers may be required. The main point is to *record any exception* to universal use of all accepted design rules. If a certain rule is found to be waived frequently to meet design objectives, clearly the rule should be altered or removed. Formal steps for revising design rules, as well as initiating them and deleting them, should be available.

System level design rules The objective of design rules applied at high levels in a system will usually be to unify design. By analogy, a top-level design rule in college campus enlargement is to maintain architectural uniformity—or at least, "harmony." Design unity for most technical systems is a largely *functional*, rather than visual, matter: the distribution of cooling liquids or air among system units, interface standards to be used in connecting major subsystems, or dimensional modularities to be applied. If subsystems are acquired from various sources, design rules may specify characteristics which should be expected of purchased subsystems. They might also provide guidance on features which, if not present in a purchased subsystem's design, would require its modification. As a simple example, electrical power source used by subsystems might be required to be 110-V, 60-cycle, or the unit would require modification.

System-level design rules should include, for example, criteria for maintenance access to subsystems or other system components. Selected *major subsystem interface standards* could be built into system-level rules, and if necessary deviated from, to meet special user requirements. Where easily misinterpreted terms such as *subsystem* or *component* are used in rules, they must of course be defined.

Architectural features imposed on a system by user requirements or during initial design may be expressed through design rules as a means of enforcement. If, for example, a feature requires that a failure warning initiated anywhere in the system be transmitted to a special maintenance display, design rules can call out necessary facilities to be implemented at major subsystem level or below.

Subsystem- and component-level design rules Other design rules may pertain to the design of subsystems or components, especially if they are to be developed by a captive design group. If they are to be developed outside, it may not be feasible, nor desirable, to attempt to dictate design aspects. Here the term *specification* will be used exclusively. Specifications are more general than design rules, and typically deal only with external characteristics of purchased parts or subsystems. Design rules required to achieve architectural unity in, for example, *maintenance features* must of course be specified for subsystem developers' use. A design rule imposed on a purchased item, unless the design is already compliant, may require costly modification or redevelopment.

At the subsystem and component level, greater reliance can be made on industry standards and industry-wide practices. Most commercial components and equipments are designed to satisfy particular industry or government standards.

When specifying system components, either for in-house or outside acquisition, a central concern is often the *ratings* of components. These should be thought of as *stress ratings*. Ratings are manufacturers' statements on the performance of the component and the conditions under which it will be attained. For example, an incandescent lamp such as is depicted by Figure 5.13 may be *rated*, at its design voltage, to produce a given amount of light while it operates for a given (expected) lifetime. If derated to a slightly lower voltage, its light output will decrease only slightly, while expected life will increase disproportionately. If it were not convenient, in a system application, to operate at a lower voltage (say,

105 V), a slightly modified lamp filament design would produce the same output and lifetime at 110 V. Manufacturers' ratings are often intended as incentives to select their product over others. Accordingly ratings must be studied carefully and, if necessary, warranted by the manufacturer. It is not uncommon for advertising matter to suggest performance parameters which cannot be achieved in combination.

Most components have ratings, which are cited in manufacturer's catalogs. Flexible hoses will be rated to withstand certain hydrostatic pressures continuously and may also be rated to sustain higher short-duration pressure. Likewise electric resistors will be rated to dissipate a certain continuous power when mounted in a specified manner (to promote cooling). Subsystems such as power supplies or pumps are rated to produce a certain current (usually at a rated voltage) or a certain pressure at a rated flow. Thus, in designing almost any system, one must determine not only the values of parameters but their stress ratings as well. An electric capacitor has as its fundamental measure *capacitance* (usually measured in microfarads), but is also rated to operate at or below some rated *voltage*. The consequence of operating above that voltage will be reduced life and increased probability of catastrophic failure. In electrolytic capacitors, the consequences of operating *below* rated voltage are not only increased lifetime but also *a lower* capacitance value. In this case and many others, a decision to use a system component well below its stress ratings may have unexpected side-effects which should be checked with the manufacturer or a qualified outside expert.

In specifying any off-the-shelf subsystem or component, a system designer will have limited choices of parameters or ratings. This means that available components may not match actual design requirements. In most cases, a component of *next larger* value or rating must be selected in order to meet requirements. Such a choice may be expected to support modest performance improvements, provide some margin for design error, or allow for system expansion. If this practice (as in brewing tea or coffee, "one for the pot") is followed by subsystem designers throughout the system, important requirements such as weight, space, and prime power required may be considerably bloated. There will be nothing concrete, in the way of improvement, to show for it. The usual way to control this sort of problem is through *budgeting* a system's extensive parameters (described in Chapter 7). Even budgeting concepts leave something to be desired, since some subsystems designers may routinely overstep budgets, while others may fail to take full advantage of available budgets for system parameters and come up short in performance.

Selecting parameter values is also at issue for system *factors of safety*, i.e., how much reserve strength or capacity should be designed into some particular aspect of the system? (Design rules are the usual media for resolving these issues.)

8.4 Special design issues

Many other important issues may influence the design of complex systems. Project and personnel management issues are taken up in Chapter 9. Here we examine several business issues.

8.4.1 Subcontracting decisions: make, or buy?

Systems organizations repeatedly face issues of whether to take on responsibility for a system component, subsystem, or service, or to contract for it with a subcontractor. Though subcontracting is a central feature of system integration organizations, it is also widely engaged in by system manufacturing firms. Few manufacturers are so vertical in organization that they produce products exclusively from raw materials taken from the earth. Most purchase cables, pipes, switches, valves, and other miscellaneous hardware from subcontractors. Many manufacturers characteristically "go outside" to purchase programming and other services requiring large quantities of special skills.

Fundamental questions which should be asked prior to a decision to subcontract or perform work in-house are those most pertinent to system performance, cost, and schedule:

1. Can both inside and external suppliers meet the necessary delivery dates consistent with the system schedule?
2. Which can most easily meet required *performance and quality standards*?
3. What will nonrecurring and recurring (per unit) *direct costs* be in each case?
4. What *indirect costs* must be incurred within our organization (such as overhead cost or cost of additional testing) for each alternative?

In principle, one might base decisions on lowest cost, summing direct and indirect costs, for suppliers which could satisfy schedule and performance needs. In practice, however there may be additional concerns:

5. How good is the "track record" (*performance history*) of each potential source?
6. Is a particular source strongly *favored by the customer*? If the customer favors a particular source, why (product quality or other reasons)?
7. Over the life cycle, considering *spare parts, reliability,* and *adaptability,* what other factors discriminate between sources?
8. Can we get *earlier warning of problems* from some sources than others?
9. Is undertaking the work consistent with *plans and resources of the firm* (such as growth into new businesses) or is this unclear?

10. In case of design, supply, or manufacturing difficulties, workload shifts, or schedule slippages, which source would be most responsive?

Most of these considerations apply broadly to subcontractor selection. Items 7 through 10 are of particular importance if internal sourcing is being considered. Many organizations which carry out system development are within corporations whose products could be used in its systems. However, *system* development organizations are expected by their customers to be objective in selection of subsystem sources. Ethics and system needs should rule against favoring particular suppliers.

Subsystem acquisition within the corporation can produce profit at both subsystem and system levels (unless denied by a government auditor, on government-purchased systems). It also provides *price leverage* if in close competition with firms lacking their own subsystem products. However, there is nothing more frustrating to a system manager to be told by a supplier within the same corporation that there will be a increase in cost, or a schedule or performance slip, and "if you don't like it, complain to the CEO!".

8.4.2 Competitive system markets

Competition in the marketplace is a reality for system engineers and developers in the United States. There are a few exceptions such as government organizations responsible for development or acquisition of publicly funded systems. The *U.S. Army Corps of Engineers* is assigned responsibility for design and construction oversight of public facilities such as dams, canals and waterways, and for some government buildings. Nonprofit systems engineering and development organizations such as the *Federally Funded Research and Development Centers* (FFRDCs) support Defense, Energy and Transportation Department system acquisitions. The only significant systems which are regularly developed *and manufactured* by noncommercial organizations are nuclear weapon warheads. While the Army and Navy once designed and manufactured in arsenals or Navy yards a large proportion of U.S. military equipment, that role has been given to industry for the past two decades.

System firms practice two principal modes of business in the competitive environment, speculative development and sale of systems, and government-funded system development and production. Huge amounts of money are required for speculative development of commercial aircraft, and (in the United States) this investment is from corporate funds. Speculatively developed systems can be sold (worldwide, with some political exceptions, especially for products containing advanced technologies) at whatever price the competitive markets will bear.

The U.S. government is the major, if not always the sole customer in government-funded developments. Both development and system production, when it follows successful developments, are performed for the customer at a negotiated profit. In most cases, they are not allowed to market freely outside the United States and have no other potential customers inside it.

Speculative system design and government-funded system development are both competitive, though *at different times in the system development and production cycle*. Once a government development contract is won, the contractor has reasonable chances of producing all, or perhaps an average of half (if forced to share the market with a second competitor) the systems produced—if the decision is made to proceed into production. Even after a production decision, continued funding depends on the support of both the current administration and Congress. (The B-1 bomber program had been initiated under Republican presidents, was canceled by a Democrat, Jimmy Carter, reinstituted by another Republican, Ronald Reagan, and its future threatened even later when performance of an important subsystem failed to meet requirements.)

When a firm develops a system speculatively, design decisions are made by the developer alone. This freedom must, however, be directed toward devising products which will have high market appeal. Where systems have few potential customers, there will probably be few serious competitors active. Losing only one or two important contract awards may lead to bankruptcy or corporate takeover, both of which are plentiful during slack periods in system acquisitions.

Business and design strategies best for system markets having their competitions *prior to development* are very different from those devised for markets where success is won by product sales. When developing systems for use by government agencies, requirements and standards are often fixed by the user. When developing systems for competitive sale, *developers* set their own requirements and select standards which best suit their own plans and capabilities. Motivation is to keep product prices low, but maintain a reasonable profit, when products are sold competitively. When a customer has no other place to buy a product, its supplier is motivated to keep both sales and profits as high as possible without losing out entirely.

8.4.3 Single-buyer markets

When there is a single customer for a certain type of system (economists refer to this customer as a *monopsony*), user system requirements must dominate design considerations. While this would suggest less difficulty in establishing requirements, it is true only if that customer is consistent about needs.

This is only one of a number of factors which may complicate system developers' activities in this sort of marketplace:

Conflicting requirements If the organization is large, it will often harbor conflicting internal views. No system design may be acceptable to a majority, and any consensus solution is weighed down by compromise.

Decision-makers may lack system skills Management of the client organization, who must make or concur in procurement decisions, often lack the knowledge required for system decisions. True decision authority may be vested in a collection of specialists who, however, lack responsibility. (This characteristic is prominent in, but by no means limited to, governments.)

Lack of precedent = uncertainty There is often little precedent for a proposed system and no experience with ones similar to it. Expectations as to cost, development schedule, and system performance are often unrealistic.

Poor requirements make for poor contractor selection Selection of a developer usually follows in-house requirements development. Since requirements uncertainty is great, contractors may be ill-chosen, with results as would be expected. Selection based on speculative proposals, before contractors have demonstrated an ability to design and produce, frequently leads to abortive development programs.

Unrealistic attitude toward value Government agencies or public utilities are purchasers of some of the most expensive system acquisitions: new military systems, utility plants, or headquarters buildings. Employees involved in acquisitions have nothing to do with generating funds required to pay for them. They may view government funds as well-deserved subsidies to their agencies and often feel no need to practice even modest economy in acquisition.

It is often recommended that system acquiring organizations invest in personnel, technology, etc., which allow them to become "smart buyers." This has long been an objective of U.S. military acquisition organizations. This limits, but cannot overcome, imbalances between authority and responsibility due to absence of rewards *and penalties*. In such environments acquisition of large or complex systems is understandably expensive and uncertain.

8.4.4 System design for speculative sale

Design *responsive to market needs* is a necessary condition for success in commercial system business. It must be accompanied by *responsiveness to competitive actions* in terms of product features and/or sales price.

The products of computer and automobile manufacturing firms are representative of speculatively designed systems. Both are designed for "general-purpose" applications, yet market differentiation takes place. This may be based more on customer perception than on fact, as in business versus scientific computers, or family cars versus "sporty" cars. (In either case the actual differences are small.) Speculative system developers go to great efforts to characterize "the market," those customers to whom they expect to sell products.

Some of the most meaningful views about any competitive market come from purchasers of earlier products in that market. Computer companies frequently interview users of their present computers to determining what they believe they need in the future. Automobile manufacturers do the same, though less directly and with less confidence of customer loyalty. It is far easier to switch auto brands than to change software to make it work on a different make of computer.

Each product or product line of the firm may be targeted against a different group of prospective buyers *if* the group is large enough and its needs or desires can be identified. Auto builders offer sedate sedans, roomy vans or wagons, "sporty" coupes, personal pickup truck transportation, or outright racing vehicles to different market segments.

In addition to market segmentation, a system developer in a commercial market must address competitive products. An important objective of the firm may be to achieve *product discrimination.* Products which look and work like those of the competition require positive points of distinction, to be selected over competition. The distinction must be sufficient to overcome other incentives offered by competitors. (*Price competition* is common. Clones of the largest personal computer manufacturer's products may be offered at 50 percent less than those which they emulate.)

Product discrimination is relatively easy for specialty products; Corvette, Porsche, BMW, and Mercedes automobiles are all considerably more expensive than average auto purchases. Each is distinctive and appeals to a slightly different market segment. Automobile firms strive for customer loyalty, though the products are nearly identical to those of competitors. New designs therefore often mimic older ones in salient visual features (traditionally the radiator grille). Lower-priced models are designed to resemble those at the top of the firm's product line.

The major U.S. computer firms each produce a relatively small number of models, software-compatible but of different performances, each with several increments of memory capacity. These are combined with a selection of disk storage and input/output devices, to be able to offer each customer a system somewhat tailored to needs.

A system industry leader enjoying a large market share has more freedom than the competitors who must follow its lead. Their development costs are not substantially smaller and must be returned via lower numbers of systems sold. They can afford less variety than a larger competitor. When a firm commands a large share of market whose size is limited, its competitors must be concerned to retain existing customers. Once lost, they are not easily replaced.

Revolutionary concepts in speculatively developed systems seldom if ever emerge into a ready-made market. Developing a new type of system for which there is *no precedent* and *no existing competition* is both exciting and frustrating. Huge, potentially fatal (to the firm) decisions must be based largely on guesswork. Resourceful firms strive to obtain some protection through patent coverage, and expand rapidly to attempt to reach a large part of the market before competition enters. Some must look on later while competitors take advantage of better appreciation in the marketplace and quickly develop *second-generation systems* more appealing than the first. Xerox and Polaroid gained valuable time for market development through market-controlling patents. Xerox recovered with difficulty from a later onslaught by competition when it was, at last, nearly free of patent limitations and able to build "plain-paper" copiers. However, it never captured the low-end copier market, if indeed it ever tried. Polaroid's "picture in a minute" products proved to fill a limited segment of the large photographic film market. The company has, however, maintained control of its special market segment.

Competitive market cost factors   As compared to designers of systems for monopsony clients, designers of speculatively developed systems certainly should have more control of system parameters. Though they do not control the size of market their product will eventually serve, they do have considerable opportunity for cost savings:

- Components and subsystems can be selected with a balance between performance and economy.
- Requirements can be made firm and will not increase throughout the development cycle. Design decisions can be made and enforced.
- Time required to complete development is limited only by the resources the firm can bring to it, not by deliberations among political bodies or users.
- Development can continue steadily, and avoid extra costs imposed by changes in funding rates or stop-and-go development.
- Systems are not required to meet standards which are unessential to their users.
- Promises difficult to keep are not made in advance. Risks are kept reasonable. If an intended feature turns out to be difficult or impossible, it can merely be omitted since no one has been promised it.
- Design features, subsystems, software, etc., developed by the firm for one design can be used in others as desired.
- Economic *substitutions* of features, purchased subsystems, parts, etc., can be made to keep costs in line.

Designers who have gained their experience in *either* commercial or government system design usually prefer not to move to the other kind of design. The supposed design *freedoms* of the speculative system designer are seen as *uncertainties*, by the designer of specific systems for specific users. The *economies* which are practiced more diligently in commercial developments are seen as *sacrifices of quality*.

Fleshing out the product line A firm's *product line* is its collection of products (or services) available at a given point in time. Few firms survive and grow for 5 years or more if they offer only a single product or service.

Arguments may be offered in support of numerous products from several points of view:

- **Marketing** Customers of one product can be sold related products (for example, supplies) with little or no extra sales effort.
- **Development** Engineering organizations will be more efficient if there is a continuing sequence of design projects requiring the specialized skills of our designers.
- **Manufacturing** Plant facilities suited for the initial product have capabilities to produce other products as well.
- **Finance** A larger business base makes better use of the firm's resources and will make available new capital.
- **Customer service** A larger product line will make possible improved customer support from additional regional offices.
- **Executives and owners** Growth and change are needed to woo investors. Profits should be larger, stock options more valuable, etc.

Each corporate function (engineering, sales, etc.) often envisions product line development in different directions. All products requiring similar design skills will not appeal to one customer. This is also true of products which make effective use of plant facilities. Accordingly a product line should be designed to meet a selected corporate objective and corporate resources adjusted accordingly. This will almost invariably lead to identifying several corporate functions which could not be put to full-time use. If the product line requires a wide range of manufacturing capabilities, perhaps most can be subcontracted. If it requires marketing presence in numerous cities, distributors may be more practical than corporate sales offices.

The cost of developing a broad line of new products and the other support that requires is often underestimated by firms flushed with success of a single good product. In some product areas it is practical to add one product at a time. However, to market successfully in many product areas, a comprehensive array of products is essential from the outset. Small firms often accomplish this by adding products made by other firms to their product lines. Many manufacturers are happy to repackage or merely relabel a product and sell, at reduced prices to other firms, what are known as OEM (*original equipment market*) products. The actual manufacturer enlarges its market in this way and thus can penetrate markets which, otherwise, it could not.

Firms having evolved very large product lines often find it useful to separate them for better penetration of several market segments. Where product lines have expanded by corporate acquisition, it is often most profitable to retain brand-name identity of acquired product lines, even if the products are merged to be almost indistinguishable and are functionally identical. This has been the evolution of two of the largest U.S. auto manufacturers, General Motors and Chrysler, and is also true of some firms in the major household appliance business.

Product-line concepts are equally applicable to firms in service businesses, as, for example, system architect-design organizations. Here the benefits must be derived from fewer sources: more efficient use of personnel expertise, expanded business opportunity from marketing effort, or distributing costs of design support tools across a larger business base.

Potential customers seldom have difficulty understanding the offerings of a product producing firm. Service producers have little difficulty if their services are in familiar areas such as medicine, law, accounting, or temporary office personnel. Even then, there may be many *conditions* which must be understood: where the services will be performed, who will perform them and what qualifications the people have, and what recourse is offered should the services prove unsatisfactory.

Systems engineering services range overall all of the types of activity discussed in this book, and more. Some firms are in business to assist system acquiring organizations in developing requirements or specifications. Others prepare requests for proposals connected with a system being acquired by a client. Others may analyze data, plan or carry out testing, or evaluate proposals for bidders or for acquiring organization. In short, any and all of the tasks we have described in this book can be among the services offered by systems engineering service firms. What most such firms *do not* do is compete against clients or prospective clients for system development. They attempt to remain disinterested, credible agents for others. As with most service industries there, is no statutory regulation or licensing of this type of businesses. A prospective client must verify, for its own protection, whether a particular firm of this kind can offer the qualifications and experience the client needs.

Timing and features of successive product generations The concept of *product generation* is applied in many speculative markets for complex products. Automobile designs in the United States are traditionally updated annually, if only superficially in most years, to give purchasers added incentive to own a newer car.

Computer product generations have traditionally been technology-driven. They have occurred typically on three- or four-year cycles, which (not coincidentally) corresponded to product cycles in the semiconductor industry which supplies much of the computer's technology advance. An aggressive market strategy was employed in the mainframe computer industry for many years to emphasize discrimination between a firm's old and new designs. Each new computer model had four times the performance of a predecessor model at about double its price. Though the computing power per dollar doubled, the marketing strategy encouraged doubling of the revenue from each customer.

Almost inevitably, the newly offered products of any generation-producing firm are announced as superior to the old—even if only lower in price. Changes in the products themselves may range from superficial to major. In some business areas, particularly those accustomed to short product cycles, cost of new product development and support represents a major part of the costs of doing business aside from production parts and labor. If new product costs are great enough, the smaller competitors in an industry may not survive. Their *nonrecurrent* product costs are a large fraction of those incurred by large producers offering similar products, while their revenues (and profits) may be far less.

Configuration control and life cycle support may become nightmares if a firm produces new products at a rapid pace but production volume for each product is small. The automobile industry handles this potentially difficult problem in a simple way. Each automotive part or subsystem is typically installed on a number of models, every year for a number of years. There are always only a few distinct models, even fewer new models or ones with major revisions, and many slight variants, distinguished from model to model and year to year by a few pieces of sheet metal or interior trim. Consequently (as any auto owner who has visited an auto supply store knows), spare parts stocking required for main-line autos is relatively small, even if one goes back 15 years.

EXERCISES

1. Assuming you are familiar with at least one *word processor*, review the previous chapters to find activities in architecture and design which might be enhanced by intelligent application of a modern word processor to design documentation. Describe any slight modifications to program operation which would improve ease of use in the design activity. (Hints: Fill in acronyms and other data in documents semi-automatically; organize searches for words or phrases in large design documents.)

2. Repeat Exercise 1 for *spreadsheets*. (Hint: many special analyses required during design can be set up on spreadsheets—think of a few in your own situation.

3. Repeat Exercise 1 for *database management programs*. (Hints: Is this a way to develop breakdowns? List parts? Cross-reference standards?)

4. Specially equipped computer graphic displays can show three-dimensional views, e.g., by creating stereo-pair images which give the illusion of depth or

by mirror-surfaced vibrating diaphragms. From your areas of design knowledge, extract one or more design problems where three-dimensional visualization might greatly improve design generation or evaluation. Describe the problem and in what way it would be aided by apparent three-dimensional design imagery.

5. Describe, in brief, each of the *computer graphic primitive operations* included as features of some particular computer graphics program, such as a "draw" or "paint" program used on Apple or IBM-compatible microcomputers or a graphics-based computer-aided design program. (A typical primitive operation might, for example, involve the creation of a rectangle.) Indicate which three operations you consider particularly useful to you and why. Also describe three or more additional features which you believe would increase the value of graphics to you, and for what purposes you would use them.

6. Describe loosely (using sketches if you like) how you might prefer a computer-graphics-based *system partitioning aid* to operate. Address these operations: defining system functions or requirements; "breaking away" a portion of the defined system functions into a separate functional block; creating additional functional blocks; defining interconnections between blocks. . . what they carry, their limitations, interface requirements posed by the block first defined; defining system parameters and relationships between them; assigning standards to an interface or functional block; checking compatibility of interface definitions, etc.

7. As a frequent system development task, you are required to verify that system dimensional clearances (in all three dimensions!) are adequate for access and easy maintainability of systems housed in a compact protective housing or inside vehicles. (One or two maintainers must be able to work inside the housing or vehicle and get access to any part of the system for maintenance.)

You have an opportunity to tailor the design of a powerful general-purpose graphics program used to create the system's physical layout. Describe the detailed control of visual objects and relationships that you would want in order to carry out each of the system maintainability assessments below.

(a) To represent a part being removed from an assembly (or replaced), with control ability to maneuver the depicted part, and a way to display the point of nearest approach to other parts, and the minimum clearance with the part in any position and orientation. (Describe the control actions and how the part would be displayed, with two-dimensional representation of three-dimensional objects.)

(b) To determine whether there is space for a technician's hands in the system, to hold and move a part while it is being removed or replaced.

(c) To find and record automatically a sequence of primitive (rotational or translational) movements appropriate to removing or replacing a part; or to explore sequences involving removing of parts not needing repair, so that a defective part can be accessed (one such is known as the "remove the air cleaner before the spark plugs" ploy).

(d) To prepare drawings illustrating certain maintenance sequences, for example, those involving complex moves needed to withdraw a fragile part from the system without damage.

8. Define a set of operational test objectives which you believe would be appropriate to:
(a) A high-priced wrist watch, guaranteed to be "accurate to 15 s/month," "waterproof to 100 m," and "shock-resistant."
(b) A personal computer shipped with memory and disk storage but without display, display adaptor or an installed operating system.
(c) A 27 in (diagonal measure) TV set equipped for cable operation and with a wireless remote control device.

9. Stress testing is intended to verify that a system design is able to withstand a special temporary environmental condition. For five products or systems you use, define one or more stress test(s) their developers should have exposed them to, based on your (good or bad) experience with them.

10. Define what you believe might represent a suitable automotive suspension stress test involving an actual vehicle. Provide a rationale for each test criterion and parameter you propose.

11. List 20 *different* types of products or systems which (in your view) must be subjected to *destructive testing* as part of a comprehensive test program during development. For each, indicate what principal parameter(s) should be examined or measured, and when (before and after or during the test).

12. Specify for a *pay telephone* design the external elements which you believe would be required for a comprehensive long-term testing *of all features*, without a human user, or any human, present during the test. How might the operating environment be simulated and what are the chief limitations in your proposed simulation?

13. You are to design a test using human subjects to compare two *word processors* designed in your company for public sale. Although some system comparison tests using human subjects attempt to expose an individual to each test system, that is made more difficult here by the software's complexity, lengthy learning period, and the prebiasing of subjects. Accordingly, you plan to install many copies of modified versions of each program, letting the modified software record technical data such as use of features, time between certain operations, and operator errors.
Describe the tests you might require the software to carry out as measures of operator performance. Outline the contents of a questionnaire which would complement the measured performance and develop some parts of it. Finally, describe how you would analyse the data to find three user-pertinent performance parameters. (Speed of correct typing, in characters per second or words per minute, could be one such parameter.)

14. Using a complex system such as a word processor familiar to you as an example, prepare a brief set of requirements for an operator training program which will include lectures or video tapes, hands-on exercises, and monitored practice. Use your familiarity of the system to emphasize areas where special attention should be paid in the training regimen.

15. Key elements of an *operational scenario* which are important for use in defining system tests are: subsequences which are repeated frequently; states of the environment, and of the system, defining stressful conditions; and operational dynamics such as average timing of peaks and lulls in activity.
Analyze your *personal work environment* in these terms. If a test designer were creating a simulated scenario to determine other people's compatibility with your work environment and work patterns what sort of scenario should be used? (That is, imagine a sophisticated employment test for someone expected to do *your* job by *your* standards.)

16. As discussed in the text, maintenance tests on systems should be tied to the *maintenance strategy* defining the system parts which are to be replaced in the operating environment.
(a) What are appropriate testing rationales if a system is to be returned as a unit for maintenance external to its operating environment?
(b) One built-in test concept involves an "I'm good!" lamp on each field-replaceable electronic module in a system. For what maintenance strategy or strategies is this well suited?
(c) A common test strategy is direct *visual examination*. Which owner-replaceable parts on a modern automobile are most often tested in this way?

17. Describe items of built-in test equipment or performance indicators (whether designed for the purpose or not) found on: (a) a personal computer, (b) a telephone set, (c) an automobile.

18. Many years ago, most standards were *physical*. The size of the king's *hand* or *foot* represented those measures. Later an inscribed metal rod defined a *meter*. Today most standards are expressed as *documents* or measured in terms of physical constants. If you have not seen standards documents, visit your library and ask to be directed to national standards in your areas of interest. First, read through several short (a few pages) standards documents in areas you understand.
Prepare a *list of standards* (perhaps better defined as *criteria*) which you believe should be applied to any standards document. It will be helpful if you begin by asking yourself *objectives* of standards, and work from there. You should have no difficulty finding good and poor examples.

19. Some government standards deal with criteria the government places on items which it purchases, while other standards describe safety protection for industrial workers or set public health criteria such as the allowed bacteria count in foods.
Investigate standards imposed on building construction in your municipality, state, and country, usually referred to as the *building code*. Compare local requirements with those imposed nearby (e.g., city versus county or neighboring states). You may in particular find differences in plumbing and electrical provisions, and in fire safety requirements. Report the significant *regional differences* you found which are respectively more, or less, constraining to a designer or builder.

20. Identify 10 or more features of similarity, such as identical sizes and shapes, which are found in everyday objects produced by different manufac-

turers. Which of these do you believe are used as *ad hoc* standards? For each, what is the rationale behind the use of standard features.

21. List all the *design rules* you personally apply in your own preparation of typed correspondence.

22. Examine carefully, inside and out, the body of your car. Try to find evidence of design rules practiced by its designers. (For example, check with a ruler the gap between abutting sheet metal parts such as door and body, hood and body, trunk lid and body.) Seek design rules in the type, size, and spacing of fasteners, size and spacing of spot welds, thickness of metals, etc. Suggestion: Look at details rather than overall appearance.

23. Create a set of design rules which you feel would be desirable in design of *user interfaces* to computer programs. (Presumably you would address amount and placement of information on the screen, number of keys required to select modes, and similar items.) Indicate which of your rules, if any, you have observed to have been violated by commercial software. Why, in your view, might the designer have departed from your standard?

24. (a) List the likely contents of a modern product line which might be developed and manufactured by an *office furniture* manufacturer.
(b) What combination of technical, resource and market factors motivated your choice of products? What period between new product generations might be followed for each product group, and why?
(c) For an office furniture manufacturer responding to opportunity as described in (b), what new products are likely to be under consideration or actively in preparation, at the present time?

BIBLIOGRAPHY

David S. Burnett, *Finite Element Analysis*, Addison Wesley, Reading, MA, 1987. Explains this broadly useful technique for reducing real-world problems correctly to finite-element forms which allow computer solution. Rather mathematical, as required.

Howard Eisner, *Computer Aided Systems Engineering*, M. Dekker, New York, 1988. A description of the processes and methods used in organizing and carrying out system engineering, mainly from a top-down viewpoint.

T. L. Kunii (ed.), *Proceedings of Computer Graphics International Conference* (annual), Springer-Verlag, New York, 1986 ff. Describes computer graphics applications including design, also computer art. (In a fast-moving field such as this, popular and trade periodicals and annual conferences are usually the most effective sources of up-to-date information.)

Personal computer software: developments evolve so rapidly that popular periodicals are one's best source of information. *Byte* magazine (McGraw-Hill, New York; monthly) is one of the most technically sophisticated popular periodicals covering this field.

(*Standards*: see references in Bibliography to Chapter 3.)

System Engineering Management Guide, (developed by Lockheed Missiles and Space Co., published by Defense Systems Management College, Ft. Belvoir, VA, 22060), 1983. Describes system analysis and design methods for Defense systems; cites numerous references and standards.

Roger T. Stevens, *Operational Test and Evaluation*, Wiley, New York, 1979 (Reprinted 1986 by Kreiger Publ. Co., Melbourne, FL.) A mostly non-mathematical work explaining the how and why of operational testing of complex systems.

Managing System Architecture and Design

Any but the smallest architecture and design projects involve management issues different from those of managing more traditional or less complex engineering projects. The qualifications and expertise of staff and other resources used for these kinds of activity are also quite different. This chapter discusses these management issues in the context of system architecture and design activities such as those described elsewhere in the book.

9.1 Organizations for system architecture and design

There may of course be enormous variations in size and constitution of systems architecture and design staffs, depending on the nature of system projects, e.g., aircraft, pipelines, manufacturing systems, computers, etc. In building architecture practice, traditionally the architect/design firm does not act as a construction contractor, but is expected to continue as an agent to represent owners during construction by a separate contractor. This stand-apart tradition has also been followed by some practices engaged in *industrial design*. Functional designs, artistic styling, and operational schemes have been developed by independent designers such as the late *Raymond Leowy*. He was the originator of automotive "streamlining" long before automobiles were routinely tested in wind tunnels. The firm whose director, *Dr. Ferdinand Porsche*, conceived the Volkswagen (*people's car*), also created many other innovative products and systems. He began automobile manufacture following World War II, at first using parts from the Volkwagen "beetle" to create the more high-spirited vehicles to which he lent his name.

In system architecture and design for modern technical systems such as computers and communications systems, ground, air, and sea vehicles, responsible architecture and design teams are usually part of the firms which produce the systems. In such enterprises, there may be substantial time-gaps between major new design activities. Where large design teams are required, as in a major aircraft design, all except a small cadre of top-ranking design personnel may be hired at start of the design effort (or receipt of a development contract), and dismissed at its completion. This has been a way of life for many journeyman engineers in the aircraft industry, where many individuals move between firms to enjoy reasonably constant employment.

A large system design activity, such as that of a new aircraft design, typically makes use of large numbers of *detail designers*, individuals assigned to design individual parts or small assemblies, but lacking the experience or ability for top-level design or developing new concepts. The more routine parts of this work may be carried out by technicians or draftsmen. A firm with as many as hundreds of engineers and technicians, however, may employ only a handful of individuals who could be characterized as *system architects*. The term *architect* will likely be used informally, sometimes as a term of respect but seldom as a job title.

Most architecture and design organizations (Figure 9.1 depicts a hypothetical example, drawn from a combination of real organizations) responsible for design of tangible systems are supported by technicians who refine design drawings, assemble and test experimental systems or subsystems, and maintain formal design records. There may be a small-scale fabrication activity, commonly referred to as the *model shop*, containing skilled model-makers who can build one- or few-of-a-kind parts and full- or subscale models from rough design drawings and descriptions. Some of this support structure may be shared with other engineering design organizations, or with the manufacturing organization if it operates on the same site. This considerably simplifies transfer of designs into production.

9.1.1 Managing design staffs

In system development organizations, the design staff will often be managed by a vigorous and productive individual, sometimes titled *chief designer*, who may also serve as chief system architect. (The chief architectual innovator will sometimes be an individual who assumes formally only an advisory role and can thus be excused from many management requirements.) The chief designer should be well supported by administrative personnel who will carry out most of the routine management tasks but take care not to interpose themselves between their boss and the team of designers.

Personality and management style of a principal designer can vary greatly, because the individual's knowledge, drive, and design experience count more heavily. At one end of the range are individuals who refuse to

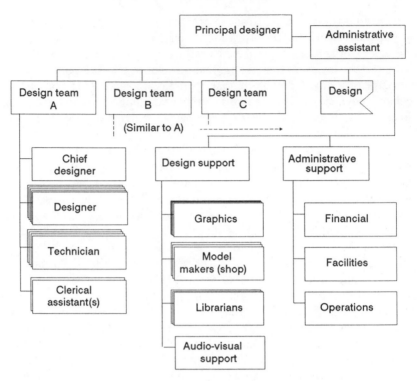

Figure 9.1. A typical structure of an organization responsible for architecture and design. Larger organizations may be matrixed, with specialist designers only temporarily assigned to a particular project. In this small organization, design groups are built around individuals. Each will direct part of the overall job.

share key design responsibility, may consider the rest of their design staff to be physical support to their own intellects, and manage the operations with little sensitivity. At the other end are fully participative managers who delegate design work, assist their designers both technically and managerially, and ask a consensus from subordinates prior to making their key decisions. Though individuals closer to this second prescription would be preferred by most subordinates and superiors as well, one can find productive examples from throughout the personality spectrum.

In a creative activity, a central management task is keeping one's organization supplied with at adequate numbers of highly capable workers. Some workers will stay longer than others, but there is little chance to rise within such an organization because it has few important managers. The amount of flow-through in such a design team, or equivalently the average tour of duty, is longer than in many other industrial positions. Not only is the work challenging and frequently changing, but the *de facto* term of apprenticeship for system designers is longer than that in traditional engineering design.

In emerging system areas, designers may be on average somwhat younger than in more mature ones. Even here, experienced system designers will be valuable for addressing long-standing design issues. Though the early designers of personal computers were youthful, many of their designs could have benefitted considerably from advice in electronic circuit packaging and mechanical design.

It is seldom effective to attempt to retain, as full-time staff, a *complete* collection of designers, able to deal competently with every aspect of a complex system design. Individual consultants or specialty firms can be used as subcontractors when internal resources are limited or absent. Even though other staff specialists may have limited experience with some special need, maintaining a uniform level of currency, quality, or innovative approach in a design usually requires seeking outside help in certain fields.

9.1.2 Activities in system technology

Much of the technology incorporated in a system will be in the form of component or subsystem technology: new structural concepts or materials, sensor elements, high-speed devices, robotic devices, and the like. There is additional technology, meaningful only in a system context. This requires activities *in advance of system design* and therefore not connected with a particular system. Otherwise it may not be available and adequately understood when needed for a system.

System technology includes, for example, techniques for large-scale system simulations which help system designers to examine effects of system design alternatives and assists in design decisions. Other system technology includes new concepts in interconnection or interfaces designed to handle system problems. Such system features as central system error-reporting must likewise be based on system technology. In short, appropriate concepts will not emerge instantaneously upon the receipt of a contract. Some system technology activities may lead to development of new system architectures and preliminary evaluations, e.g., demonstrating feasibility or system characteristics, prior to application in an actual system.

Any given design organization will deal with certain classes of systems and system design problems. Likewise, particular kinds of system technology will be most useful. An organization with intent to excel in system architecture and design should pursue system technology activities in a separate technology activity, or in the design activity itself, if it can afford to take advantage of periods of reduced activity. A major difficulty, if using only designers' "spare time" to carry out this work, is that no reasonable schedule of objectives can be planned or will be met. A combination of a small, dedicated system technology staff, with practicing designers temporarily away from system design, is an effective way to pursue system technology.

In the custom-system marketplace, including governmental system development, activities anticipating future user needs are often carried out using the firm's own funds, rather than a separate contract. Both technology development and future-requirements-based R&D are funded in the defense industry, via an overhead allowance termed IR&D (*independent research and development*). A certain percentage of overhead charges collected from the government on ongoing defense work may be spent on IR&D activity. The principal shortcoming of IR&D, of course, is that it is not available if the firm has not yet received substantial government contract work.

9.2 System life-cycle architecture and design activity

Architecture and design organizations are involved most heavily in early parts of the life cycle (Figure 9.2): Their work may begin with requirement development, usually tops out as specification preparation reaches its peak, but may peak again during system integration and developmental testing. In many phases of the life cycle there is relatively little system-level design work.

One design organization may build its expertise in architecture or initial design work while another specializes in updating existing systems. In either case, expertise will be developed through a limited number or variety of systems. It would be unlikely to find a design team, expert in small private aircraft design, which would also be competent and competitive for designing business jets. This correctly suggests that systems incorporating major new elements into conventional systems, or combining distinct classes of old elements, will require combinations of design teams or their personnel.

Intermittent involvement during a typical design cycle means that, to be kept reasonably busy, a designer may move through one or perhaps several projects during a year, depending on project size and outcome. If this pace of work cannot be sustained, some designers may be moved between high-level and detail design. Designers expected to be the organization's innovators should be assigned, during such intervals, to advanced projects. As construction architects often do, they may create system designs not intended to be carried beyond the concept stage. Unless they are able to spend continuous effort of months or up to several years on these advanced projects, results will be too fragmentary to add usefully to the individual's (hence, the organization's) well-developed skills.

In commercial firms which plan their product line updates, in-house design organizations can be assigned projects by which they can effective-

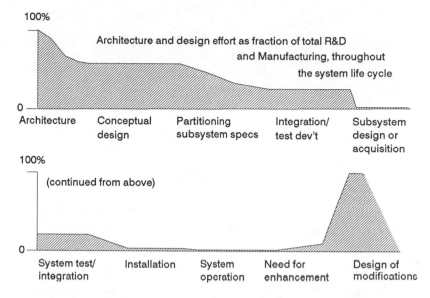

Figure 9.2. Typical proportion of total effort requiring system architecture and design work during each major phase of system development. Because most of the design work is in the front end, it is desirable to maintain involvement of key design personnel in later phases of the project.

ly apply their talents throughout the product development cycle. Firms which survive by responding to proposal requests arriving at uncontrollable times must deal with irregular and overlapping peaks of activity. Design managers will recognize that creative organizations do not thrive well in an environment of consistently overtime workloads and repeated contract losses. If there is no way to handle several major proposals concurrently, the usual decision is to work on some while ignoring others.

9.2.1 Processes and process models for design

In planning the execution of architecture or design work, it is useful to base plans on some generic *process model*, which helps guide initiation, execution and revision of the plans. The simplest, most frequently used high-level design model is sometimes referred to as the "waterfall model" (Figure 9.3). It is fully understandable almost at a glance, even by non-technical individuals. In the example of Figure 9.3a, the model suggests graphically that, at the conclusion of step 1 (which may, for example, be architectural design), step 2 (perhaps system decomposition) will be begun. The drawing implies that the results of step 1 somehow flow down to step 2. Since there is no overlap, at least as depicted here, the same personnel might carry out both steps.

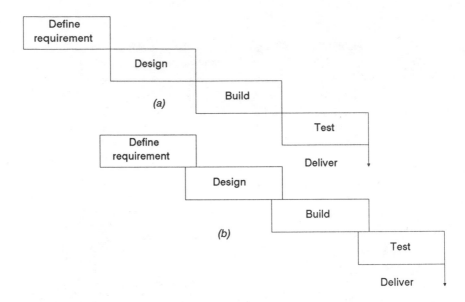

Figure 9.3. The "waterfall" development process model. In *(a)*, each phase is assumed completed before the next is begun. In *(b)*, there is some time allowed for transfer of results from one activity to the next.

Practically speaking, it seldom makes sense to plan to bring one phase of activity to a full stop prior to beginning the next. A useful modification to the "waterfall" is shown in Figure 9.3b. This suggests a transition period, during which step 1 is being completed and step 2 being started. If the two were performed by *different personnel*, an overlap would be very desirable to effect the transfer of results. The amount of time actually needed would of course vary with the circumstances.

The waterfall model is acceptable to general management in the sense that it is specific and suggests no uncertainty. It is characterized by a never-look-back approach. It suggests to many practical designers that, if step 2 were to get into trouble, a way out must be found within the kinds of activities and resources involved in step-2 work. Most experienced designers realize that *serious enough* snags arising in step 2 are signals that some parts of step 1 *must* be revisited. For example, an architectural change may be needed to overcome a critical problem exposed during detail design.

If one were to consider all possibilities imposed by needs to revisit earlier activities, in the waterfall model, one might arrive at the *fully iterative model* of Figure 9.4. Difficulty in any part of this process can invoke a decision to return to any of the earlier steps.

To most *designers*, this iterative model appears more realistic than the waterfall model. They see it as a logical way to describe dealing with uncertainties, present at the start, or discovered only later in the process.

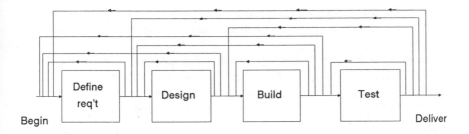

Figure 9.4. A possible but intolerable result of making a project fully iterative. One might presumably return to the start, from the end of any phase. Each phase of a sound development must have feasible objectives which are consonant with reasonable requirements.

From a manager's point of view, however, it is a horror. It appears to commit to a process which could absorb infinite time and money without ever being completed. What's wrong with Figure 9.4 is that a manager looks at the diagram as a schedule of activities, whereas it is intended only to depict a *process of iteration*. It is not acceptable, in any event. A process which can return to its beginning point whenever some result suggests doing so is *out of control*. Unless a project can be brought to a specific state of completion in a finite time, completion schedules will be arbitrary. If there are uncertainties, *each phase* must contain contingency effort (time, money, other resources) for revisiting earlier decisions which may have to be changed. This risk should be identified and contingency time/funds applied. We return to Figures 9.3*a* or 9.3*b* for schedule, *and use* Figure 9.4 *for process*, understanding that each schedule block represents *all of the activity* necessary to complete the system development to some particular checkpoint. *Any* project which repeatedly turns back from its final test phase to initial architectural design is *not* a development project. It might better be described as a succession of research activities directed toward a goal which is unreachable with the resources available.

Requirements uncertainty In developing a system for a military agency, one of the main sources of uncertainty is system requirements. However, requirements uncertainties may arise in development for any client which does not have complete control over its requirements projected to the time the system could be fielded. Unless most uncertainty can be removed from requirements, chances of satisfactory system development or of system acquisition are poor.

There are two recognized ways to plan and proceed with design programs for which significant requirement uncertainty exists:

1. Set up and carry out a preliminary program whose objectives are to resolve uncertainties, prior to proceeding with system design (Figure 9.5).
2. Establish initial program objectives which are consistent with those requirements least likely to change. Develop a system meeting these more limited requirements, but make provisions for future evolution *incrementally*, as requirements continue to evolve.

Both of these are management techniques rather than design approaches. The first is valid only if uncertainties could be removed by a *funded requirements study*, or perhaps by the development, test and demonstration of a *prototype system*. If the second is pursued, the user should be able to field a system meeting some or most requirements. That system must evolve incrementally since the user cannot articulate requirements far enough ahead.

Problems with requirements appear frequently in military system acquisition, largely because of uncertainties in international affairs, national politics or operation of potentially hostile military systems. These problems are notable in connection with software which, within limits, can be altered quickly, and can alter the personality and functions of a

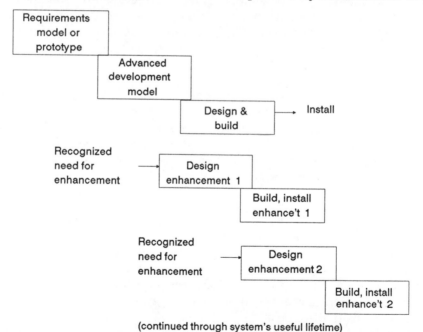

(continued through system's useful lifetime)

Figure 9.5. A more feasible *incremental* development plan than those represented by Figures 9.3 or 9.4. One or even several initial phases are carried out to reduce uncertainty in requirement or feasibility. If the initial system development cannot meet all objectives, planned product improvements do so after initial fielding of the system.

hardware system. Either of the approaches mentioned above for dealing with requirements uncertainty is fully applicable to software.

9.2.2 Organization of architectural and design efforts

The structure of any system design organization will depend very much on how *many* individuals are involved. Small organizations involving a handful of designers may operate very informally, with the manager becoming a participant as well.

Building and retaining a permanent staff is essential. There is, at least with current information management concepts, no other good way to build experience and recall the lessons learned. Even so, top-level designers and design managers may spend as little as a portion of the design cycle for one major system in the organization, before moving on.

Support staff are also essential. Personal computers and workstations, as never before, have reduced dependence of engineering designers on secretaries. Model shop and other support facilities require seasoned individuals. They are well worth their cost, but for cost-effectiveness must be assigned some independent activity to carry out as well as sharing their services between different designers.

Although small design organizations tend to be informal, in larger ones two basic types of organization are found: *project-based* organizations and discipline-based or *matrix* organizations. A project-based organization anticipates long continuation of projects or a succession of related projects and sets up a permanent organization to deal with them. In a matrix setup there are also permanent organizations. These are identified by discipline or function: electronic design, mechanical testing, packaging, materials.

The matrix organization responds to a new design requirement by assigning a manager to be responsible for a project and temporarily attaching specialists as needed. Matrix organization is most useful in large organizations which have little continuity in design activities from year to year. Some firms have compromised on mixed organizations, wherein permanent design groups are assigned to major or long-term projects. The smaller matrix structure there supplies only part of the projects' personnel requirements.

Personnel acquisition and management Where does a systems engineering organization find personnel for architecture and design activities? College graduates in traditional engineering, or from those few schools offering systems engineering degrees, are not productive for interdiscipinary system work until they have years of experience.

Organizations of any size consider their main employee market as consisting of employed, experienced designers. There is thus good, usually upward mobility available to competent system designers. Some of the larger organizations which plan personnel requirements hire new college graduates and "grow their own" system designers. They may have formal training regimens which bring young designers into the company of a number of experienced hands. System design work in a given organization often employs a set of design aids and methodologies different from those used by other system organizations. New design employees may be given on-the-job training in the use of these processes, to improve their productivity.

Managing an interdiscplinary group of innovative individuals *itself* requires innovation. Today's young employees often see no reason why they should be required to work an 8:30 a.m. to 5 p.m. schedule if they perceive their most productive period to occur at other times. This has been accommodated by some employers through the *flextime* concept. In this, employees may be required to include a certain set of midday hours in their daily schedules but may elect to begin and end work at times they select—corresponding to at least a full workday. The principal problem with broadly flexible work hours in systems work is that opportunities for interaction can be greatly reduced. If technicians are required to work by a traditional schedule, professionals working a flexible schedule may not have support when they need it. Time management of personnel involves many other challenging problems for supervisors. It is merely one part of the resource manager's challenge. Sharing limited and expensive test facilities between several designers or teams, all in immediate and urgent need, is another.

Reward for design professionals often takes the form of generous salary increases. Opportunity to attend a far-distant conference may be used as incentive or reward. Some large organizations, too dogmatic in establishing uniform salary scales based on age or experience, find themselves unable to attract or retain high achievers. Flexibility in rewards administration is helpful. Where a firm finds it impractical to compete with outside salary offers, superior design and test facilities or location in an area rich in avocational opportunity can sometimes compensate.

Providing a separate, attractive, and stimulating work environment for a design organization is another way to attract and retain employees. A generous budget for support equipment (e.g., computers, workstations, or special test equipment) will also be appreciated.

System architecture and design in the corporate organization   The system architecture and design activities of a system-manufacturing organization are only a minor part of the operation. Engineering activity may also take place in advanced development, manufacturing, test, and field engineering (support of installed systems) departments. The prominence of system design activities within the organization depends on the business of the overall firm and attitude of top management. More simply put: If the corporation is headed by a former system designer, there could be a single central systems architecture activity reporting near the top of the firm. If headed by a marketing or financial officer, system design activities will more likely be distributed across product units. In this case central oversight will be limited to a staff Director or vice president, to act as a cross-pollinator.

Good access to high-level management by a system design activity has both positive and negative aspects. While designers' resource needs may receive top-level attention quickly, the work may be disturbed and redirected too often, receive generous praise when it meets management objectives but harsh treatment if not.

The best measure of effectiveness of a system architecture and design group is how often their work is translated into useful and profitable systems. An idependent architecture group not tightly coupled to design activity is not only a waste, it is a disaster. Architectural proposals have to be "thrown over the wall." The results please neither the architects, who are not recognized as participants, nor the designers, who usually accept such contributions less happily than ones from outside the firm.

Decision making   Even where requirements are fully and clearly stated, system architecture and design involves formulating, analyzing, and completing vast numbers of system decisions. These decisions are of enormous variety: system structure and layout, modularity measures, types of connectivity, priority of operations, control concepts, display formats, etc. To do this well, design personnel must approach it in a disciplined way. Once having realized a need to take a position, (1) determine pertinent alternatives, (2) assess those within the system requirements and other constraints, (3) select from the alternatives, (4) record conclusion and rationale, and (5) distribute them to those who need to know.

Dealing with optimization and compromise in large organizations is, as a rule, much more complicated and slightly less successful, than in smaller ones. This is simply because there are more players and points of view. Past some point in deliberation, little or nothing can be gained by gathering more opinions, but time and project momentum can be lost.

Large organizations whose managers perceive this recurring problem may establish rapid, formal processes to bring important decisions to a conclusion quickly. The main features of such a process:

1. Prepare a "white paper" which states the situation clearly but succinctly, naming those parts of the organizations, or individuals, who will be involved with the outcome. Preferred direction(s) and rationale(s) should also be presented. The individual who is responsible to make the decision should sign the paper.
2. Distribute this to all organizations involved, at all levels, simultaneously.
3. Allow a few days (or less, if extemely urgent) for review.

4. The decision maker holds a meeting to receive opinions and recommendations from all interested and involved parties (high-level executives may send a staff member).
5. The decision maker decides, and (perhaps after ratification by a superior, if a major decision) notifies all parties. The decision should surprise no one.

A frequent weakness of large and long-established organizations is that individuals or groups which lost some previous decision may be allowed "in fairness" to win the next, sometimes even if offering an uncompetitive proposal. Other rewards may be distributed over long time periods in such a way that no individual or group feels left out. Smaller enterprises are usually less egalitarian, especially if ruled over by a reasonable, competent, and productive despot. Large organizations have traditionally enjoyed the potential to recover from a few mistakes. Smaller ones lead a more precarious existence and can expire because of a single faulty decision. With the current trend in large organizations to decentralization into small *profit centers*, managers of those smaller units may lead just as precarious an existence as those in independent firms.

In the United States, the most successful system development and acquisition management programs place major design decisions for a given system project in the hands of one responsible individual, who may be titled *system manager*. This individual may fearlessly accept or reject subordinates' counsel and even that of superiors (perhaps more carefully, in that case). It is not uncommon for such individuals to report directly to an executive at a high level in the organization so that a minimum of intermediate managers need to be involved or informed. Notwithstanding the almost dictatorial power implied in this management arrangement, it can be extremely effective.

In designing a complex system, though assistance from a wide variety of specialists is essential, major system trade-off decisions *must* be made from a central perspective. This is probably the most important of the many responsibilities of the system manager. This person must be forceful and determined, but also intelligent and open-minded. Even granting the authority which goes with the job, he or she will benefit greatly from the respect and good will of those both above and below in the organization.

9.2.3 System design work breakdowns and task schedules

Systems architecture design or traditional engineering design are very similar, in terms of planning and scheduling development work. They have much in common with organizing any complex activity sequence. The general process requires:

1. Subdivide the problem or system into aspects (or, subsystems and parts) which *can* be separately designed or specified. (In the case of

systems, this implies that partitioning has been carried out, or will be done so as part of this activity.) Then, for each item *define* the kinds and amounts of work which will be required to define, design, build, and test it. The collective result, when organized into a single document, is similar to what is referred to in federal procurement documents as a *work breakdown*. If certain parts are to be the responsibility of others, only definition and perhaps testing work items will represent them here. Estimated work hours required for each item in the breakdown can be collected for use in budget estimation or budget update. Facility use can also be estimated from this data.

2. Determine the most appropriate *sequence* in which the work should be done, based on need for completion of certain work as required input to further work. It is usually easiest to develop this kind of schedule in reverse time sequence, i.e., from finish to start. This corresponds closely to the work breakdown and partitioning. The result is often referred to as a *task schedule*. It may be put in graphic form, if desired, as a "PERT chart" (discussed below).

3. Associate each work activity with the *individuals and/or organizations* which must perform it. Then determine the expected elapsed time, based on personnel and other resource limitations (test facilities, model making, etc.).

4. If the time required exceeds that allocated (which is not uncommon), reexamine and, as feasible, adjust the task schedule. If that fails to bring the schedule into reason, reexamine the assumptions made in the work breakdown.

Management support tools are available which can assist in routine aspects of this activity. Database management is probably the most helpful as a means of collecting and arranging large amounts of data. PERT (*program evaluation and review technique*) is a simple procedure, best understood in a graphic form where *nodes* represent *states of completion* of system or its parts, and *arcs* represent *activities* required to get from one state to the next (see Figure 9.6). Given a specific time required, or a maximum/minimum estimate for each activity, it is a relatively simple computation to determine which start-finish track(s) are of longest duration. (This procedure has been termed CPM, or *critical path method*.) The longest, as illustrated in the figure, is termed the *critical path*.

One objective in PERT/CPM schedule manipulation might be to reduce the duration of the critical path. The usual result is recognition of another sequence as the longest path. As this process continues, a larger fraction of activity sequences may be on critical paths. This usually makes possible easier management of resources and *slack time* than where all else hinges on one long sequence.

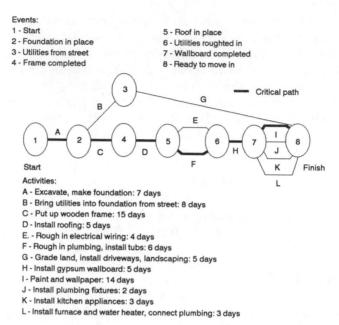

Events:
1 - Start
2 - Foundation in place
3 - Utilities from street
4 - Frame completed

5 - Roof in place
6 - Utilities roughted in
7 - Wallboard completed
8 - Ready to move in

Activities:
A - Excavate, make foundation: 7 days
B - Bring utilities into foundation from street: 8 days
C - Put up wooden frame: 15 days
D - Install roofing: 5 days
E. - Rough in electrical wiring: 4 days
F - Rough in plumbing, install tubs: 6 days
G - Grade land, install driveways, landscaping: 5 days
H - Install gypsum wallboard: 5 days
I - Paint and wallpaper: 14 days
J - Install plumbing fixtures: 2 days
K - Install kitchen appliances: 3 days
L - Install furnace and water heater, connect plumbing: 3 days

Figure 9.6. A PERT chart illustrating the main steps in construction of a wooden residence in the United States. The *critical path* is shown as a heavier line. (In this simplification, the typical delay times required to get certain construction crews on the job have been ignored.)

In practice, time is not the only limited resource. The development schedule may be limited by resources such as design specialists, facilities, or computer time. When this happens it does not constitute a sufficient excuse to leave planning to luck. If one is using a PERT/CPM framework, one can determine if concurrent activities must share a resource. If so they must be shifted if possible to be non-concurrent by changes in other parts of the schedule.

PERT/CPM was once the darling of the U.S. Department of Defense and mandated for most contracts. It fell from favor because it was improperly used. PERT charts containing hundreds or thousands of activities, prepared using computer plotters, were far too complex for humans to work with. And, once a project began, typically the diagram and its defining data were not updated and quickly became meaningless.

There exist a variety of project planning tools, the simplest of which, sometimes referred to as a *Gantt chart*, is a form of calendar which displays activities of an individual or small group (Figure 9.7), usually in the form of a large wall-hanging board. A week or month of days, or sometimes a year of months or weeks, is laid out horizontally. Each schedulable task or subtask is assigned a row of the chart and may be described by an entry on the left side of the row. Provision is made in each task-date cell to write the work description or other data.

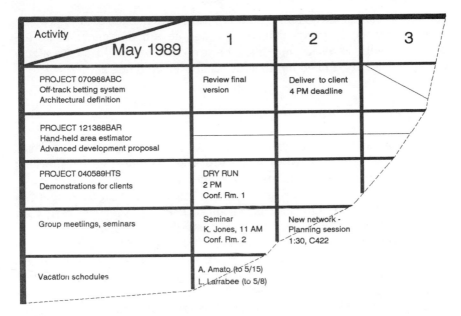

Activity May 1989	1	2	3
PROJECT 070988ABC Off-track betting system Architectural definition	Review final version	Deliver to client 4 PM deadline	
PROJECT 121388BAR Hand-held area estimator Advanced development proposal			
PROJECT 040589HTS Demonstrations for clients	DRY RUN 2 PM Conf. Rm. 1		
Group meetiings, seminars	Seminar K. Jones, 11 AM Conf. Rm. 2	New network - Planning session 1:30, C422	
Vacation schedules	A. Amato (to 5/15) L. Larrabee (to 5/8)		

Figure 9.7. A type of chart often referred to as a *Gantt chart.* This one is in essence a calendar of activities for each of a number of projects or other scheduled activities.

Beyond this basic format (which can be laid out on plain paper if need be), a vast variety of wall charts is available. Some use small cards which can be slipped into tracks. Some have colored strings, on pegs which can be pulled over from the left edge to quickly display project duration. Others use magnetic markers and still others have a wide variety of accessories.

Like racks of personnel time cards surrounding a time clock, wall-hung schedules seem more appropriate to a machine shop than a sophisticated design operation. Some system design operations designate a particular conference room as a "control center," where such schedules and other project information are stored and can be displayed for internal meetings and reviews with clients.

Any organization carrying out time-critical work of high complexity deserves a flexible, easy-to-use, computer-based planning and scheduling system. A wide variety of special-purpose and generic software programs of this kind are available for use on personal computers, minicomputers, or mainframes. A simple system adapted to the particular needs of a design organization can also be put together, based on spreadsheet or database-management software, and made instantly available via computer networking. Computer-based planning systems have the advantages of being able to present a wide variety of display and print formats, and ability to arrange the same information in ways which fit specific needs. Effective project use of any such system demands that project personnel early define and structure data needed for the project, with assistance from specialists or other design personnel more experienced with its use. They must be prepared to readjust data structures, repeatedly if need be, to address unpredicted needs.

9.3 Managing design progress

The typical manager of a system architecture and design project takes a major role in most system decisions. He or she is responsible for enforcing a planned schedule and altering it when necessary.

Many creative individuals are convinced that additional time spent in generating and assessing alternative approaches, or otherwise refining a design, will pay off in system improvement. This is true, but only up to a point. A representative "results versus effort" relationship applicable to system design (and to many other activities) is depicted in Figure 9.8. An alert and sensitive manager who asks the right questions can prevent overworking of design problems. Even if the quality of the results does not actually deteriorate, as depicted, the extra effort can often be better spent attacking more important and more difficult problems.

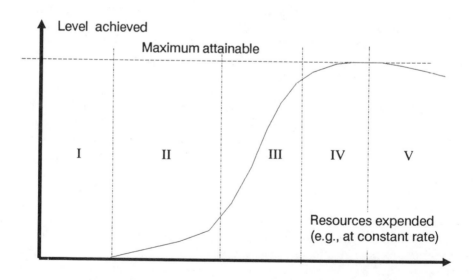

Figure 9.8. A universal plot of utility attained vs expense or effort of development. I: Some minimum investment is required before any results occur. II: Below a critical expenditure level little is accomplished. III: There is often a high-productivity per dollar region. IV: Only marginal improvements obtained. V: The design has been overoptimized.

Management style is a term referring to the way in which a manager carries out the task of achieving results by expending time and money. Managers in architecture and design are not unlike those found in other management functions. Below are four management styles of significantly different effectiveness:

Management by objective *implies the preliminary setting of objectives, then assessment of progress in terms of meeting those objectives. Under one title or another, it has been widely approved for business management for several decades. In seasoned organizations, a determined effort is made to practice it. Although recommended for management of creative work and better than some alternatives there, it is probably less effective than it is found to be in managing more routine work. Most qualified designers are inherently objective-oriented. The main role of the manager is keeping them from wandering too far off on side tracks, without discouraging their initiative.*

Management by leadership *(or by example) emphasizes the drive or personality of the manager, who presumes this enthusiasm to be infective and thus must be the greatest workaholic of all. Some such managers are innovative, others merely energetic. They are not rare in management and appear often in design management positions. An associated weakness is too strong a focus on personal accomplishment, while paying too little attention to employees' needs.*

Management by memo *is a dreary management style characterized by total reliance on documentation, with little attention to what has been accomplished. This style may be countenanced at upper levels in large organizations, if face-to-face management becomes impossible because a manager must spend all available time reacting to higher management. Notwithstanding the need to document activities, challenging system developments are too dynamic to allow their being handled by mail. A slight modification, using electronic mail, may speed the process somewhat, but results are almost as poor.*

Management by committee *needs no explanation. When rationalized by a desire for "broad participation," it usually produces huge waste of both managers' and designers' time. The larger the group, the more wasted. Decisions are obfuscated; in fact, there is probably no one in charge. This is not to say that group meetings must be avoided, since for some purposes they are the most effective means of communication. But they should be used in moderation.*

Japanese businesses typically practice a *consensus management* style which to an outsider to their culture, at least, looks much like committee management. The inefficiency of U.S. committee activity has a lot to do with poor agenda preparation and time control by chairpersons, along with the propensity of some individuals to discourse at great length, about very little.

9.3.1 The resource manager

Many employees perceive management as being largely a matter of personnel supervision. In fact, managers are expected to be *resource managers*, of whatever resources pertain to the circumstances. At the lowest levels of management, the resources to be managed *are* mainly people. At higher levels, they are, almost exclusively, *money*. At levels in between they may include buildings, land, and production machinery as well, all of which require managing.

Frequent shortfalls in upper management's attention to needs of employees is natural. Employees are not, in a real sense, investments of the firm. They cannot be depreciated or sold. In most organizations, upper-level managers learn not to expect sensitive handling by the higher-level managers above them.

Dynamics of resource management What then does resource management really require? There are almost as many schools of thought as there are successful managers (i.e., those who are able to survive, and thrive). Some managers might be classified as "real-time" managers, concerned about up-to-the-minute project status, the facial expressions of clients during presentations, and *weekly* financial reports (when the norm is monthly).

While all managers must frequently react to immediate problems, some managers attempt to understand *trends*. This means that they must study business data of both past *and* present. They look into the future as well, often using as vehicles *strategic plans*, whose time horizons typically extend 5 to 7 years.

Depending on the type of business activity, the real-time manager has a place. However, a competent manager of system architecture and design needs a broad time-line. Past design work represents a jumping-off point for new design, while projection of future changes in technologies and requirements enables the designer to create systems with long useful lifetimes. However, design activities take place in the present and must be checked frequently lest they wander off track.

When focus is on several time scales, different sources of information are required. Current designs often need to be based on earlier ones still in users' hands. A design manager requires a reliable source of "corporate memory," either through personal experience or derived from experienced designers on the staff. Managing in real time requires direct contact with what is happening, meetings with employees in their offices or labs, phone calls to subcontractors, with an efficient assistant to coordinate calls and travel. Managing from the predictive point of view requires spending time at conferences or symposia, visits at universities, and personal reading. The skilled system design manager should do *all* of these well. Clearly, an adaptable personality is desirable.

Schedule and cost control Schedule and cost control are especially important in complex system development, since there are so large a number of tasks, procurements, and fabrication operations to be done. Good schedule and cost control requires a combination of intelligent planning, responsive and accurate data collection, quick understanding of financial data sheets, and the determination to correct problems *before* they get out of hand.

The first component in management control of costs and schedules is *information*. Financial and technical information arrive separately, and in different forms.

Most industrial firms accrue *financial data* over three time periods: monthly, quarterly, and annually. Monthly data is used for project control, quarterly and annual data for tax purposes, business analysis and stockholder relations. Some managers would prefer to review financial status more frequently than monthly. This is usually of no real value, because the most important financial activities (billings) are almost universally carried out on monthly cycles. Even on a monthly basis, a firm's accountants, presenting fiscal data, may be required to "adjust" it. For example, if payrolls are distributed biweekly, in some accounting months there will be three and in others only two payrolls. Certain payments may be made just before or just after the monthly closing of records. Without explanation or adjustment, this could give management an incorrect view of financial status.

Although top management is concerned about income versus expense, a design manager's view of financial records is almost solely one of *cost of operations*. Monthly cost data is usually presented subdivided *organizationally*, then by *project* and *expense category*. Categories will include wages, equipment, lab supplies, office supplies, telephone, and sometimes items such as rent, electricity, or overnight delivery services.

The objective is to provide the manager with a level of breakdown sufficient to permit identification of trends, but without so much detail that the report would be hard to interpret. In some organizations with especially responsive computer support, a manager can request a special financial report layout in addition to the standard one. As an aid in assessing financial progress, the monthly report may include plots of cumulative expense (or income) for certain account categories. These may be displayed as segmented curves or bar charts, and will show monthly totals or changes, usually from the beginning of the present *fiscal* year

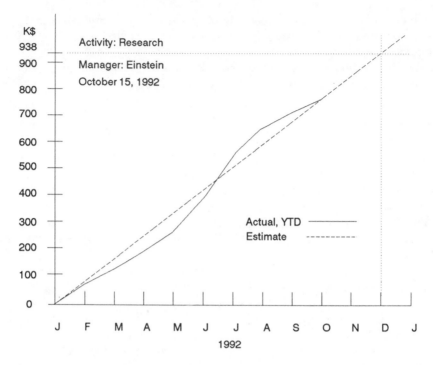

Figure 9.9. A typical accounting plot describing rates of expense of a project over a current year. (Similar plots might be provided for salaries, expenses, capital equipment, communications, and other accounting categories.) Such plots quickly reveal trends which would not be as apparent in data tabulations.

(Figure 9.9). In strongly project-oriented organizations, monthly project data may be shown from the start of each project.

In most design organizations, *progress* is also documented formally, though briefly, on a monthly basis. Documented progress may include estimates of percent completion if that is meaningful. It more often deals with bad news, punctuated occasionally by some good news such as a project completion or award.

Short progress meetings of the members of project teams may be held weekly. Reports on completion of scheduled project activities may be made at such meetings and noted by the manager. When meetings are held on planned schedules at the different management levels, it allows important but nonpriority information to be passed between employee and manager and up the chain, in no more than a week. Priority information of immediate interest can be passed along in person or by telephone.

Financial and technical information documented, reported, or otherwise obtained by a design manager is the principal basis for management control at working levels. At highest levels in most organizations, and lower levels in some, activity *outside the enterprise* may be the most important component of management data. Managers may obtain external data about clients and competitors in various ways. Periodicals, industry newsletters, reports from sales and marketing organizations inside or outside the firm, consultants, and occasionally new employees who previously worked for competing firms—all are potential sources of information.

Management attempts to influence technical directions or control expenditures take forms depending on management style and perceptions of importance and urgency. A manager may decide to shift more personnel to a project which requires more than planned effort. An individual whose budget has gotten out of bounds, or who is spending far *below* what was planned can expect to be asked to give the reasons. This may occur privately. However, the manager may choose to bring it up at a group meeting so others can also get the point.

In some cases it will be necessary for a manager to alert higher management when planned performance, cost, or schedule targets will not be reached. With thoughtful executives, the sooner they are alerted, the better. However, "real-time" managers may react so sharply at such news that their subordinates will go to almost any efforts to avoid bad news. Transfer, demotion, or even dismissal may be the reward of the manager who brings such news repeatedly. Young managers quickly learn that they no longer enjoy the forgiveness they may have been accorded as nonmanagers.

Dealing with program uncertainties In system architecture and design, uncertainties enter a program's path from many sources: clients, subcontractors, program personnel, immaturity of technology, economic inflation, or competitors' pricing—to name merely a few.

System development managers must accordingly deal with uncertainty, as a *major* part, if not *the* most important part of their task. The best-laid plans seldom work out without snags along the way. Some managers focus *solely* on variances from business as usual in a style known as *exception management*. For some this works well, while others seem to gather more problems day by day. One key to successful exception management is for the manager to remain in close enough contact with the organization, that bad news can be picked up far in advance and acted on thoughfully and without sense of crisis. The ". . .don't bother me" manager doesn't usually learn bad news until too late to fix the problem. When exception management is overdone, it becomes "crisis management." The true crisis manager needs several cups of strong coffee and a serious business or technical crisis merely to get going in the morning. . . not much fun to work for!

The development of a plan or schedule, establishment of an architecture, or layout of a design is a satisfying thing. It often leads its developers to the sensation, at least temporarily (and until new problems arise), that uncertainty has been eliminated. The consequences are that many plans, schedules, or designs make few or no provisions for unforeseen difficulties. The prudent manager is not paranoid but insists that every plan, schedule, or design contain recognition that uncertainty has been assessed and responded to. The response should take a form appropriate to the nature of the uncertainty. Schedules should allow time for unforeseen activities or iteration, cost breakdowns should have contingency funds (even if they must be "hidden" from accountants or clients), designs must respond to the possibility that demanding performance criteria may not be met.

The prudent manager also recognizes that if too much "insurance" against uncertainty is built into the organization's plans, it may not produce competitive results. Careful attention must be paid to avoiding *compounding* of uncertainty allowances, as by expanding every estimate at every level. (This is natural for a designer accustomed to including a factor of safety at each level in a design.)

Once equipped with a rational protection against downside risks, the prudent manager must fight to prevent worst-case predictions from being fulfilled. If employees perceive that a worst-case plan (taking advantage of all risk-removing contingencies) is acceptable to management, they may pursue it too literally, and actually expend the contingency allowances. Hence, it is desirable that schedule, cost, and performance progress be charted against *reasonable but challenging* norms. A fitting technique involves planning estimates of worst-case *and* best-case performance. Actual performance should lie between, but with the challenge to move it nearer the best-case estimates.

Just as an efficient helmsman does not attempt to steer a course which requires continuous correction, a shrewd manager usually does not apply "hard rudder" when first aware of a possible slip in schedule or cost overrun. However, some sort of precautionary action or plan for future action should be considered whenever ominous signs appear. A manager new to a particular organization, or managing a new organization, may not understand the dynamics of his or her information sources and the effectiveness of controls, for several measurement cycles (i.e., months) or longer. Though it is important to get into position to execute available control as quickly as possible, most smart managers do so by immediate insistent nudgings of their controls, rather than by hard-over steering.

9.3.2 Managing external interfaces

The external entities most important to a system architecture and design activity are often clients, subcontractors, and competitors, and industry and standards organizations. Occasionally, stockholders, legislative bodies, communication media (e.g., the general press, trade press, or broadcast media), or even the general public must be dealt with. In large firms supporting marketing and public relations organizations, those in design activities may be almost totally shielded from direct external involvement. In smaller ones, however, these contacts must often be faced at first hand.

Client interfaces Where there is a marketing organization, the design organization's contacts with clients will usually be indirectly, through marketing, or at least with participation by marketing personnel. At the least, engineering personnel will be asked to document client contacts, so that marketing will be made aware of any requests or agreements. The marketing representative assigned to a client may set up and be present at meetings with that client.

The most common reason for meetings between design personnel and clients is to obtain more specific or complete data on requirements not well, or fully, documented. Some firms encourage controlled interaction between important or prospective clients and the firm's system design personnel, with the intent of impressing clients with the firm's technical know-how. This may indeed have positive value. However, results can be disastrous if design personnel get involved in technical arguments with client personnel or fail to treat them with ordinary courtesy and respect. Personality screening and informal preparation is advised for *any* personnel who will interact with clients or prospects.

Professional publications and meetings Technical personnel in any competent design organization should be encouraged to participate in professional activities through local and national meetings, membership and receipt of publications, and where possible submission of publications. The benefits include becoming more aware of competitive designs and of applicable new technologies or methologies. The negative side, especially for firms which lack self-image, is that employees may receive hints of job offers, or even real ones. Professional employees and managers alike must remember that the firm's proprietary "edge" resides with these employees. The firm's financial support of conference attendance, both the direct cost and that of the employee's time (with overhead!) represent a sizeable contribution. The employee is expected to remember who paid the bill, and act in the employer's best interests. It is not unreasonable to point this out.

Industrial *researchers* expect, and are usually expected by their management, to publish and present their own work in professional publications and meetings. The same is less often true for design personnel, whose work usually remains proprietary until near the time at which the system is ready for sale. More progressive system firms encourage their design employees to publish, after it ceases to be of immediate benefit to competition, but before it becomes stale. A large fraction of technical society publication editors are academics having little interest in design principles or experiences. Accordingly, most design articles appear in the trade press, or in a "house organ"—a periodical published by a firm to enhance its own image.

Market evaluation for systems developers Market evaluation, though important for both custom systems development and commercial developments, takes considerably different forms in those market areas. In custom system development, emphasis is placed on understanding the needs of a particular client, or a potential client. In speculative development of commercial systems, the developer needs to estimate and weigh needs and desires of many clients, though certain major clients may be selected as being representative of their kind. Understandably, in the custom system environment, interaction between system developers and clients may be closer than in commercial developments.

Information must be gathered from various sources and in many ways. Both plans for its continual collection *and* processes for its timely analysis must also be developed. Otherwise, information may be fragmentary, poorly distributed, and sloppily arranged. But, this won't matter because it cannot be put to good use without thorough analysis. Many firms without continuous need for analysis of some market area will contract for this work from specialty marketing organizations.

For highly specialized system applications typical of small systems firms, there will often exist no competent outside marketing organization. The system developer should recognize this and lay out marketing activities. These may, for example, involve a small in-house organization to collect market data from "clipping services" which screen newspapers or other information sources. Outside marketing consultants should be called in when required. They will expect a retainer fee, for their contracts should proscribe similar work for others. Consultants in custom-system market analysis are often retired or former employees of key clients who maintain their contacts within the client organization (typical examples are retired military personnel). Consultants in *commercial system* market analysis usually obtained their experience as employees of a system user or of a leading system development firm.

No matter what sources of market information are available, and how extensive they appear to be, analysis is essential to relate the information to the particular needs of a system firm. The same collection of market data relating to midsized computers should be analyzed differently for use by two computer makers whose business is centered in mainframes and minicomputers, respectively. The two firms would approach the needs of that gross market differently because their system offerings are different. If an outside marketing firm is to go farther than merely organizing collected market data, it must know considerable about the capabilities *and* plans of its clients. If the firm is reluctant to release details of its system developments to outsiders, it *must* analyze markets internally. Internal marketing is the customary practice for established firms in their main markets. If such a firm is contemplating entering a new market, it will use outside analysts. They will have a double hurdle to overcome: first, evaluating the market; then, convincing their clients in a course of action responsive to that market.

Market analysis for new systems Market analysis addressing a new product or product line normally attempts to matching its characteristics, *including cost*, to estimated needs or desires of *identifiable groups* of potential customers. Firms will be identified and their personnel interviewed, where feasible. Counts will be taken of similar firms in the local area, region, nation or world (depending on the breadth of marketing intended). The number who might conceivably purchase the proposed system, multiplied by the average numbers of systems each might purchase, is estimated and accumulated, to define the *potential market.*

Thus there might exist within the United States or internationally, 1500 organizations, each of which might be able to use *one* of the proposed systems, 300 which might be *able to use* two, and a potential market for 400 more units to clients which could use three or more. This would constitute a potential market of 2200 units. To this must be applied a realistic factor estimating what fraction of this potential market would be *likely to purchase* the systems. This factor will depend on the existence and market position of competitors and on the vigor employed in selling the new product. It is, however, unlikely to be much above 10 percent, and even less if well-financed and technically capable competition is already in the market. In our example, that factor suggests about 220 sales of the system, over its sales life. While this estimate should be discouraging to the developer of an automobile, it would not be, to the developer of a containership. Nor might it discourage an individual who planned to sell computer systems and special software to physicians.

To be most effective, market data gathering and analysis should be carried out continuously or, at least, regularly. As with any other supporting activity, one may have too little or too much. Most large firms evolve marketing practices which will be applied to any new product or system. Though a mature marketing organization can relieve the system design manager of much of the tedium of marketing, it requires guidance if it is to serve special needs of the system area. In turn, the system manager must be alert to whether the marketing organization seems to be headed in reasonable directions, and ask the same objective types of questions which would ordinarily be directed to designers.

EXERCISES

1. Design a hypothetical system design organization, for a specific type of system with which you are familiar. First, identify those system specialties so important that they should be represented by in-house expertise. Identify the essential areas of building/testing (or simulation) activity required to prove out the system designs, and how many individuals will be required to perform and support this work. Note the facilities critical to this support. Consider the kinds and amounts of technical support personnel which the group will require: technicians, document and standards librarian, technical writers, artists, or others. Consider need for administrative assistance for correspondence, meeting arrangements, personnel matters, equipment and supply management, and project planning. Lay out the results as a proposed organization. Then examine and rebalance creative, technical support, and administrative personnel in a way most consistent with the needs of the organization.

2. Define a research program in system technology. For a system area of which you are knowledgeable, (a) characterize the nature of the systems and (b) describe the *principal systems design issues* which you believe to be currently important in this area of system design. These may relate to: recent availability of pertinent new technology; need to react to new requirements or changes in typical system environments; or to availability of new design tools or production technology. (c) Examine these issues, their complexity and character, then describe theoretical or experimental work which could permit progress to be made. (d) Based on your response to (c), describe appropriate *objectives* for research or methodology development which could shed light on the major issues, through experiment, simulation, conceptual system design, or some other specific activity.

3. For each of these requirements or objectives: (1) identify *uncertainties* which it appears to embody from the "requirement" statement shown. (2) Suggest likely source(s) and plausible reasons for that uncertainty. (3) Recommend a course to follow which might resolve the uncertainty or demonstrate that it is irresolvable.
(a) "A system which will halve the response time of city ambulances for answering emergency calls"
(b) "A system through which you will be able to do all of your banking by a single phone call"
(c) "A system which will permit federal crime-fighters to keep track of every convicted felon on a 24-h basis"
(d) "A dieting system by which one can lose 40 lb or more in only a week"

4. "Requirements creep" refers to continuing change of system requirements occasioned by matters outside the control of the system user. A concern is that system design based on a current requirement will be obsolete by the time of system delivery and installation. The effect of requirements creep can in some cases be reduced by (1) minimizing system development time to put systems in the field faster, (2) addressing the requirements farther ahead in time, or (3) fielding initial systems which, though not fully responsive at installation, can be incrementally upgraded.

Find and describe some system area where system requirements have been moving rapidly for several years or more. (Among other areas, integrated circuit manufacture or test systems would qualify at time of publication.) By studying systems being delivered currently or over a period of time, determine in which of the three ways these rapidly moving requirements are being addressed. If none of the three applies, what *is* being done?

5. By library study of recent business or trade publications within a selected system industry, identify two sets of firms which are cooperating with one another to produce systems beyond the capabilities of one firm. Describe the objective(s) of cooperation (what will they design or produce), means or mode of cooperation (how is their separate and joint activity controlled), what functions each will perform, and why they engaged in cooperation.

6. Lay out, in PERT diagram form, the sequence of tasks required to prepare some favorite meal, including the preparation of three of more recipes. (Use recipes from a cookbook, if you wish, to define specific steps.) Assume you have as many helpers, pans, ovens, etc., as needed, to perform all the tasks which can be carried out simultaneously. Assume all components needed are immediately at hand. For simplicity, include in time required for a following step, transfer of food prepared in a previous step. Determine the time which should elapse from start to finish and mark the critical path(s) requiring that time.

7. For the exercise above, assume that *one individual* must do the entire job, equipped with sufficient equipment to hold and cook all recipes in parallel. Sketch the activity schedule which you believe to produce the shortest completion time, and describe the strategy you employed to find it.

8. Characterize the management style of one, several, or all the managers you have worked for or known well. Use specific anecdotes as a basis for your description, including any evidence suggesting that the individual used different styles under different circumstances.

9. Perform a market analysis exercise to attempt to estimate the *potential U.S. market* for a stenographic system which would deliver, directly from dictated material, a draft letter or memorandum both in computer-readable and printed form. Typical performance in the hands of an average dictator is 1 misinterpreted and 1 unrecognizable word in every 10.

BIBLIOGRAPHY

Norman R. Augustine, *Augustine's Laws*, Viking Press, New York, 1986. This witty top system manager, industrial executive and high Defense Department appointee looks at U.S. business and management practices and finds some important lessons.

Harold Kerzner, *Project Management – a Systems Approach to Planning, Scheduling and Controlling*, (2nd edn.), Van Nostrand Reinhold, 1984 . A highly readable modern treatment written by a former military and industrial manager.

J. J. Moder, C. R. Phillips and E. W. Davis, *Project Management with CPM, PERT, and Precedence Diagramming*, (3rd edition), Reinhold, New York, 1983 . An updating which takes into account resource limitations and other constraints.

Albert Shapiro, *Managing Professional People*, Macmillan/The Free Press, New York, 1985. This book focuses on creative employee selection, development and motivation.

SHOPLINK—A CASE STUDY

A.1 Background

This appendix discusses the requirements, architecture, and many of the design concepts employed in "Shoplink," a special-purpose information distribution system. It is classified, by its application, as a Distributed Numerical Control (DNC) system. The system was developed in 1984-85 by the author, Stephen Silverman, and several detail designers in his firm, which currently produces and markets the system. Its principal function is to distribute *programs* describing the tool path for the machining of mechanical parts, to the numerically controlled machine tools which will use the programs, and to store modified programs transmitted from the machine-tool controls.

This appendix is a pragmatic insider's look at architectural and design evolution, a viewpoint seldom found in professional or trade literature. Its organization follows closely the plan of the book itself. Only a fraction of the issues discussed in the book were pertinent in Shoplink. There were no special system or user safety considerations, for example. Despite these understandable limitations, it is hoped that the reader will find the treatment a useful reinforcement to the main body of material.

A.1.1 Application area

The system is a data storage and transmission network specially designed for application in machine shops using numerically controlled machine tools. Machine shops, for the most part, are in the business of "machining"—removing metal in chips or strips from solid blocks or rods, using sharp-edged hardened metal tools, to create parts for machinery, automobiles, home appliances, and other industrial products. A considerable fraction of these shop operations are owned and operated independently of product manufacturing firms and therefore referred to as *job shops.* Both independent job shops and their counterparts in final-product manufacturing firms range in size from as few as a single computer controlled machine tool to as many as 50 or more.

The market is, for the most part, a cost-conscious one. Though there are a few such shop operations which have been set up by huge corporations as showplaces, most are busy, disorganized-looking operations closely supervised by their owners or foremen.

Numerically controlled machine tools Numerically controlled machine tools have been produced since the 1950s, initially for very high volume precision machining. Their use has grown continually until today they are used for both high- and low-volume production. Though they require certain skill to load and set up, actual operation is a matter of pushing a few buttons and monitoring the machine while it does the rest.

The latest such tools are controlled by integrated circuits and may contain large memory capacity—the same kinds of components as used in Shoplink. However, the majority of the machines in use in the United States are older models without communications ports for inputting the data needed to operate them. Most of these machines produced up until the early 1980s were equipped with a *paper tape reader* as an input device. The industry was built up around paper tape, and some of it still operates using that medium.

The data, *part programs* as they are called, are almost always alphanumeric in content (rather than binary). There is nothing about the data which limits it to the paper tape medium. Older *numerical control (NC)* machine tools employ slight variants of a 7-bit-coded character set identified with Electronic Industries Association (EIA). Newer, *computer numerical control* (CNC) machine tools principally employ the same ASCII standard character set (7- or 8-bit codes) used in personal computers and most data communication.

Newer machine tools may still be ordered with paper tape reader and punch unit, but are usually equipped with a *serial signal interface* conforming closely to the EIA *RS-232C* interface standard. This, interface is used in personal computers and modems for telephone transmission of digital data.

CNC machines of intermediate ages usually include both paper tape reader and punch. Memory of a only few thousand characters is not uncommon because of its high cost when these machines were designed. A user can usually *edit* a part program in the machine tool's memory, using alphanumeric keys on the control console of the machine tool. While this was acceptable while very simple programs were used, large mold-cutting may require programs as large as a million characters. These program sequences are very complex and problems cannot quickly be interpreted by inspection. Design correction in the shop is thus becoming less common.

The machines behind the tape readers and RS-232C interfaces are varied. To transmit a part program (in computer storage, a *file*) serially, it is necessary to carry on a digital conversation called a *protocol*. Many unique protocols have been devised by CNC machine tool designers, though standard protocols have been available. A representative protocol referred to as Xon/Xoff calls for the receiving party to transmit a control character (usually an ASCII character 17 = 11 hex) when it is ready to receive. The sending party transmits file characters until an "Xoff" (usually character 19 = 13 hex) is heard from the recipient. Thenceforth transmission occurs after an Xon until the next Xoff. Is is usually completed by the sending party adding an ASCII end-of-transmission character (EOT = 04 hex).

As elementary as this signaling plan seems, many variants are found: some CNCs expect the sender to announce readiness to send, while others must first announce willingness to receive. Some may *abort* if a final EOT arrives, while others aren't happy without one. There are many other protocols built into CNC machines: a few simpler than this but most more complex. This variety is the main reason why a special data system such as Shoplink was needed.

Machine shop operations In the typical shop operation of the late 1980s, part programs are laid out using graphically aided computer based manufacturing programs by a few designers, for shops having 10 times as many machine operators. In few instances do these individuals have formal training in computers or software development. The design programs do not require it.

Many shops using computer-aided manufacturing are not equipped with a DNC system. In these operations, small rolls of paper tape (at 120 characters per foot, 20 ft is a common length) fill numerous cabinets. When an order is received for a quantity of parts of a design previously produced, a supervisor or administrator locates the correct tape in a storage cabinet and gives it and any instructions needed to a machine operator. If the tape cannot be found, it must be punched from a file stored on the CAM computer. Of late, this is often a personal computer.

If changes must be made as a result of discrepancies in the part(s) produced, a new tape will be punched. Clearly the frequent handling of many different versions of cryptically marked tapes, by personnel who have more useful things to do, is another reason why a better method of data handling was needed. The tapes did, however, make it possible to ignore the variety of transmission protocols. Since the tapes' dimensions and their (ASCII) codes were standardized, the machine control had complete control over input of the tape data. Furthermore, tapes served as a cheap and simple memory, for the older NC machines. If a tape is too long for its contents to fit in memory, even CNC machines can usually be operated on data directly from the tape reader.

A.1.2 Initial objectives

This application should quickly be recognized as having a very limited market size. Few prospective purchasers other than a machine shop owner are likely to be interested in a data handling system suitable for use with a GE, Cincinnati Milachron, Bridgeport, or Hitachi machine tool. One exception might be the builders of the machine tools themselves. However, their own DNC products seldom address needs of a shop including several CNC manufacturers' products.

The tools themselves cost $100,000 and up, and the materials they machine are often quite costly. There was considerable incentive—especially in shops using 15 or more tools—to overcome the errors and inefficiencies to which paper tape was prone.

The principal in the Shoplink project, Steve Silverman, had earlier experience in this market. His first DNC product appeared about 1980 and was operating in more than 300 machine shops when Shoplink's development began. A brief discussion of this first-generation DNC product will set the stage for discussion of Shoplink.

The N.C.H.Q.   *N.C.H.Q.* ("Numerical Control HeadQuarters"), marketed in the early 1980s, was a stand-alone data-transfer unit similar in appearance and internal operation to early personal computers. Based on the CP/M operating system and an 8-bit Z80 microprocessor, it stored files totalling up to about 90K characters on a "floppy disk" and contained a simple line editor for creating or editing short part programs. On its front panel were eight push-button switches allowing it to be electrically connected by cables to the inputs of up to eight machine tools. A commercially available cathode-ray-tube terminal served for display console and keyboard. Either 32 or 64 kilobytes of memory could be installed, and one or two 5 1/4-in diameter diskette drives.

N.C.H.Q. was equipped with about a dozen file-transfer protocols, suitable for the limited numbers of tools which had serial inputs when it was designed. In its operation, a diskette containing the desired part program (as one of multiple files on the disk) was taken from a diskette library and placed in the drive. The operator then pressed one of the eight output selector buttons, then keyed a command to the computer which identified the file name, protocol, and parameters to be used. The situation was complicated by need to mount the disk, start the program, and then coordinate it with the control operations on the machine tool.

Because many of the older machine tools had no serial digital interface, a variety of serial-to-parallel "adaptor boards" were designed, each adapted to having the parallel signal end connected to tape reader/punch data and control signals in a particular machine-tool control.

N.C.H.Q. was a popular product at the time Shoplink was begun, but it represented late-1970s design, in a computer field which, at the time, doubled raw performance every 3 years. Something better was needed, and soon.

Recognized needs: The competition   In 1984, advances in small digital systems were represented by growing acceptance of 16-bit microprocessors for personal computers. While the Apple, Osborne and other 8-bit computers were struggling and would not survive long, the 16-bit machines—represented at that time by the slow *IBM Personal Computer*—were no faster than 8-bit ones.

More importantly, there were growing trends toward connecting numbers of microcomputers, printers, and other devices together by serial data networks. Though this had no technical barriers and had been done for years with larger machines, the difference was that low memory costs (a minimum was reached about that time) and microcomputer controlled devices could be put together to handle very large amounts of data at very low costs.

New personal computers such as the Apple Macintosh featured more sophisticated user interfaces which made the earlier ones, and even the PC, look clumsy by comparison.

What was needed, Silverman determined, was a DNC system which could make use of low-cost personal computer technology (for memory, storage, and user interfacing), coupled with easily expandable communication networks.

Competition was in view. As is often the case, it came in two forms, respectively more and less expensive than one's own projected system. The more costly DNC systems used a separate terminal and a computer well endowed with memory, for each machine tool. The low-cost solutions were simply computer programs which read a stored file and transmitted it one of a limited variety of CNCs. Because machine tools may read in a new file only once or a few times a day, and have no need for a DNC otherwise, equipment committed permanently to a machine appeared wasteful. Although the type of push-button *multiplexing* represented by N.C.H.Q. would not be acceptable, it was clear that minimizing the necessary investment *per machine* should be a key design objective.

As for the competitors who wrote software for personal computers, they faced two limitations: Nothing prevented a user from copying the program or giving it to a colleague. There was no good way to charge enough for a program to provide the heavy customer support needed in non-computer-literate customer establishments.

A.2 Architectural considerations

Architectural studies went on for 2 months, beginning in August 1984. A few architectural issues had not been settled when system development

began with the HUB design. We present below the flavor of the deliberations, without getting involved with too many details.

A.2.1 System concept

The system needed to be able to support shops containing at least 30 NC
or CNC machine tools with common part-program storage. Since personal
computer disk files were already available which could store large
libraries of part programs, it was decided to use a personal computer, as
what came to be called the *Library* but would colloquially be termed a *file
server*. At that time the IBM PC and compatible "clones" had not captured
the PC market, so it was originally intended to be able to use any of
several popular personal computers (Apple II and CP/M based 8-bit
machines, and the IBM PC). This anchored one end of the system, but the
other end (which was assumed to be at the machine tools) called for 30 or
more terminal points. Since no personal computer available at the time
was able to handle that many directly connected serial ports, it was
realized that some intermediate level was needed in the connecting
network.

This level evolved in hardware form into what was called a *HUB*,
Shoplink's keystone element.

System structure Figure A.1 depicts the complete network architecture
of Shoplink, in a format used for diagrams of networks having variable
numbers of branches.

The HUB is in some ways an offspring of N.C.H.Q. It supports eight
(or four, in a low-end version) machine tools. In addition it has two serial
communications ports for control terminals (called *Shop Terminals* to
distinguish them), and a connection to the PC-Library, which also has the
alias of *Master Terminal*. Terminals may also be connected on the serial
lines which serve the machine tools, where they are referred to as *Line
Terminals*.

The key to this system structure is that the HUB also is the controlling
device in the structure. The HUB responds to specific keyed input
sequences (always initiated by a carriage return character) from any of
its 11 input ports. In case of inputs which request information held in the
Library PC, the HUB formats a command and initiates an interchange
with the PC which causes the desired action (directory list, file transfer,
queued file list, etc.). The PC-Library thus has two roles to perform: as
a "dumb" terminal, and as a file server. Communication between a HUB
and the PC is controlled so that terminal operation is suspended when a
command is being responded to. A user is unaware of this complexity.

The HUB serves, in addition, as a file converter and protocol adaptor,
changing ASCII files to EIA and the converse on the fly if required, while
carrying on a preselected protocol to permit communication with each

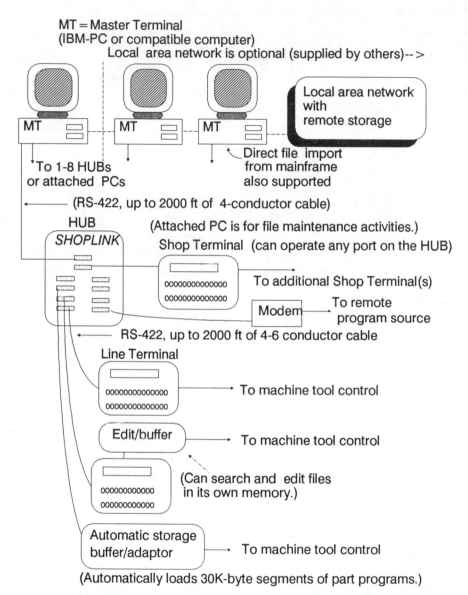

MT = Master Terminal
(IBM-PC or compatible computer)
    Local area network is optional (supplied by others)-- >

Local area network with remote storage

MT   MT   MT

To 1-8 HUBs
or attached PCs

Direct file import
from mainframe
also supported

◄── (RS-422, up to 2000 ft of 4-conductor cable)
    HUB   (Attached PC is for file maintenance activities.)
    SHOPLINK   Shop Terminal (can operate any port on the HUB)

    00000000000000   To additional Shop Terminal(s)
    00000000000000
                Modem ── To remote
                        program source
    ◄── RS-422, up to 2000 ft of 4-6 conductor cable
    Line Terminal

    00000000000000   → To machine tool control
    00000000000000

    Edit/buffer ───── → To machine tool control

    00000000000000   (Can search and edit files
    00000000000000   in its own memory.)

    Automatic storage
    buffer/adaptor ── → To machine tool control
    (Automatically loads 30K-byte segments of part programs.)

MINIMUM SYSTEM: 1 PC w/floppy drive(s), 1 HUB, 1 to 8 machine tools
MAXIMUM SYSTEM: network with 24 PCs, up to 24 drives & 192 HUBs,
1536 machine tools or other devices.

Figure A.1. A network representation of the Shoplink data system, described in the text.
Each type of equipment which can be connected into the system is illustrated along with
applicable limitations on terminal counts and connecting-line lengths.

machine tool (or other device) on its ports. The HUB has a storage buffer adequate in capacity for most small part programs on a *download* or transfer *to* a machine tool. In *upload* transmissions, it isolates the machine tool when the PC needs to store the contents of its own storage buffer.

A necessity for the design concept is that the memory buffer in the HUB be a completely dynamic *ring buffer*. That is, it continually transfers characters in at one end while others are being transferred out at the other and will not accept more characters when full.

User interfaces A principal aim was an easy-to-use interface. Except for the Master Terminal (a PC) all other terminals used in the system are simple ones, unable to support graphic operations. In that environment, it was decided that *user-friendly* meant *single-key commands* (N.C.H.Q. required a user to enter a command string including several function selecting characters and a file name). Filenames required some user-friendliness, too, hence users were prevented from entering a name whose format or any of characters were unacceptable. If an incorrect character is entered it does not display. A "dot" cannot be entered without first entering one or more filename characters; but if eight filename characters have been entered, the dot appears automatically.

Other user-friendly features include automatic generation numbering of file extensions (.001 for the original, then .002 for the first revision, etc. to .999) unless a "rename to .BAK" option is selected. A reminder of the one-character selection required next, is given at each step in a command selection process.

Dozens of port parameters (communication rate, protocol selections, character modes, etc.) are selected by a HUB command which can be executed only from the Library or an authorized Shop Terminal. Though this usually is done only at initial installation, it is simplified by a feature which allows another port to be parameterized like one previously set up. The port personalization data can be saved in a file in the Library, and if need be restored to the original, or a replacement, HUB.

External interfaces Remote access to or from Shoplink was expected to be required by some users to permit transfer of files between a CAD/CAM system and the Library. Both the distance and type of computer could be variables, however. A compromise was required. Shoplink either had to perform the transfer on-line (i.e., as it would for machine-tool file transfers), or if offline the operation would have to be coordinated by an operator.

In the initial design it had been intended to permit a HUB to be connected directly to a minicomputer or possibly to a mainframe computer, rather than a dedicated personal computer Library. This option was carried into early production. It was never used because the minicomputers and mainframes could not be accessed remotely without special software or coordination between two operators.

As an alternative, a port was provided on the PC-Library which could *receive files* (only) through attachment to another computer. Since the Shoplink Library PC was commonly located in a CAD/CAM room, this made the needed coordination much easier. (However, Local Area Networks later provided a much better solution.)

Early on it was realized that some users would need to transfer files to or from Shoplink at great distances. Since this *had to be* controlled from the remote site, it was necessary to connect the remote line to a HUB machine-tool port. An optional protocol set (download, upload) in the HUB was designed to provide XMODEM protocol capability (a public domain ad-hoc-standard serial communication protocol, developed originally by Ward Christensen). This allows error-corrected communication with a microprocessor using almost any communication package, through modems and the public telephone system.

Extendibility considerations In this architecture, extendibility means one of several things: adding new functional features, growing additional structure at the top (Library) or bottom (HUB port) ends, or expanding the numbers of connections provided at other points in the network (growing "sideways").

The chief system feature which assists addition of new functions was the command structure by which HUBs ask the Library for its services. A number of commands were added to the original set, and existing commands were assigned additional parameters, with very little incompatibility. If the Library does not recognize a new HUB command, it sends a message to that effect. An earlier-version HUB lacking some new command does not cause problems either.

The meaning of growth "at the top" only became clear after original design. The growth of personal computer LANs (*local area networks*) began in earnest *after* Shoplink was fielded. Thanks to the "plain vanilla" software design of the Library, LANs allowed part-program files to be distributed elsewhere than on the Library. A minor change later permitted up to 24 different Shoplink systems to operate simultaneously on any PC network.

Growth at the bottom was allowed for by supporting existing serial-parallel adaptor board designs. The HUB design permitted new devices attached to these ports to communicate as if they were Line Terminals and control file transfers. This ultimately allowed Shoplink to be operated directly from the console of modern CNC controls supporting complex protocols. Other bottom-end growth was represented by terminals which could use Shoplink's message-communication features to transmit formatted data such as shop activity reports.

Standards considerations Selection of the IBM PC, and its highly compatible successors and clones, all using compatible versions of IBM's PC-DOS or Microsoft's near-identical MS-DOS operating system was an important and beneficial standardization step. While it meant that Shoplink's builders could not sell a slightly modified PC as a proprietary product, that was never part of the business strategy.

Serial data interfaces between Library, HUB and adaptors were designed to support either RS-232 or RS422 (for greater distances) standards. However, the protocol by which HUB and Library communicate is proprietary and designed expressly to meet (what were thought to be) the communications needs. This protocol, which was fairly complex to implement in software, is used solely *within Shoplink*. Its proprietary character ensures that Library software and HUBs, both controlled by Shoplink's supplier, will work together.

A.2.2 Important uncertainties

In 1984, there were uncertainties both with respect to the components and tools which would be most appropriate to build the system, and in the future requirements of the computer-aided manufacturing application area. At time of design, pertinent capabilities such as local area networks, very large capacity personal computer disk storage devices, read only memory (ROM) chips, and programming-language development tools were primitive but experiencing rapid improvement in availability, performance, quality and price. Although conventional wisdom requires a designor to plan for the future, systems must be built with the tools and components available in the present.

In the application area, there were murmurings that a factory-control netword protocol called MAP, identified with General Motors Corporation, would rapidly replace the polyglot industrial networks. As of late 1989, this has not occurred and does not appear imminent.

The presence of market competition, of both highly hardware-intensive and software-only form, did not give any comfort *at the time*. (What did they know, which we didn't?) It later became clear that Shoplink had a reasonable market niche of its own. It could also attack both the high end of the low-cost (software only) solution and the low end of the "Cadillac" DNC market.

A.3 Initial system design

The first marketable system was ready for operational testing in May 1985. Development was pushed as rapidly as resources permitted, while the remaining N.C.H.Q. inventory continued to be assembled and sold by Silverman.

By the time of the first Shoplink installation, any thought of using as Library any other computer than an IBM PC (or compatible) machine had disappeared. PC clones had at last arrived in quantity, making it clear that Shoplink's users would have no trouble selecting from a range of Library computers. More important for Shoplink's builders, users could afford backup Library systems, since the Library PC (a high-volume product) was now considerably cheaper than the lower-volume HUB and related products.

Economy and practicality dictated an 8-bit processor (a Z80) in the HUB. Few 16-bit choices were then available; what's more, the main job of the HUB was to handle characters— one at a time. Its 64K byte memory address space, together with the uncertain performance of high-level language compilers for the Z80, dictated use of assembly language as the available programming vehicle.

A Library program originally prepared for operation on a class of 8-bit machines was translated for use on the PC without difficulty. It too was programmed in assembly language. At the time, the "C" language did not appear to offer significant advantages, though it was the only high-level programming language efficient enough to operate at the high data transmission rates needed. (Later, when advanced debugging tools became available, C was used for Shoplink offline utilities and special programs such as a remote PC program which emulates a HUB for program maintenance.)

Initially, machine-tool protocols were included in the 32K bytes of HUB read-only memory. One of the two 16K byte ROM chips could be replaced, to add new protocols.

The first delivered system was set up for operational testing in a large nearby machine shop which had previously used the N.C.H.Q. It was selected because it needed the extra capacity Shoplink offered, had a large but well-known variety of machine tools, was viewed as progressive by both clients and competition, and its management and personnel were cooperative. During development it had not been feasible to simulate actual operation of the HUB with more than two simulated machine tools connected, or the Library with more than two HUBs. Thus the field test posed real uncertainties. It was successful, however, and the test installation became a permanent one.

A.3.1 Design criteria and issues

In this section we examine *design objectives* which dictated Shoplink's architecture and design features. As indicated in Chapter 5, many system users have a combination of requirements of near-equal priority. Even system characteristics for which there are no special concerns are usually expected to be of competitive quality.

Reliability In a busy job-shop environment, any delay is revenue lost. A DNC system must not hold up its users more than seconds or very few minutes. If there is any delay from the DNC, it must be for a sound reason such as completing service to another user. Gaps in operation of several days are totally unacceptable. A single character error in transmission of a part program can cause a part to be ruined. Single parts worth up to $1 million (an entire airplane wing) have been worked on using Shoplink-delivered programs.

Several potential causes of unreliability may haunt a machine shop DNC system.

- **Problem:** The shop environment collects tons of metal chips, oil and other byproducts which could damage unprotected electronic equipment.
- **Response:** The PC need not be installed in the shop itself. (Sealed and ruggedized PC compatibles are available if absolutely necessary.) The HUB is a closed unit with no user-accessed parts. A small, rugged, sealed two-line liquid crystal display (LCD) terminal was supplied for installation as a Shop or Line Terminal.
- **Problem:** The electromagnetic environment is often stressful because of operation of welding equipment in the immediate vicinity.
- **Response:** Both RS-232 and the better-protected RS-422 standard serial digital connections are standard on both HUB and Library, the latter to be used for longer cable runs. However, should a file character be received incorrectly by the HUB, transfer will be automatically aborted by the HUB-Library protocol.
- **Problem:** Personnel in such an environment have dirty hands and cannot be expected always to be gentle of touch.
- **Response:** The OEM terminal is the only component which foremen or operators need to touch. It has a flat, sealed, pressure-sensitive keypad which can be wiped clean.

Availability Availability requirements are closely related to the reliability requirements discussed just above. Shoplink was originally designed to service one user at a time.

- **Problem:** Failure of a HUB might cause up to eight machines and their operators to be idled.
- **Response:** A HUB can be replaced by transferring no more than 11 plug-in cables to a replacement HUB. After this reconnection, the replacement can be personalized identically to the original by transfer of a single file from the Library. This could be done in little more than 5 minutes, assuming a spare HUB is available.

- **Problem:** Failure of the Library could cause loss of part programs stored there.
- **Response:** This sword hangs over the head of any personal computer user. Though disks continue to improve in reliability, the only reliable solution is periodic backup to tape or diskette storage. However, if Shoplink is connected to a local area network for data access the problem can be handled in a variety of ways, none of which are related to Shoplink itself.
- **Problem:** What happens to other users when one user transfers a million-byte file at, say, 1200 bits per second?
- **Response:** The market envisioned originally had no such applications, and the problem was not addressed initially. Later enhancements addressed the large-file user market. Very large files are transmitted by the HUB in large, error-corrected segments—first to a two-segment buffer then on to the machine tool.)

Compatibility with other systems The primary concern here was compatibility with a wide variety of NC and CNC machine tools. A secondary one was the ability to move N.C.H.Q. users over to Shoplink with minimal inconvenience and delay.

- **Problem:** How to ensure compatibility with both existing machine tools and ones which system users would purchase after they bought Shoplink.
- **Response:** Devising protocols which will operate properly is made difficult by poor protocol design and even poorer descriptions. In some cases, it seemed possible that tool manufacturers which sold proprietary DNC systems for use with their own tools prepared spurious descriptions of their protocols. More than one first-line manufacturer has outright protocol faults which must be "worked around." The N.C.H.Q. protocol code, most of which was well tested, could be transcribed almost line for line into the HUB firmware. Protocol routines were intended from the outset to be easily added, initially by replacement of a ROM chip.
- **Problem:** How to move N.C.H.Q. users to Shoplink.
- **Response:** This was made difficult by the incompatibility of the N.C.H.Q. diskette format with any other. It was thus not possible merely to copy the contents of diskettes into the Shoplink Library. Since N.C.H.Q.'s design was understood, a program was written for use on it. This transferred all or any selected group of files on a diskette automatically to Shoplink, though a machine-tool port on a HUB.

Ease of use Ease of use was a concern because most machine tool operators, like people, are not computer-trained. We have already mentioned the use of single-key function selection and no-error filename keying. Commands in the HUB provide terminal-type-unique menus on request. For internal functions, the Library uses a selection mechanism somewhat like that in Lotus 1-2-3.

- **Problem:** How to provide common user interface for file operations at all three types of terminals (Master, Shop, and Line).
- **Response:** This in part dictated the architecture. Since the HUB can communicate directly with all three terminal types (Master, Shop, and Line) it provides a common interface without duplicating system control functions.

Enhancement capability During 4 years of field operation, Shoplink's software or firmware was added to or modified over 300 times, its hardware on less than a dozen occasions. Initially there had been no intention of such an incremental evolution of capability. However, the system's command structure and separation of machine tool protocols from other functions made it relatively easy to change. Most enhancements could be added without affecting existing system features or requiring hardware changes in installed systems.

The common user interface at all terminals allowed operational control to be exercised on any of those inputs to the HUB. This made possible a wide variety of add-ons at the "bottom" end, since they could make use of this same interface. Likewise, variants of the Library could use the "Master Terminal" interface to ask the HUB, to ask the Library, to carry out a command. At the "top end," the Library program has been modified and extended to do almost every function possible on a PC, including communication with other applications. One additional HUB and Library feature proved to have important extendibility value: A *terminal-to-terminal message* feature provided an independent means to transfer data and control information around the system, whether or not the Library is immediately available.

Eventually, each of these extension points was used to expand basic capabilities or provide special features required in unusual applications.

A.3.2 Design trade-offs

When a small design team is involved in a race against exhaustion of its resources, trade-offs often receive less formal attention than if a client's resources are supporting the work. Many trade-offs were informal. When there are a great many decisions to be made, some are made subjectively, to make way to consider those believed more important.

The factors which dominated most trade-offs in this design were economic. The principal objective was *to minimize the prices at which DNC systems of high reliability and adequate functionality, simple enough for typical machine operators to use could be marketed.* This led, most importantly, to selection of "plain vanilla" HUB hardware which would be thoroughly exercised, rather than more elaborate hardware which would remain largely idle. It also dictated the use of a conventional PC as Library.

Software and firmware development were initiated, and a software prototype of user interface firmware was demonstrated before HUB hardware design was completed. *After* hardware design was fixed, thousands of detailed software and firmware design trade-offs were required. These had to be dealt with swiftly. Errors and less-than-ideal decisions occurred and some were recognized only after fielding. Corrections were made, but some less-than-ideal features were simply tolerated. Shoplink has a few of these, of which most users are unaware.

HUB-Library interface The most worrisome area of trade-off, in retrospect, was the protocol used to pass commands or files between HUB and Library, and the data rate used. The protocol detects errors in transmission by echoing each character received back to the sender where it is compared with the original character. This performs a very effective error check.

In simple versions of such protocols, the sender waits for a character to return before sending the next. The original design anticipated that many HUBs would be connected to the Library through a *tree* of two-way or four-way multiplexers resembling the tree structures shown in Figure 2.13. These multiplexers, which are commercially available products, would allow a HUB at the bottom of the tree's "roots" to find a path to the Library. The added tiers of multiplexers were to be computer controlled; at any instant, each might be holding 2 to 3 characters. Without any tiers of multiplexers, if the protocol waited for the echo before sending the next character, the data rate achieved would be slightly less than *one-half* that at which communication takes place. With one layer of multiplexers, instead of two characters in transit, there are usually four (one in each link, in each direction). The effective data rate will drop by an additional factor of two. To offset this, the Shoplink protocol was originally designed to allow 16 characters to be out in the HUB-Library "echo loop" before waiting for a character to return.

The multi-tiered multiplexer trees were replaced simply by multiple HUB ports on the Library, when it was determined in a trade-off that users *should not be allowed to* connect more than 8 to 10 HUBs to a single Library PC. The original design assumed that when the HUB-Library transfer was under way, the Library would be doing *nothing else*. Although this was a reasonable assumption initially, it was not true of certain PC variants (typically, lap-top computers) which spend a large fraction of available time refreshing their displays. This meant that some of the (up to 16) characters in transmission at any instant might be "lost." The HUB-Library protocol was then revised to limit it to a single character in the loop, while the communication rate was doubled (from

19,200 to 38,400 bits/second). This maintained actual transmission capability near 19,200 bits/second. (Later, refinements were made to optionally increase speed of transfer of large files.)

Machine-tool protocols Machine-tool protocol expansion was required from the start—users needed to be able to add protocols for new machine tools. Logistically (and for user appeal) it was desirable that added protocols become part of the standard HUB product. Finite memory to contain the protocols required that they be compact. Many protocols differed from others only slightly, suggesting the value of building several protocols in a single routine.

An early trade-off contemplated building each protocol as a modular sequence, built from a set of simple sub-sequences. This scheme is common in general purpose serial communications programs, which send selected character sequences, after receiving a carriage return, etc. This modular concept seemed to have promise for simplifying implementation of additional protocols. But when the collection of needed protocols was studied, it became clear that they were far too different to make a modular scheme practical. The practical choice, then, was to design each protocol explicitly. (However, similar protocols share much of the same firmware, achieving the same sort of memory economy for which the modular concept was intended.) CNC protocol firmware routines were kept very compact by eliminating need to invoke, explictly, a routine transferring data between HUB and Library; instead, this was handled as a background activity, whenever the HUB is waiting to communicate with a CNC.

As a result of this design, more than forty distinct protocols fit into little more than 8K of read-only memory, leaving generous expansion room for growth.

A.4 Integration and testing

Despite its relative simplicity as compared to some of the examples mentioned in the text, the initial integration of Shoplink was difficult and challenging. This took place on three levels which will be separately described.

A.4.1 HUB testing

Although the HUB hardware design was conventional, it was unusual in having a total of 10 serial interfaces (8 machine-tool ports, Shop Terminal and Library). Since the HUB would have no direct access to disk drive, program loaders or operating system features software developers have come to expect, initial tests on the HUB would be relatively primitive. A short program was prepared, and stored in a ROM chip. It first set up a uniform known data rate and other parameters on each of the 10 ports, then wrote to and read from each memory location. It finally went into a loop which caused each port to echo any 8-bit code (e.g., ASCII character), sent to it at the correct data rate. With the aid of a hardware *emulator* which replaced the Z80 and allowed it to be stepped one instruction at a time, the program could be stepped through the instructions, and demonstrated to operate correctly at all ports. The HUB hardware was then ready for its operational tests.

The complex firmware for the HUB could not be tested and debugged efficiently on the actual hardware. A simulator was used, in the form of a complete Z80 computer (Ampro Corporation's "Little Board") with disk storage and a (CP/M) operating system. This could test the HUB firmware in *software* form. Since the I/O design of the Ampro board was similar to that of the HUB (though with 2 serial ports instead of 10), few modifications were required to test HUB firmware. A software debugger was used, where necessary, to step through the program and examine results.

Firmware integration and testing began with hardware test and port setup initialization operations. The input-output routines were then tested, followed by parts of the firmware which allow the HUB parameters to be altered by users.

A.4.2 HUB and Library integration

The Library, being software-only, was integrated in a manner similar to the HUB: first checking hardware initialization, then input/output, and finally file transfers. Before the HUB-to-Library command functions could be tested it was necessary to test and correct the HUB-Library protocol. This proved to be a difficult challenge, since two different computers had to run through their paces. As an initial step, each was tested separately while a keyboard/display terminal was used to emulate the other, while all timeouts in the software were disabled to permit slow step-by-step operations. The protocol sequence as it responded to keyed inputs was compared with that expected, and discrepancies fixed.

By careful isolated testing of both HUB and Library it was possible to correct all important errors prior to connecting them. Integration then expanded into demonstrating that the Library could actually transfer commands or short files to the HUB; during this time the HUB-CNC interface was disabled. Thus, step by step over a period of several weeks the complex HUB-Library communication routines were tested and corrected. Subsequently, timeout conditions were restored and additional testing carried out. Until all this was completed, it was not possible to begin testing transfers to a (simulated) machine tool.

A.4.3 Protocol simulators

The simplest machine-tool (CNC) protocols are those which simply send a file non-stop or receive a file, timing out when nothing more has been received for a certain period. These were also easiest to test, since they could be tested using a "dumb" terminal as a surrogate for the machine tool.

Most machine-tool protocols, however, involve *two-way* sequences. It is necessary to simulate the tool's control using a terminal or another computer. At this point, one was back to the coordination problems encountered in HUB-Library integration, with the additional complication that the HUB-Library transfer must now be operated.

Once the basic characteristics of the HUB and Library became stable and well understood, it proved most efficient to write (or modify) a *simple* simulator program for each supported machine-tool protocol. (These were written in assembly language, out of convenience rather than necessity.) They would transmit a previously stored file or receive a file and store it for comparison with the original. Often, as a convenience, simulators included messages to indicate how far a protocol had proceeded (before bombing!). Real-time hardware/software/firmware testing and debugging is challenging. Under these particular circumstances, this test-based approach proved time- and cost-effective. Organizations of larger size, having more personnel and more generous budgets, might have relied more heavily on analytical tests of correctness, before entering trial-and-error experimental phases.

A.5 Design evolution

The author has emphasized in the earlier chapters the importance of a good original system concept and architecture in making possible extensions to the original design. Shoplink has been given numerous functional extensions since its initial design. All were desirable from some application standpoint, but most could not have been anticipated at the outset. The groundwork had been well laid for some, but not for others. Each of the extensions discussed below contains a lesson for would-be system architects.

A.5.1 Binary file transfer capability

The original HUB-Library protocol used, as error or completion signals, certain ASCII characters found in neither EIA or ASCII machine-tool part programs. A major revision was required, when it was learned that some users wanted to transfer *binary computer programs* for use in their CNC machines from Shoplink's Library. Since such programs could contain *any* 8-bit characters, the original protocol had to be modified to handle binary data.

Lesson: It is best to provide, in features of a system, the greatest generality which may be needed, especially when (as here) there are some bounds on what may be required.

A.5.2 Access by part number (indexed Library)

In the original design, access to files required entering the file name. This is limited to certain characters and can contain no more than 8 characters, plus an optional 3-character "extension" which Shoplink users as a rule employ to designate *revision number* of the file.

A number of users called for ways to specify parts by a *part number* assigned to each part. Though the user-friendly name entry was no longer pertinent, being able to use a user-assigned part number was essential in some installations. This was implemented by associating each part number (20-character maximum length, 1 character minimum) with a file name assigned when the part number is introduced to the Library. Automatic search of a sorted list ("Master Index") of part numbers is used to locate the actual file name. The user may independently select a three-character extension, hence many files can be associated with a single part number if desired. The user of an authorized terminal can enter a new part number, or part numbers can be imbedded in a file "header" which is removed from the file before it is sent to the CNC.

This part-number access feature is incorporated in a variant of the Library. If that version is not present, the special HUB commands are not responded to. Files can be accessed in the original way, even in the modified Library, unless their names begin with "S," which in this Library version is reserved for part-number-accessed files.

- **Lesson:** This was not a simple change. The part number/filename cross-index and its management required entirely new program functions. One feature of the original HUB design greatly aided implementation: When a filename is set up for download (i.e., for transfer to a machine tool), or again when actual download is initiated, the Library is queried about whether it knows of the file. If so, it returns the name of the file to the HUB, and this determines the file which is actually transferred. In this modification, the HUB did not need to "remember" the part number, since the Library returned the correct filename to it. However, the fact that indexed files are an addition is reflected by the appearance of the filename at the user terminal. If the original Shoplink design had used only part numbers, rather than file names, a file name would probably never be displayed at the HUB, where it can occasionally confuse some user who wonders what the name S0001233.001 really refers to.

A.5.3 Library interface change

After about three successful years in the market, Shoplink received the ultimate compliment of being imitated by a competitor. The imitation was a limited one, but the physical appearance of the HUB was closely copied and the terms *HUB* and *Library* used. The competitor implemented Library control commands in color, and used cursor keys to move a highlight for selecting Library functions.

Few users operate the system from the Library. They would ordinarily not purchase a color display for the Library PC. However, in sales demonstrations the color-display menu feature proved very effective.

A similar capability was quickly added to Shoplink. The minimum number of keystrokes to select a function increased with this change, since selecting one out of a dozen options became a two-level, two-key process. However, on balance it is less forbidding to a novice user while asking slightly more effort from an experienced one.

Lesson: There are three lessons here. One is not to underestimate user appeal which may help sell a product, even though product utility may not be improved. A second is that ease of use is partly a matter of familiarization. The third lesson is that departure from development tools and methods used in original development of a system often makes modification difficult or impossible. (This realization came when it was attempted to add new menu routines written in the C language to the highly mature assembly-language Library.)

A.5.4 Downward extension

The existence of "adaptor boards" for the precursor product should have given warning that additional adaptation devices would be called for to fit the special needs of certain CNCs or provide other features. *Most of these represent compensation for inability of Shoplink (or any similar system) to adapt to all possible interfaces.*

Some machine tools require very long part program files, and read them in so slowly that they would tie up both HUB and Library. One Shoplink adaptor provides 60K bytes of buffer memory, and is served the next 30K-byte segment of the file whenever it is less than half full. A second "edit buffer" adaptor was provided for use with machine tools having no memory and accordingly no editing capability. An additional pair of HUB-CNC protocols was provided for error-free upload and download from, or to, this adaptor. A port is provided on it for a terminal, to control file transfer and for editing the file before final transfer to the machine tool.

Even more complex adaptors permit certain CNC machine tools to access files by entering a filename using their own controls (rather than by controlling Shoplink via a separate Line Terminal). These are complicated by requiring a HUB-like double-ended buffer, to permit data to flow in from the HUB while it simultaneously flows to the CNC. Integration is extremely difficult since these devices may have no means of communicating problems to the operator through the CNC's control. With the machine tool control simulated by one computer, the adaptor simulated by a second, and a real HUB and Library all operating in real time, debugging is a true challenge.

Lesson: Provision of Line Terminal control capability (control signals carried by the same line as data transfer) in the original, opened the way for simple implementation of a variety of adaptors. These can map Shoplink's operating sequences to those of systems having very different design philosophies. (But nobody said it would be *easy*!)

A.5.5 User-unique enhancements

Because the Library was implemented in software, it was early decided to implement some special requirements of users entirely within the Library wherever possible. This left the HUB, which has the more challenging tasks and very limited resources, free of user-unique modifications. (Additions to or changes of protocols does not affect the HUB's operational program, since protocols were eventually collected into a single file in the Library, easily updated.) Some of the special user requirements have related to file listing formats, others to access to file-comparison records. Library variants could be implemented in many cases by changing only one of its dozen source program modules.

Some special requirements demanded new commands between HUB and Library. When this was recognized, the HUB and Library were equipped with two generic, normally inoperative single-key commands which could address special needs. Each of these responds by soliciting from the user an input string which is forwarded to the Library. If no command has been implemented, the response "Bad Command" is returned, indistinguishable from any other nonexistent or disallowed commands. If the Library has been modified, however, it acknowledges the command to the HUB. Both HUB and Library transfer into a conversational mode in which the HUB echoes each character back to the Library after it has transferred it to the inquiring terminal. The system will time-out if no key is pressed within a given period, or will abort if a key on the terminal is struck while a stream of characters is being transmitted to the HUB.

Some users had very complex needs. One needed to verify that a file transmitted to a CNC was received correctly, before proceeding. A million-dollar part hung in the balance. The HUB had a standard feature requiring enforced upload of a program after its download, and this was invoked. The modified Library carried out an *offline* comparison of the uploaded file with the original one. The results were placed in a uniquely named file for the particular HUB and port. Once the terminal operator observed that the Library was once again online, execution of the "W" command displayed result of the comparison on the terminal.

Lesson: The quality of the Library structure and its design decomposition was testified to, by ability to make many major within limited portions of the complex software. Unpredicted control- and information-flow adaptations were possible, and many of these later became standard.

The Library program, a simple file server program which required only 4K bytes in its initial form, grew in four years to exceed 60K bytes of program length. This is typical of what can be done with software, which is currently the most flexible medium for engineering design. It is also often the most complex: there may be many thousands of design decisions in a 50K-byte program.

A.6 Final comment

The author hopes that this discussion of an actual system design cycle will bring together for interested readers the concepts taught in earlier chapters. The scale of system activities here was small, by comparison to the "large and complex systems" mentioned often in the book. However, the size has permitted more detailed and more pragmatic reporting of key steps in creation of the system. In the author's experience, following much larger system developments the perspectives of those at working levels are much narrower, and those at management levels shallower, than in this example.

BIBLIOGRAPHY

B. Leatham-Jones, *Introduction to Computer Numerical Control*, Wiley, New York, 1986. This describes the nature of digitally controlled manufacturing machinery and the procedures for programming the controls.

ABOUT THE AUTHOR

Walter R. Beam is an electronics engineer by training. While doing teaching and graduate study at the University of Maryland, industrial summer work led him into mechanical design when he was given responsibility for the development of a new radiosonde weather instrument able to operate to three times higher altitudes then standard radiosondes of that period. That experience was his first in interdisciplinary engineering.

After receiving the doctorate in electrical engineering at Maryland, he spent the next seven years as a research staff member and later a manager at David Sarnoff Research Lab in Princeton, NJ, where his doctoral work in microwave electronics broadened into work on other microwave electron beam devices, television color tubes, and parametric amplifiers. He next served five years as professor of electrical engineering at Rensselear Polytechnical Institute; there he initiated research in magnetic thin film computer memories and served for part of the period as head of the Electrical Engineering Department.

Industrial research called him again and he became manager of exploratory memory at IBM Research Center, and later, director of engineering technology in IBM's Systems Development Division. The variety of technical work in these experiences turned his interests increasingly toward system-level concerns in the late 1960s. His interests have been cross-disciplinary since that time.

After a four-year period as a consultant in computer-related work, he joined the U.S. government where he served for more than six years as deputy for advanced technology in the Office of the Secretary, U.S. Air Force. He has since that time served as vice-president for research and development in the defense operations of Sperry Corporation (now Unisys), spent additional time consulting, and served four years as professor of systems engineering at George Mason University, Fairfax, Va. His services have been in demand since 1981 for high-level Defense Department advisory boards. He was, at time of publication, an adjunct professor of systems engineering at George Mason. The project described in the appendix was a five-year consulting activity.

Dr. Beam is a registered professional engineer and a licensed glider pilot who has built and flown his own high-performance sailplane.